# Emergence, Complexity and Computation

Volume 46

The Emergence, Complexity and Computation (ECC) series publishes new developments, advancements and selected topics in the fields of complexity, computation and emergence. The series focuses on all aspects of reality-based computation approaches from an interdisciplinary point of view especially from applied sciences, biology, physics, or chemistry. It presents new ideas and interdisciplinary insight on the mutual intersection of subareas of computation, complexity and emergence and its impact and limits to any computing based on physical limits (thermodynamic and quantum limits, Bremermann's limit, Seth Lloyd limits...) as well as algorithmic limits (Gödel's proof and its impact on calculation, algorithmic complexity, the Chaitin's Omega number and Kolmogorov complexity, non-traditional calculations like Turing machine process and its consequences,...) and limitations arising in artificial intelligence. The topics are (but not limited to) membrane computing, DNA computing, immune computing, quantum computing, swarm computing, analogic computing, chaos computing and computing on the edge of chaos, computational aspects of dynamics of complex systems (systems with self-organization, multiagent systems, cellular automata, artificial life,...), emergence of complex systems and its computational aspects, and agent based computation. The main aim of this series is to discuss the above mentioned topics from an interdisciplinary point of view and present new ideas coming from mutual intersection of classical as well as modern methods of computation. Within the scope of the series are monographs, lecture notes, selected contributions from specialized conferences and workshops, special contribution from international experts.

Indexed by zbMATH.

More information about this series at https://link.springer.com/bookseries/10624

Igor Balaz · Andrew Adamatzky
Editors

# Cancer, Complexity, Computation

Springer

*Editors*
Igor Balaz
University of Novi Sad
Novi Sad, Serbia

Andrew Adamatzky
Unconventional Computing Laboratory
University of the West of England
Bristol, UK

ISSN 2194-7287 ISSN 2194-7295 (electronic)
Emergence, Complexity and Computation
ISBN 978-3-031-04378-9 ISBN 978-3-031-04379-6 (eBook)
https://doi.org/10.1007/978-3-031-04379-6



This Springer imprint is published by the registered company Springer Nature Switzerland AG
The registered company address is: Gewerbestrasse 11, 6330 Cham, Switzerland

# Preface

Cancer is the most enigmatic health disorder humanity ever experienced. The last decade witnessed rethinking and reunderstanding of cancer. Cancer is not an evil creature. It is just a creature who wants to live, as all other living beings do. The book brings together biomedical scientists, biophysicists, mathematicians and computer scientists. They share their unorthodox views on origins, progress and treatment of cancer. J. James Frost opens the book with his chapter uncovering two key properties of cancer: symmetry-breaking and computational intelligence. Then Przemyslaw Waliszewski discusses cancer in the light of complementarity, complexity and fractal dynamics. A chapter from Thomas E. Yankeelov's laboratory analysis of how medical imaging technologies can be used to obtain patient-specific parameters for mathematical models of cancer is given. The effects of over-feeding in computational models of tumour growth are discussed by Pan Pantziarka and colleagues. Conway's Game of Life cellular automata are modified with global coupling in the chapter by Vladimir García-Morales and colleagues to model tumour growth. Oscar J. Suarez and colleagues discuss the potential of controlling genetic regulatory networks to tune cellular response to cancer. Milos Savic and colleagues present implementations of heterogeneous tumour modelling with PhyssiCell software and the implication of the heterogeneity in precision medicine. Sensitivity analysis of the cancer stem cells dynamical simulation is studied by Branislava Lalic and Igor Balaz. Marina Kovacevic presents her pioneer results on the molecular multiscale simulation of nanocarriers for cancer treatment. Michail-Antisthenis Tsompanas and colleagues present a haploid-diploid evolutionary algorithm for the Optimization of nanoparticles for cancer treatment. Perspectives of nanomedicine for cancer treatment are overviewed by Petra Gener and colleagues. A chapter from the Sabine Hauer's group outlines the future of cancer medicine based on swarms of nanorobots. The book ends with a chapter on modelling of angiogenesis with differential equations and cellular automata authored by Ioannis Karafyllidis and colleagues. Benedetta Casadei with co-authors overview strategies and mathematical models to study the complexity of drug resistance in cancer. Hector Zenil and colleagues discuss how to use computational analysis to unravel the complexity of the immune system. The book is mostly self-contained and well-illustrated, and chapters are written in a lively style. The book is accessible to readers from all walks of life and level of education, from laymen and high school students to experienced researchers.

Novi Sad, Serbia
Bristol, UK
March 2022

Igor Balaz
Andrew Adamatzky

# Contents

# What Cancer Is

**J. James Frost**

> *...it seems to me that I must consider as contained in the individual concept of myself only that which is such that I should not longer be me if it were not in me: and that all that is to the contrary such that it could be or not be in me without my ceasing to be me, cannot be considered in my individual concept.*
> *Letter from Antoine Arnauld to Gottfried Leibniz* [1].

**Abstract** The problem of cancer is examined from the metaphysical standpoint of essence and ground. An essentialist definition of cancer is assumed that would be valid in all possible worlds in which cancer could logically exist. The grounds of cancer are then examined and elucidated. Two grounding cancer properties are identified and discussed: symmetry-breaking and computational intelligence. Each examination leads to concrete conclusions for novel therapeutic approaches and a more fundamental understanding of what cancer is at bottom. Other possible cancer grounding properties related to evolution, adaptability and stochastic features are identified for future work. This approach is novel and offers new solutions to the problem of cancer.

**Keywords** Cancer · Intelligence · Symmetry-breaking · Computation · Ground

## 1 Introduction

Identifying a solution to the unsolved problem of cancer is currently based on a program of scientific reductionism, in which each new solution attempt provides temporary patient benefit until the oncologist and the field of oncology itself reaches the end of yet another box canyon.[1] In 1990 5.7 million people died from cancer worldwide; in 2017 the number of deaths increased to 9.6 million. Each new

[1] A canyon with vertical walls and closed upstream with a similar vertical wall, that is, a dead end.

J. J. Frost (✉)
Professor emeritus, BioMolecular Imaging, Baltimore, MD 21218, USA
e-mail: jfrost@jhmi.edu; jjamesfrost@comcast.net



I. Balaz and A. Adamatzky (eds.), *Cancer, Complexity, Computation*,
Emergence, Complexity and Computation 46,
https://doi.org/10.1007/978-3-031-04379-6_1

approach, from deep DNA sequencing, to exquisitely-designed molecular targeting, to the intricacies of epigenetics follows a well-worn path of near certain scientific justification and overwhelming enthusiasm to modest gains in some cancers in select patient populations—all at a worldwide cancer research expenditure of $100 billion in 2019.

This chapter examines an alternative to the regime of scientific reductionism. It is concerned with an essentialist definition of cancer and what then grounds cancer 'at bottom'. The approach strives to strip cancer to its bare bones, and then to examine the necessary properties cancer requires to manifest the characteristics observed everyday in cancer patients. It then leads to a consideration of concrete inferences that, in turn, result in greater insight into cancer and therapeutic strategies that have not previously been considered. Two aspects of this analysis are cancer symmetry breaking and computational intelligence. Each can be derived from a fundamentality approach presented here and can be considered as two of cancer's grounds. Each additionally leads to material conclusions for novel solutions to the cancer problem.

## 2 Cancer's Essence and Ground

Essence is what it is for an object to be the very thing that it is. It is the fundamental definition of an object—its definition at bottom. Essence is a metaphysical concept of what it is to be that thing in all possible worlds that are logically realizable [2–4]. Essence is thus a necessary feature of an object [5–8]. It is a property an object could not lack and still be that object. An essence definition should be independent of the conceptual apparatus that observers bring to bear on it [9]. It may be singular (monadic) or plural [10, 11].

To be human is the essence of Socrates and that would be true in all possible worlds in which Socrates could exist. A possible world is an instantiation of reality in which the world's accidental or contingent properties are different than in our actual world [9], for example, a world in which I myself, giraffes, Harvard, the US Constitution, Samuel Clemens, oak trees, and Chevrolets don't exist. In their place are other objects, concepts, and relationships. For cancer, a possible world is one in which biochemistry, some physical laws, and organizational properties are different than in our actual world. For example, a world in which life is based on RNA or some other coded information, where transforming growth factor, tumour angiogenesis, or any of the cancer hallmarks [12] don't exist. A statement of essence will not include the properties of the object itself to avoid problems of self-reference and circularity [13, 14]. Essence statements may extend from coarse to fine-grained [15, 16]—fine-grained essence statements are most relevant to scientific explanations of natural phenomena.

What might be an essence statement for cancer that would be necessary in all possible worlds in which cancer could logically exist? Logically, existence means that properties cannot violate first order logical principles, such has having and simultaneously not having a certain property, or violating implication principles. We can try on the following essence statement and see if it works as we move forward: *Cancer expands physically and biochemically to overcome and destroy its host and, thereby, the cancer itself*. One could construct other essence statements in this spirit. That is, statements that do not reference the actual properties of cancer, thus avoiding circularity. Conversely, the statement could encompass other processes, like man's destruction of the environment as "a cancer on the planet" [17–19]. The cancer essence statement incorporates none of the traditional cancer hallmarks [12, 20], which can be viewed as accidental features of cancer in our world. Certainly, we have to deal with the cancer we know, including its mutable properties. But here, examination of matters of fundamentality is the goal. Again, the actual properties of cancer in our world are not features of the essence statement above. Can we learn something new about cancer by considering its essence?

Each term in this essence statement has meaning. Expansion encompasses physical extent and scope. The biochemical expansion of cancer and subversion of normal homeostatic physiology and biochemistry is a key aspect of cancer. It is as much responsible for cancer's deadly features as is physical expansion in space. Metastasis is omitted since cancer can expand in possible metaphysical scenarios without metastasizing, e.g., by the expansion of a solitary mass, as is the case in some malignant cancers. Cancer must have a host with a preexisting structure, mode of energy production, and biochemical machinery that can be co-opted for the cancer's new purpose.[2] Cancer is a progressive process that first overcomes the host's normal biochemistry by overthrowing and subverting it for a new functional purpose. It eventually destroys the host and thereby the cancer itself—how it accomplishes this is left out of the essence statement. The nature and period of time of the subversion and destruction process is variable, from months to years. Indeed, a current treatment approach is to lengthen this period in order to prolong the life of the host—the patient [21–23]. Space, energy, and information are key background features of this essence statement that require examination.

Leaving essence for now, the concept of "ground" is examined. There are a number of meanings of ground in the philosophical literature, but a useful one for scientific purposes in describing natural phenomena is: ground is a relationship between facts [15, 24–27]. A fact is grounded by grounds or ground claims. For example, that this drink is a cocktail is grounded by the drink being alcoholic and mixed [15]. Other examples include a physicalist asserting that nonphysical facts are grounded by physical processes; natural laws are grounded in patterns of local, qualitative matters of fact; and a person in pain is grounded in the firing of c-fibers [14, 15]. Thus, ground has explanatory and deterministic features that create a hierarchy of stratified facts of reality. Ground provides an explanatory or "because of" characterization of facts, creating a level structure that ends with a foundational fact in the explanatory chain. In this construct there may be ungrounded facts at bottom that are primitive or pre-rational [28]. Ground is typically a many-to-one relationship, in that many properties can jointly ground an item [29, 30]. Grounds portray

[2] Viral and other external (e.g., radiation) modifications of the host's molecular machinery that lead to cancer are a subcategory.

an antecedent-consequent relationship, in that the antecedent grounds the grounded consequent.

Grounding can be differentiated from causality on two fronts. Causality is typically diachronic whereas grounding is synchronic [31]. Second, causality is a relationship between events whereas ground is a relationship between facts [25, 31–33]. This distinction will deserve further examination for cancer. A ground may be full or partial. For example, being a mixed drink and being alcoholic are both partial grounds for being a cocktail. Partial grounds will be important for cancer. Ground obeys transitivity: if A grounds B and B grounds C, then A grounds C. The hierarchy of grounding relationships in cancer will be important to examine, as discussed below.

Ground and essence are related: when a connection of grounded facts exists, the truth of the grounded connection will be based in the essence of the constituents of the grounded fact [24, 25, 27, 34, 35]). In fact, the relationship of ground to essence, causality, and necessity is a current area of active investigation. Some argue that ground reduces to essence, and others the converse [36–39]—avoiding circularity is key here. Essence statements can be divided into *objectual, generic* and *factual* categories [11, 38–40]. Essence statements as definitions are equivalence or identity statements, such as: water is $H_2O$. Full essence statements are symmetric and imply necessity. The proposed essence statement for cancer above is viewed as generic essence since it refers to a process or concept, rather than a pure object, such as Socrates. Whether it is full or partial is explored subsequently.

Ground, in contrast to essence, reflects an asymmetric, irreflexive relationship between the grounded and the grounds [29, 41, 42]. In scientific analysis, in particular, ground offers an explanation between the *explanandum* and the *explanans*, the phenomenon to be explained and the sentences offered to explain that phenomenon [35, 38]. Ground is a "because of" or "makes it the case" type of statement. Ground is a *constitutive* relationship between one or more grounds for the grounded object or phenomenon that results in a *sufficiency* relationship. In the essence or definitional framework, the relationship is of *necessity* and is reflexive [36, 43, 44]. That a ball is red and round is grounded by the fact that it is red and that it is round, but not the converse direction. Another example: the laws of nature are grounded in patterns of local, qualitative matters of fact. A grounded fact is less fundamental than the fact that grounds it. Ground is a type of deterministic relationship, the strongest form of metaphysical explanation [44–46]. Hence, determination implies metaphysical sufficiency for what is grounded, and metaphysical sufficiency then suffices for necessitation [14]. For cancer, both necessary and contingent grounds exist, the former providing the explanatory role in all possible worlds and the contingent in this.[3]

Of the many varieties of grounding, "grounding explanation" is most useful for natural phenomena and scientific accounts and laws [35, 45]. An explanation has a fact that is explained, the *explanandum*, and the *explanans* fact. As addressed above,

[3] Cancer has indeterministic properties as well, for example, random mutations related to its origin and evolution, and the stochastic binding of transcription factors to their genomic sites. The overlap and interplay of deterministic grounds and non-deterministic features await further research.

the *expanandum-explanans* is a "because of" relation: the fact that the window shattered, the *explanandum*, and Suzy threw a rock at it, the *explanans*. This grounding explanation is synchronic, whereas the equivalent causal explanation is diachronic. Grounding explanations are irreflexive, asymmetric and transitive [41, 42, 47]. To clarify the overlap with causation, one considers constitutive grounding explanations. These are a hierarchy of chaining facts that constitute an object: there is a table here in virtue of there being wood arranged here in a certain 'table-like way. Lest this become too abstract for the present purposes, one can posit the question of whether cancer is explained by the arrangement of genes, proteins, signaling receptors, histones, etc. in a 'cancer-like way'. Is there something beneath the facts of cancer that provides a grounding explanation for cancer?

An important grounding category, particularly for scientific explanation, is that of *essentialist grounding mechanisms*, which link back to the essences of the objects in a hierarchical chain [32, 33, 35, 48] In this view, object essences require certain determination relations to exist. A grounding mechanism explanation then delineates how the connections run between the grounding facts and the fact they ground [49]. This can be a fine or coarse-grained description that encompasses quantum mechanics, covalent and ionic bonds, molecules, DNA, and transcription to a functioning gene. Different levels of explanation and their corresponding scientific disciplines will bottom out at different levels; molecular biologists will bottom far before quantum mechanics, for example. Here we address where cancer explanations may bottom out in the domain of necessary facts that would be true in all possible worlds.

In the case of cancer we will examine the relation of the essential definition of cancer, above, to its grounding mechanisms. This view overlaps with law-like natural phenomena and, indeed, the concept of a law itself [46]. Are there any laws of cancer that can be discovered? The Cancer Hallmarks were an early attempt at describing the recurring features or properties of cancer that might give rise to law-like explanations. The Hallmarks are, however, the "accidental" features of cancer [50]. They are, accordingly, not necessary and cannot be the basis of "cancer laws". Here, we seek laws that are not based on nor incorporate the contingent cancer facts, but rather ground these facts. The detailed relationships among grounds, laws, events, and causality is a complex area of active research [46, 49, 51, 52], but is beyond the scope of this work.

## 3 Cancer Hallmarks as Contingent Properties

The Hallmarks of Cancer was a seminal advance in the conceptual view of cancer [12]. The hallmarks delineate the common features or traits of cancers across organ origin and pathology. The initial list of hallmarks includes: self-sufficiency in growth signals; evasion of apoptosis; insensitivity to anti-growth signals; sustained angiogenesis; tissue invasion and metastasis; and limitless replicative potential [12].

In an updated account, two hallmarks were added: deregulating of cellular energetics, and avoiding immune destruction [20]. Two additional enabling characteristics are: tumour-promoting inflammation, and genome instability and mutation. While the hallmarks provided needed insight into common cancer traits and have been applied to new therapeutic developments, the problem of cancer has remained largely unsolved.

The centrality of the hallmarks has been questioned on several accounts. One category is the emphasis on proliferation and motility in the context of a similar importance in development and in benign tumours. Several analyses have pointed out the overlap in feature of benign and malignant tumours [53–57]. Endometriosis is example of an abnormal cell migratory process that is not cancer. Another category of critique is the reliance on the reductive regime of somatic mutation theory (SMT) [57–60]. This approach has recently given rise to the non-deterministic "bad luck" theory of cancer neogenesis based on copying mistakes proportional to the normal cellular replication rate [61, 62]. Further, the hallmark traits shift in their hierarchical importance at different times in a cancer's evolution and during treatment [20, 56]. SMT can be contrasted with the multilevel-based tissue organization field theory [54, 63]. This viewpoint avoids a purely reductionist theory and examines cancer as a multilevel process that features upward and downward causation.

The Hallmarks may also be questioned as grounds of cancer that would be necessary in all possible worlds in which cancer exists—clearly they are not. In other worlds, other Hallmarks would exist, some possibly overlapping with those in this world, some eliminated, and new ones added. The central question is: what grounds the Hallmarks at all in any possible world, including this one?

## 4 Cancer's Broken Symmetry

Symmetry pervades the world, from nature to human art and construction to the fundamental particles of physics [64, 65]. It is, however, uncommonly a perfect symmetry, like a sphere, but most often an imperfect or broken symmetry [66–68]. Broken symmetries appear and resolve continuously over time, for example, as water freezes and thaws again to its more symmetric liquid state. Perfect symmetry achieves little in nature nor in art or human construction. Only with a judicious level of broken symmetry does functional structure arise. Conversely, states of maximal or fully broken symmetry are chaotic and generate nothing. Symmetry and symmetry breaking is a foundational property of the universe and is grounded at bottom in space and logic itself. This can be most directly seen in the Noether theorems that demonstrate an underlying symmetry in the key conservation laws of physics, including the conservation of energy and angular momentum [69, 70]. Cancer is an interesting example: its *sine qua non* is broken symmetry from the cancer cell to the chromatin itself. A useful consequence of this fact is that all cancer is diagnosed and graded based on the degree of cellular disorganization.

The categories of cancer's broken symmetry have been examined in a previous work [71]. Three essential questions are addressed: what features or mechanisms ground the observed macroscopic broken cellular symmetries; what classes of broken symmetry exist in cancer cells; and how can improved knowledge of broken symmetry in cancer be used to advance patient treatment. The three categories of broken cancer symmetry are: combinatorial; geometric; and functional symmetry breaking [71]. One example of combinatorial symmetry breaking is the broken symmetry of the epithelial-mesenchymal hybrid cell state to the pure epithelial or mesenchymal type. Since it is the mesenchymal cell type that metastasizes, this broken symmetry is a key determinant in cancer progression. The molecular constituents underlying this symmetry change include ZEB, SLUG, TWIST, miR-200, miR-34, and many others [72–75].

The most widely-recognized feature of broken symmetry in cancer is that of the cell shape. Shape progresses from uniformity within different organs and functional cell types to an increasingly disorganized structure that correlates with malignant potential and patient prognosis. What grounds this broken symmetry? The internal cellular matrix, consisting of microtubular structural filaments, largely determines the cell shape [76–78]. These microtubules comprise a tensegrity-based supporting structure that has its own internal symmetries. Microtubules are also thought to transmit information from the environment to the internal cell structures [78–80]. Recent investigations have even shown that microtubule and actin networks can compute [81–83]. An outstanding question for future research is the degree to which broken symmetry in microtubule networks grounds intrinsic cellular computation, a topic addressed in the next section. Does cell asymmetry and coincident intracellular actin filament asymmetry confer a computational advantage for cancer cells?

In the category of functional symmetry breaking, the symmetry of gene, protein and other cellular networks is key. All networks can be classified according to the symmetry features of the connected network components [84–87]. Graph theory grounds the network symmetries and is a powerful tool for investigating cellular networks [88–91]. The automorphism group, Aut(G), is a key parameter [92–95]. The Komogorov complexity, K(G), is another that is inversely related to Aut(G) and correlated positively with information processing capacity [96–100]. Function networks must be stabilizable and controllable to store and process information required for survival of the cell or organism [101–104]. They must be resistant to disruption due to a loss of functionality at one or more points in the network [71]. Network homeostasis is closely related to network symmetry. Conversely, network vulnerability is directly related to focal network asymmetries. The vulnerability of electric power grids, for example, is related to the number of hubs where multiple connections are concentrated. The broken subgroup symmetries in cancer cell networks are related to the level of attack tolerance at the sites of reduced symmetry in cancer networks [105–111]. Identifying subgroup cancer network broken symmetries is a viable strategy to direct targeted therapeutics to these sites [84, 89, 90]. This is one of the many concrete examples of knowledge of symmetry breaking in cancer potentially leading to improved therapeutic strategies, and perhaps more importantly, the development of new, targeted drugs for these sites. In parallel, could this knowledge be used to repair cancer networks to reestablish homeostasis? Other directions for cancer symmetry-breaking research are listed elsewhere [71].

Symmetry and symmetry breaking are thus one possible ground for the essence of cancer statement above. Since symmetry exists in all possible worlds in which hosts and cancer can logically exist, it would satisfy necessity. It is, however, only a partial cancer ground, so satisfaction of the sufficiency criterion remains a question. The next section examines computational intelligence and information transfer in cancer. A question of where symmetry and symmetry breaking lie in the explanatory hierarchy with computation and information emerges: a question of fundamentality.

## 5 Cancer's Intelligence

Intelligent oncologists wage a valiant battle against cancer every day. They nevertheless have limited success in curing cancer or even arresting its progression in the majority of cases: there are still some 10 million cancer deaths per year worldwide. Cancer computes inventive solutions to the oncologist's various strategies, and eventually wins the game in the majority of cases. It is indeed a gameplay: the intelligence of the oncologist vs. the intelligent cancer, as has been described in a recent article [112]. This view is grounded in the definition of intelligence and what can be intelligent. The definition of intelligence for these purposes reduces to computational intelligence and excludes conscious awareness. It is most immediately concerned with the ability to learn, problem solve and adapt. Deception, bluffing, and prior computation of actions for an array of possible future events are additional features of intelligence.

Daniel Dennett has examined intelligence in terms of biological competency. He describes four levels: Darwinian, Skinnerian, Popperian and Gregorian, in order of increasing competency [113]. Each level can be linked to features of cancer, thus viewing cancer as possessing many properties of intelligence [112]. For example, does cancer learn from its earliest encounters with host defenses, store this information in memory, preemptively compute solutions to future chemotherapeutic actions by the oncologist, and at the right time implement these defense measures? Or, does cancer generate real-time intelligent strategies as new survival problems arise? The strategic balance of memory and active computation is a feature of intelligence; both are limited as a function of the computing architecture and the supply of energy. *Cancer's Intelligence* examines these and related questions [112].

What can be intelligent is broad and extends from animals to bacterial swarms [113–120] (ref from paper 9,10). In short, anything that can compute can be intelligent. It is well known that pebbles, knots on a string, a slide rule, and silicon-based computers themselves can compute. Chemical matter can compute, as well, including the Belousov-Zhabotinsky reaction, liquid crystals, carbon nanotubes and conductive foams [121–123]. Biologic matter also performs computations, including with gene networks, DNA, actin filaments and microtubules, and the famous slime mold [124–128]. In fact, biological matter can solve some of the most challenging computational problems in existence, such as the traveling salesman problem [129]. Biological computation in organisms can be classified as intrinsic computation, that is, self-computation, as opposed to external computation by the desktop computer in front of us [130–132]. Cancer, with its varied cellular composition and extended networks, eventually computes its spread throughout the body. A co-opted microenvironment and relentless evolutionary adaption presents a formidable computational adversary for the most intelligent oncologists.

The intrinsic computation of cancer can be described by its ε-machine [133–136]. The ε-machine is an inferential model of the cell's or multi-cell system's capturing and processing of information "that permits the system to read the environment's information and rate of change; store and process that information; create an internal efficient model representation of the environment; and use the model for future decisions and actions". As a function of evolutionary change, the caner ε-machine can be updated to a new computational architecture in order to better outplay the oncologist's strategies, or to develop improved resistance, such as by adopting a quiescent state. Energy, space, and other resources limit this adaptive process. Although ε-machine theory has been applied to many computational systems, it has not yet been developed for cancer. Further discussion of the ε-machine, the role of Shannon entropy-complexity descriptions, and the outlook for its application to cancer are covered elsewhere [112]. Boolean network theory provides a tractable approach to defining the cancer ε-machine(s) [137–143].

Once the caner ε-machine structure has been elucidated, the question then becomes: how can it be out-computed by the oncologist? In the concept of gameplay between the two adversaries, one can consider the most advanced types of human gameplay, including chess, go and poker. How can new artificial intelligence (AI) algorithms for human gameplay be applied to the cancer problem? One of the most exciting recent AI developments in human gameplay has been for limit and no-limit Texas hold'em poker [144–147]. Two-player poker is a vastly more complex game that chess or go, with a state space of $10^{160}$. The new poker AI algorithms use counterfactual regret minimization and other techniques to make the problem solution tractable—it can now beat the best professional poker players [145, 146]. The algorithms also incorporate the bluffing plays that are central to poker and mimic the deception observed throughout nature [112, 147]. In the early development stages of game theory with imperfect information, von Neumann recognized the importance of deception in nature: "Real life consists of bluffing, of little tactics of deception, of asking yourself what is the other man going to think I mean to do. And that is what games are about in my theory." Further: "An organism that has no poker face, that communicates its state directly to all hearers, is a sitting duck, and will soon be extinct" [148]. *Cancer's Intelligence* explores the question of whether cancer bluffs and how the new poker AI algorithms can be used as future AI tools for the oncologist [112].

Limit Texas hold'em is easier to solve compared to no-limit Texas hold'em since the limit on betting places significant restraints on the number of possible game plays. The degree to which cancer gameplay is a limit or no-limit game is an interesting question for future investigation—here the limits are resource expenditures or allocations to the game by the cancer and the oncologist's time and actual monetary expenditures, or bets, for the therapy actions. Multiplayer poker is even more complex and possesses a non-computable Nash equilibrium [146]. Cancer's game against the oncologist is most certainly a multiplayer game on cancer's part since it plays multiple strategies simultaneously, and even shares cards from the hands of its multiple players within the molecular defense armamentarium it brings to game. Perhaps that is an insurmountable game for oncologist, even with improved AI tools. Only further investigation will provide the answer. Several signposts for further research are provided in *Cancer's Intelligence* [112].

Since information and information transfer grounds computation, and computation can be regarded as a partial cancer ground, we can ask what grounds information? Information theory is derived from work in the nineteenth and twentieth centuries from Gibbs, Boltzmann, Turing, Nyquist, von Neumann, Shannon, Landauer and others. Landauer showed the important relationships between information, entropy and energy and Shannon derived the measure of information as the Shannon entropy. Landuuer showed that the storage of 1 bit of information requires work equal to kTln2, k being Boltzmann's constant and T temperature [149]. Landauer also originated that concept that "all information is physical." That is, to be information it must be physically represented or encoded [150]. In other words, information is a physical entity. In fact, the universe can be regarded a having computed itself from its very beginning through the information transfer among its physical entities, essentially a self or intrinsic computation [151–154].

Further examination of these important concepts is far beyond the scope of this article, but current consideration do demonstrate the point that grounding will eventually reach its bottom at some primitive or irreducible fact that is itself ungrounded, some of which are energy and entropy, and even symmetry, as addressed above. For the purposes of understanding foundational aspects of cancer, we do not need to reach a bottom ground to achieve a useful and fundamental of explanation of cancer. Rather we only need to understand cancer at a level where we can attack and eradicate it using computation, symmetry breaking, and other foundational concepts yet to be examined. We are not concerned with what happens to the information cancer formerly possessed once it is destroyed.

## 6 Conclusion

This examination of the fundamental nature of cancer by employing essence and ground concepts yields new insights. Two aspects of cancer's ground have been highlighted here and in more detail previously [71, 112]. Cancer's broken symmetry is a central feature of cancer diagnosis, monitoring, and treatment employed every day in medical practice. More abstractly, the broken symmetry of the cancer gene and other regulatory networks provides guidance for directed therapies at network sites of reduced Aut(G), since these sites have greater attack vulnerability. Other practical applications of broken symmetry in cancer are presented elsewhere [71].

The approach of drilling down to cancer's fundamentality is not just a metaphysical exercise, but one that can yield new practical insights for cancer treatment. Indeed, the science we practice today is based on foundational philosophical concepts.

A second result of the cancer ground analysis is that computational intelligence is a necessary cancer property. Each cancer move in space and time is a computed decision within a vast possibility of alternatives, whether in response to innate body defenses or to the oncologist's use of lethal measures to eradicate the cancer. There are many incompletely-explored questions in this category, including: the structure of cancer's intrinsic biochemical computer, how information is stored, and how cancer's computation can be quantified in calculations per second as for external computers (such as the Summit computer, at 200 quadrillion calculations per second). Recently-developed AI algorithms for the most complex human gameplay of poker could be extended and adapted for better play against cancer once intrinsic cancer computation is better understood. Bluffing, self-aware computing, and pre-computation of future moves are other intriguing aspects of computational intelligence that merit further examination in cancer [112].

Evolution and adaptation is another ground property of cancer, but is beyond the scope of this article. It requires examination, as has been accomplished for the cases of symmetry breaking and computational intelligence. A central question will be to determine the extent to which evolution is grounded by broken symmetry and computational intelligence, or grounded by other phenomena. There is a limit to how much information can be stored by cancer cells and networks. Evolution therefore seems to be a necessary cancer property. Improved understanding of the grounds of cancer evolution could result in improved approaches to disrupting it, thereby benefiting cancer patients. This has been the approach in the analysis of symmetry breaking and computational intelligence, that is, to take the abstract analysis to concrete measures for clinical care.

The stochastic or indeterministic nature of cancer is another category requiring further ground analysis. Random fluctuations extend from gene mutations to small numbers of transcription factors stochastically binding to gene regulatory sites. How do deterministic features of ground overlay and interact with the random properties of cancer? Some signposts exist in the ground literature, but more work is needed. Use of the Langevin equation has proven helpful in examining the behavior of chromosomal system dynamics—including in VDJ recombination—and could therefore find application in systems with deterministic and stochastic features [31, 46, 155–158].

This chapter has examined the application of essence and ground analysis to cancer. Some conclusions have been reached for symmetry breaking and computational intelligence, but more work is needed, as is commonly the case in introducing new ideas to solve old problems. This must be a novel multidisciplinary effort, including the familiar cancer disciplines of molecular biology, evolution and genetics, and adding back philosophy, from which all conceptual knowledge flows.

**Acknowledgements** I thank the many individuals who brought me to the field of cancer and its manifestations in information science, physics, complexity and computation. They are Donald Coffey, Eshel Ben-Jacob, Kenneth Pienta, Shawn Lupod, William Issacs and others from The Johns

Hopkins University and the Brady Urological Institute. Finally, I thank Ms. Nina Frost for her expert editing work.

## References


1. Mason, H.: The Leibniz-Arnauld Correspondence. Manchester University Press, Barnes & Noble, New York (1967)
2. Kovacs, D.M.: Modality (2020)
3. Stalnaker, R.C.: Possible worlds. Noûs, 65–75 (1976)
4. Divers, J.: Possible Worlds. Routledge (2006)
5. Correia, F.: On the reduction of necessity to essence. Philos. Phenomenol. Res. **84**(3), 639–653 (2012)
6. Ditter, A.: The reduction of necessity to essence. Mind **129**(514), 351–380 (2020). https://doi.org/10.1093/mind/fzz045
7. Leech, J.: Essence and mere necessity. R. Inst. Philos. Suppl. **82**, 309–332 (2018). https://doi.org/10.1017/s1358246118000139
8. Zylstra, J.: Essence, necessity, and definition. Philos. Stud. **176**(2), 339–350 (2017). https://doi.org/10.1007/s11098-017-1018-y
9. Yablo, S.: Identity, essence, and indiscernibility. J. Philos. **84**(6), 293–314 (1987)
10. Shumener, E.: Identity. In: The Routledge Handbook of Metaphysical Grounding. Routledge (2020)
11. Zylstra, J.: Essence. In: The Routledge Handbook of Metaphysical Grounding, pp. 324–335. Routledge (2020)
12. Hanahan, D., Weinberg, R.A.: The hallmarks of cancer. Cell **100**(1), 57–70 (2000)
13. Hicks, M.T., van Elswyk, P.: Humean laws and circular explanation. Philos. Stud. **172**(2), 433–443 (2015)
14. Wilsch, T.: Laws of metaphysics. In: The Routledge Handbook of Metaphysical Grounding, pp. 425–436. Routledge (2020)
15. Raven, M.J.: Introduction. In: The Routledge Handbook of Metaphysical Grounding. Routledge (2020)
16. Zalta, E.N.: Essence and modality. Mind **115**(459), 659–694 (2006)
17. Hern, W.M.: Has the human species become a cancer on the planet? A theoretical view of population growth as a sign of pathology. Curr. World Lead. **36**(6), 1089–1124 (1993)
18. MacDougall, A.K.: Humans as cancer. Wild Earth **6**, 81–88 (1996)
19. Pauly, D.: Homo sapiens: cancer or parasite? Ethics Sci. Environ. Polit. **14**(1), 7–10 (2014)
20. Hanahan, D., Weinberg, R.A.: Hallmarks of cancer: the next generation. Cell **144**(5), 646–674 (2011). https://doi.org/10.1016/j.cell.2011.02.013
21. Cunningham, J.J., Brown, J.S., Gatenby, R.A., Staňková, K.: Optimal control to develop therapeutic strategies for metastatic castrate resistant prostate cancer. J. Theor. Biol. **459**, 67–78 (2018). https://doi.org/10.1016/j.jtbi.2018.09.022
22. Gatenby, R.A., Silva, A.S., Gillies, R.J., Frieden, B.R.: Adaptive therapy. Cancer Res. **69**(11), 4894–4903 (2009). https://doi.org/10.1158/0008-5472.CAN-08-3658
23. Zhang, J., Cunningham, J.J., Brown, J.S., Gatenby, R.A.: Integrating evolutionary dynamics into treatment of metastatic castrate-resistant prostate cancer. Nat. Commun. **8**(1), 1–9 (2017). https://doi.org/10.1038/s41467-017-01968-5
24. Rosen, G.: Real definition. Anal. Philos. **56**(3), 189–209 (2015)
25. Raven, M.J.: Ground. Philos. Compass **10**(5), 322–333 (2015)
26. McSweeney, M.M.: Logic. In: The Routledge Handbook of Metaphysical Grounding, pp. 324–335. Routledge (2020)
27. Schnieder, B.: Dependence. In: The Routledge Handbook of Metaphysical Grounding, pp. 107–120. Routledge (2020)

28. Wittgenstein, L., Anscombe, G.E.M., von Wright, G.H., Paul, D., Anscombe, G.E.M.: On certainty, vol. 174. Oxford, Blackwell (1969)
29. Correia, F.: Granularity. In: Routledge Handbook of Metaphysical Grounding. Routledge (2020)
30. Correia, F.: Grounding and truth-functions. Logique et Anal. **53**(211), 251–279 (2010)
31. Wang, J.: Cause. In: The Routledge Handbook of Metaphysical Grounding. Routledge (2020)
32. Trogdon, K.: Grounding-mechanical explanation. Philos. Stud. **175**(6), 1289–1309 (2017). https://doi.org/10.1007/s11098-017-0911-8
33. Bliss, R., Trogdon, K.: Metaphysical grounding (2014)
34. Rosen, G.: Metaphysical dependence: grounding and reduction. Modality: Metaphys. Logic, Epistemol. 109–136 (2010)
35. Glazier, M.: Explanation. In: The Routledge Handbook of Metaphysical Grounding. Routledge (2020)
36. Carnino, P.: On the reduction of grounding to essence. Stud. Philos. Est. 56–71 (2015). https://doi.org/10.12697/spe.2014.7.2.04
37. Zylstra, J.: The essence of grounding. Synthese **196**(12), 5137–5152 (2018). https://doi.org/10.1007/s11229-018-1701-3
38. Fine, K.I.T.: Unified foundations for essence and ground. J> Am. Philos. Assoc. **1**(2), 296–311 (2015). https://doi.org/10.1017/apa.2014.26
39. Correia, F., Skiles, A.: Grounding, essence, and identity. Philos. Phenomenol. Res. **98**(3), 642–670 (2017). https://doi.org/10.1111/phpr.12468
40. Fine, K.: Identity criteria and ground. Philos. Stud. **173**(1), 1–19 (2015). https://doi.org/10.1007/s11098-014-0440-7
41. Thompson, N.: Strict partial order. In: The Routledge Handbook of Metaphysical Grounding, pp. 259–270. Routledge (2020)
42. Correia, F.: Real definitions. Philos. Issues **27**(1), 52–73 (2017). https://doi.org/10.1111/phis.12091
43. Zylstra, J.: Constitutive and consequentialist essence. Thought: J. Philos. **8**(3), 190–199 (2019). https://doi.org/10.1002/tht3.419
44. Lewis, D.: On the Plurality of Worlds, vol. 322. Oxford Blackwell (1986)
45. Maurin, A.-S.: Grounding and metaphysical explanation: it's complicated. Philos. Stud. **176**(6), 1573–1594 (2018). https://doi.org/10.1007/s11098-018-1080-0
46. Emery, N.: Laws of nature. In: Routledge Handbook of Metaphysical Grounding. Routledge New York, NY (2020)
47. Dixon, T.S.: Infinite descent (2020)
48. Trogdon, K.: Truthmaking (2020)
49. Trogdon, K.: Grounding-mechanical explanation. Philos. Stud. **175**(6), 1289–1309 (2018)
50. Robertson, T., Atkins, P.: Essential vs. accidental properties. Stanford Encyclopedia of Philosophy (2008)
51. Korman, D.Z.: Debunking arguments. Philos Compass **14**(12), e12638 (2019)
52. Barker, J.: Debunking arguments and metaphysical laws. Philos. Stud. **177**(7), 1829–1855 (2019). https://doi.org/10.1007/s11098-019-01287-z
53. Lazebnik, Y.: What are the hallmarks of cancer? Nat. Rev. Cancer **10**(4), 232–233 (2010)
54. Sonnenschein, C., Soto, A.M.: The aging of the 2000 and 2011 hallmarks of cancer reviews: a critique. J. Biosci. **38**(3), 651–663 (2013)
55. Sonnenschein, C., Soto, A.M., Rangarajan, A., Kulkarni, P.: Competing views on cancer. J. Biosci. **39**(2), 281–302 (2014)
56. Fouad, Y.A., Aanei, C.: Revisiting the hallmarks of cancer. Am. J. Cancer Res. **7**(5), 1016 (2017)
57. Baker, S.G.: Paradoxes in carcinogenesis should spur new avenues of research: an historical perspective. Disruptive Sci. Technol. **1**(2), 100–107 (2012)
58. Sonnenschein, C., Soto, A.M.: Cancer metastases: so close and so far. J. Natl. Cancer Inst. **107**(11), djv236 (2015)

59. Sonnenschein, C., Soto, A.M.: Carcinogenesis explained within the context of a theory of organisms. Prog. Biophys. Mol. Biol. **122**(1), 70–76 (2016)
60. Soto, A.M., Sonnenschein, C.: Paradoxes in carcinogenesis: there is light at the end of that tunnel! Disruptive Sci. Technol. **1**(3), 154–156 (2013)
61. Weinberg, C.R., Zaykin, D.: Is bad luck the main cause of cancer? JNCI J. Natl. Cancer Inst. **107**(7), djv125-djv125 (2015). https://doi.org/10.1093/jnci/djv125
62. Tomasetti, C., Vogelstein, B.: Variation in cancer risk among tissues can be explained by the number of stem cell divisions. Science **347**(6217), 78–81 (2015)
63. Sonnenschein, C., Soto, A.M.: Over a century of cancer research: inconvenient truths and promising leads. PLoS Biol. **18**(4), e3000670 (2020). https://doi.org/10.1371/journal.pbio.3000670
64. Weyl, H.: Symmetry. Princeton University Press, Princeton (1952)
65. Feynman, R.: Symmetry in physical laws. In: The Feynman Lectures on Physics, vol. 1 (1963)
66. Brading, K., Castellani, E.: Symmetry and symmetry breaking. In: The Stanford Encyclopedia of Philosophy (2013)
67. Collier, J.: Information originates in symmetry breaking. Symmetry: Cult. & Sci. **7**, 247–256 (1996)
68. Strocchi, F.: Symmetry Breaking. Springer, Berlin (2010)
69. Kosmann-Schwarzbach, Y.: The Noether theorems. In: The Noether Theorems, pp. 55–64. Springer (2011)
70. Neuenschwander, D.E.: Emmy Noether's wonderful theorem. JHU Press, (2017)
71. Frost, J.J., Pienta, K.J., Coffey, D.S.: Symmetry and symmetry breaking in cancer: a foundational approach to the cancer problem. Oncotarget **9**(14), 11429–11440 (2018). https://doi.org/10.18632/oncotarget.22939
72. Jolly, M.K., Jia, D., Boareto, M., Mani, S.A., Pienta, K.J., Ben-Jacob, E., Levine, H.: Coupling the modules of EMT and stemness: a tunable "stemness window" model. Oncotarget **6**(28), 25161–25174 (2015). https://doi.org/10.18632/oncotarget.4629
73. Lu, M., Jolly, M.K., Levine, H., Onuchic, J.N., Ben-Jacob, E.: MicroRNA-based regulation of epithelial-hybrid-mesenchymal fate determination. Proc. Natl. Acad. Sci. USA **110**(45), 18144–18149 (2013). https://doi.org/10.1073/pnas.1318192110
74. Lu, M., Jolly, M.K., Onuchic, J., Ben-Jacob, E.: Toward decoding the principles of cancer metastasis circuits. Cancer Res. **74**(17), 4574–4587 (2014). https://doi.org/10.1158/0008-5472.can-13-3367
75. Jolly, M.K., Tripathi, S.C., Jia, D., Mooney, S.M., Celiktas, M., Hanash, S.M., Mani, S.A., Pienta, K.J., Ben-Jacob, E., Levine, H.: Stability of the hybrid epithelial/mesenchymal phenotype. Oncotarget **7**(19), 27067–27084 (2016). https://doi.org/10.18632/oncotarget.8166
76. Tadeo, I., Berbegall, A.P., Escudero, L.M., Álvaro, T., Noguera, R.: Biotensegrity of the extracellular matrix: physiology, dynamic mechanical balance, and implications in oncology and mechanotherapy. Front. Oncol. **4**, 39 (2014)
77. Pienta, K., Coffey, D.: Cellular harmonic information transfer through a tissue tensegrity-matrix system. Med. Hypotheses **34**(1), 88–95 (1991)
78. Ingber, D.E., Wang, N., Stamenović, D.: Tensegrity, cellular biophysics, and the mechanics of living systems. Rep. Prog. Phys. **77**(4), 046603 (2014)
79. Goehring, N.W., Grill, S.W.: Cell polarity: mechanochemical patterning. Trends Cell Biol. **23**(2), 72–80 (2013)
80. Schiffhauer, E.S., Robinson, D.N.: Mechanochemical signaling directs cell-shape change. Biophys. J. **112**(2), 207–214 (2017)
81. Adamatzky, A., Huber, F., Schnauß, J.: Computing on actin bundles network. Sci. Rep. **9**(1), 1–10 (2019)
82. Mayne, R., Adamatkzy, A.: Cellular automata modelling of slime mould actin network signalling. Nat. Comput. **18**(1), 5–12 (2019)
83. Siccardi, S., Adamatzky, A., Tuszyński, J., Huber, F., Schnauß, J.: Actin networks voltage circuits. Phys. Rev. E **101**(5), 052314 (2020)

84. Dehmer, M., Emmert-Streib, F.: Quantitative Graph Theory: Mathematical Foundations and Applications. CRC Press, Boca Raton, London, New York (2015)
85. Lauwerier, H.A.: Applied Graph Theory (2014)
86. Rietman, E.A., Colt, J.Z., Tuszynski, J.A.: Interactomes, manufacturomes and relational biology: analogies between systems biology and manufacturing systems. Theor. Biol. Med. Model. **8**(1), 19 (2011)
87. Rietman, E.A., Karp, R.L., Tuszynski, J.A.: Review and application of group theory to molecular systems biology. Theor. Biol. Med. Model. **8**(1), 21 (2011)
88. Garlaschelli, D., Ruzzenenti, F., Basosi, R.: Complex networks and symmetry I: a review. Symmetry **2**(3), 1683–1709 (2010)
89. MacArthur, B.D., Sánchez-García, R.J., Anderson, J.W.: On automorphism groups of networks (2007). arXiv:0705.3215
90. MacArthur, B.D., Sánchez-, R.J., Anderson, J.W.: Symmetry in complex networks. Discret. Appl. Math. **156**(18), 3525–3531 (2008)
91. Rodriguez, L.: Automorphism groups of simple graphs (2014)
92. Babai, L.: Graph isomorphism in quasipolynomial time (2016). arXiv:1512.03547v2
93. Yaveroglu, O.N., Malod-Dognin, N., Davis, D., Levnajic, Z., Janjic, V., Karapandza, R., Stojmirovic, A., Przulj, N.: Revealing the hidden language of complex networks. Sci. Rep. **4**, 4547 (2014). https://doi.org/10.1038/srep04547
94. Fortnow, L.: The golden ticket: P, NP, and the search for the impossible. Princeton University Press, Princeton (2013)
95. Weisstein, E.: Graph automorphism (2008). http://mathworld.wolfram.com/GraphAutomorphism.html
96. Rietman, E.A., Platig, J., Tuszynski, J.A., Klement, G.L.: Thermodynamic measures of cancer: Gibbs free energy and entropy of protein–protein interactions. J. Biol. Phys. **42**(3), 339–350 (2016)
97. Zenil, H.: Algorithmic Data Analytics, Small Data Matters and Correlation versus Causation (2013). arXiv:1309.1418
98. Zenil, H., Kiani, N.A., Tegnér, J.: Methods of information theory and algorithmic complexity for network biology. In: Seminars in Cell & Developmental Biology, pp. 32–43. Elsevier (2016)
99. Zenil, H., Kiani, N.A., Tegnér, J.: Quantifying loss of information in network-based dimensionality reduction techniques. J. Complex Netw. **4**(3), 342–362 (2016)
100. Zenil, H., Soler-Toscano, F., Dingle, K., Louis, A.A.: Correlation of automorphism group size and topological properties with program-size complexity evaluations of graphs and complex networks. Phys. A **404**, 341–358 (2014)
101. Chapman, A.: Semi-Autonomous Networks Effective Control of Networked Systems Through Protocols, Design, and Modeling. Springer International Publishing, Cham (2015)
102. Chapman, A.: Controllability and observability of Cartesian product networks. In: Semi-Autonomous Networks. Springer (2015)
103. Chapman, A., Mesbahi, M.: State controllability, output controllability and stabilizability of networks: a symmetry perspective. In: 2015 IEEE 54th Annual Conference on Decision and Control (CDC), pp. 4776–4781. IEEE (2015)
104. de Badyn, M.H., Chapman, A., Mesbahi, M.: Network entropy: a system-theoretic perspective. In: 2015 IEEE 54th Annual Conference on Decision and Control (CDC), pp. 5512–5517. IEEE (2015)
105. Albert, R., Jeong, H., Barabási, A.-L.: Error and attack tolerance of complex networks. Nature **406**(6794), 378–382 (2000)
106. Bak, J.H.: Error and attack tolerance of scale-free networks: effects of geometry. Korea Advanced Institute of Science and Technology (2010)
107. Crucitti, P., Latora, V., Marchiori, M., Rapisarda, A.: Error and attack tolerance of complex networks. Phys. A **340**(1), 388–394 (2004)
108. Matta, J., Borwey, J., Ercal, G.: Comparative resilience notions and vertex attack tolerance of scale-free networks (2014). arXiv:1404.0103

109. Schieber, T.A., Carpi, L., Frery, A.C., Rosso, O.A., Pardalos, P.M., Ravetti, M.G.: Information theory perspective on network robustness. Phys. Lett. A **380**(3), 359–364 (2016)
110. Nie, T., Guo, Z., Zhao, K., Lu, Z.-M.: The dynamic correlation between degree and betweenness of complex network under attack. Phys. A **457**, 129–137 (2016). https://doi.org/10.1016/j.physa.2016.03.075
111. Manzano, M., Sahneh, F., Scoglio, C., Calle, E., Marzo, J.L.: Robustness surfaces of complex networks. Sci. Rep. **4**, 6133 (2014). https://doi.org/10.1038/srep06133
112. Frost, J.J.: Cancer's intelligence. Int. J. Unconv. Comput. **16**(1), 41–78 (2021)
113. Dennett, D.: From Bacteria to Bach and Back: The Evolution of Minds. Springer Science & Business Media (2018)
114. Wissner-Gross, A.D.: A new equation for intelligence (2013)
115. Wissner, A.D., Freer, C.E.: Causal entropic forces. Phys. Rev. Lett. **110**(16), 1–5 (2013). https://doi.org/10.1103/PhysRevLett.110.168702
116. Gardner, H.: Multiple Intelligences, vol. 5. Minnesota Center for Arts Education (1992)
117. Ben-Jacob, E.: My encounters with bacteria–learning about communication, cooperation and choice. Phys. Biol. **11**(5), 053009 (2014). https://doi.org/10.1088/1478-3975/11/5/053009
118. Ben-Jacob, E., Coffey, D.S., Levine, H.: Bacterial survival strategies suggest rethinking cancer cooperativity. Trends Microbiol. **20**(9), 403–410 (2012). https://doi.org/10.1016/j.tim.2012.06.001
119. Ben-Jacob, E., Cohen, I.I., Shochet, O., Tenenbaum, A., Czirok, A., Vicsek, T.: Cooperative formation of chiral patterns during growth of bacterial colonies. Phys. Rev. Lett. **75**(15), 2899–2902 (1995). https://doi.org/10.1103/PhysRevLett.75.2899
120. Kauffman, S.: At home in the Universe. Oxford University Press, New York, Oxford (1995)
121. Zhabotinsky, A.M.: Belousov-Zhabotinsky reaction. Scholarpedia (2007)
122. Adamatzky, A., Akl, S., Burgin, M., Calude, C.S., Costa, J.F., Dehshibi, M.M., Gunji, Y.P., Konkoli, Z., MacLennan, B., Marchal, B., Margenstern, M., Martínez, G.J., Mayne, R., Morita, K., Schumann, A., Sergeyev, Y.D., Sirakoulis, G.C., Stepney, S., Svozil, K., Zenil, H.: East-west paths to unconventional computing. Prog. Biophys. Mol. Biol. **131**, 469–493 (2017). https://doi.org/10.1016/j.pbiomolbio.2017.08.004
123. Stepney, S., Rasmussen, S.: Computational Matter (2018)
124. Harding, S., Koutník, J., Schmidhuber, J., Adamatzky, A.: Discovering Boolean gates in slime mould. In: Stepney, S., Adamatzky, A. (eds.) Inspired by Nature: Essays Presented to Julian F. Miller on the Occasion of his 60th Birthday, pp. 323–337. Springer International Publishing, Cham (2018)
125. Vallverdú, J., Castro, O., Mayne, R., Talanov, M., Levin, M., Baluška, F., Gunji, Y., Dussutour, A., Zenil, H., Adamatzky, A.: Slime mould: the fundamental mechanisms of biological cognition. Biosystems **165**, 57–70 (2018). https://doi.org/10.1016/j.biosystems.2017.12.011
126. Adleman, L.: Molecular computation of solutions to combinatorial problems. Science **266**(5187), 1021–1024 (1994)
127. Ignatova, Z., Martínez-Pérez, I., Zimmermann, K.-H.: DNA Computing Models. Springer (2008)
128. Nicolau, D.V., Jr., Lard, M., Korten, T., van Delft, F.C., Persson, M., Bengtsson, E., Mansson, A., Diez, S., Linke, H., Nicolau, D.V.: Parallel computation with molecular-motor-propelled agents in nanofabricated networks. Proc. Natl. Acad. Sci. USA **113**(10), 2591–2596 (2016). https://doi.org/10.1073/pnas.1510825113
129. Baumgardner, J., Acker, K., Adefuye, O., Crowley, S.T., Deloache, W., Dickson, J.O., Heard, L., Martens, A.T., Morton, N., Ritter, M., Shoecraft, A., Treece, J., Unzicker, M., Valencia, A., Waters, M., Campbell, A.M., Heyer, L.J., Poet, J.L., Eckdahl, T.T.: Solving a Hamiltonian path problem with a bacterial computer. J. Biol. Eng. **3**, 11 (2009). https://doi.org/10.1186/1754-1611-3-11
130. Crutchfield, J.P.: The origins of computational mechanics: a brief intellectual history and several clarifications (2017). arXiv:1710.06832
131. Crutchfield, J.P., Ditto, W.L., Sinha, S.: Introduction to focus issue: intrinsic and designed computation: information processing in dynamical systems—beyond the digital hegemony. Chaos **20**(3), 037101 (2010). https://doi.org/10.1063/1.3492712

132. Crutchfield, J.P., Ellison, C.J., Riechers, P.M.: Exact complexity: The spectral decomposition of intrinsic computation. Phys. Lett. Sect. A: Gen. At. Solid State Phys. **380**(9–10), 998–1002 (2016). https://doi.org/10.1016/j.physleta.2016.01.008
133. Crutchfield, J.P.: Between order and chaos. Nat. Phys. **8**(1), 17–24 (2011). https://doi.org/10.1038/nphys2190
134. Görnerup, O., Crutchfield, J.P.: Primordial evolution in the finitary process soup. In: Physics of Emergence and Organization< pp. 297–311. World Scientific (2008)
135. Roli, A., Villani, M., Filisetti, A., Serra, R.: Dynamical criticality: overview and open questions. J. Syst. Sci. Complex. **31**(3), 647–663 (2018). https://doi.org/10.1007/s11424-017-6117-5
136. Shalizi, C.R., Crutchfield, J.P.: Computational mechanics: pattern and prediction, structure and simplicity. J. Stat. Phys. **104**(3–4), 817–879 (2001)
137. Fumiã, H.F., Martins, M.L.: Boolean network model for cancer pathways: predicting carcinogenesis and targeted therapy outcomes. PLoS ONE **8**(7), e69008 (2013). https://doi.org/10.1371/journal.pone.0069008
138. Hu, Y., Gu, Y., Wang, H., Huang, Y., Zou, Y.M.: Integrated network model provides new insights into castration-resistant prostate cancer. Sci. Rep. **5**, 1–12 (2015). https://doi.org/10.1038/srep17280
139. Poret, A., Guziolowski, C.: Therapeutic target discovery using Boolean network attractors: improvements of kali R. Soc. Open Sci. **5** (2017)
140. Nagaraj, S.H., Reverter, A.: A Boolean-based systems biology approach to predict novel genes associated with cancer: application to colorectal cancer. BMC Syst. Biol. **5**(1), 35 (2011)
141. Stratmann, A.T., Fecher, D., Wangorsch, G., Göttlich, C., Walles, T., Walles, H., Dandekar, T., Dandekar, G., Nietzer, S.L.: Establishment of a human 3D lung cancer model based on a biological tissue matrix combined with a Boolean in silico model. Mol. Oncol. **8**(2), 351–365 (2014)
142. Srihari, S., Raman, V., Leong, H.W., Ragan, M.A.: Evolution and controllability of cancer networks: a boolean perspective. IEEE/ACM Trans. Comput. Biol. Bioinf. **11**(1), 83–94 (2013)
143. Von der Heyde, S., Bender, C., Henjes, F., Sonntag, J., Korf, U., Beissbarth, T.: Boolean ErbB network reconstructions and perturbation simulations reveal individual drug response in different breast cancer cell lines. BMC Syst. Biol. **8**(1), 75 (2014)
144. Bowling, M., Burch, N., Johanson, M., Tammelin, O.: Heads-up limit hold'em poker is solved. Science **347**(6218), 145–149 (2015). https://doi.org/10.1126/science.1259433
145. Brown, N., Sandholm, T.: Superhuman AI for heads-up no-limit poker: libratus beats top professionals. Science **359**(6374), 418–424 (2018). https://doi.org/10.1126/science.aao1733
146. Brown, N., Sandholm, T.: Superhuman AI for multiplayer poker. Science **365**(6456), 885–890 (2019)
147. Moravčík, M., Schmid, M., Burch, N., Lisý, V., Morrill, D., Bard, N., Davis, T., Waugh, K., Johanson, M., Bowling, M.: DeepStack: expert-level artificial intelligence in heads-up no-limit poker. Science **356**(6337), 508–513 (2017). https://doi.org/10.1126/science.aam6960
148. Morgenstern, O., Von Neumann, J.: Theory of Games and Economic Behavior. Princeton university press, Princeton (1953)
149. Landauer, R.: Irreversibility and heat generation in the computing process. IBM J. Res. Dev. **5**(3) (1961). https://doi.org/10.1147/rd.441.0261
150. Landauer, R.: Information is physical. Phys. Today **44**(5), 23–29 (1991)
151. Lloyd, S.: Ultimate physical limits to computation. Nature **406**(6799), 1047–1054 (2000). https://doi.org/10.1038/35023282
152. Lloyd, S.: Computational capacity of the universe. Phys. Rev. Lett. **88**(23), 4 (2002). https://doi.org/10.1103/PhysRevLett.88.237901
153. Lloyd, S.: Programming the universe: a quantum computer scientist takes on the cosmos. Vintage (2006)
154. Lloyd, S.: The universe as quantum computer (2013). arXiv [quant-ph], https://doi.org/10.1111/tbj.12461

155. Di Pierro, M., Potoyan, D.A., Wolynes, P.G., Onuchic, J.N.: Anomalous diffusion, spatial coherence, and viscoelasticity from the energy landscape of human chromosomes. Proc. Natl. Acad. Sci. **115**(30), 7753–7758 (2018)
156. Khanna, N., Zhang, Y., Lucas, J.S., Dudko, O.K., Murre, C.: Chromosome dynamics near the sol-gel phase transition dictate the timing of remote genomic interactions. Nat. Commun. **10**(1), 1–13 (2019)
157. Lucas, J.S., Zhang, Y., Dudko, O.K., Murre, C.: 3D trajectories adopted by coding and regulatory DNA elements: first-passage times for genomic interactions. Cell **158**(2), 339–352 (2014)
158. Skiles, A.: Necessity. In: The Routledge Handbook of Metaphysical Grounding, pp. 148–163. Routledge (2020)

# Complementarity, Complexity and the Fokker–Planck Equation; from the Microscale Quantum Stochastic Events to Fractal Dynamics of Cancer

**Przemyslaw Waliszewski**

**Abstract** Tumourigenesis possesses no equivalent among known physical phenomena. It is initiated at the quantum level by thermodynamic fluctuations of macromolecules. Accumulation of non-lethal alterations in quasi-deterministic dynamic cellular network of genes and their regulatory protein elements facilitated by changes in microenvironment results in a weak emergence of non-complementary, malignant phenotype. The Weibull distribution of cancer incidence suggests that neuro-immuno-hormonal network modifies that process. Eucaryotic cells are supramolecular objects. They make use of quantum entanglement, quantum tunneling, coherence, and chirality in formation of novel molecular couplings with both multiple feedbacks, synergy, and hysteresis. Complementarity at each integration level and non-ergodicity are their distinguishing features. Quantum effects may contribute to the conjugated appearance of cancer mutations. Connectivity, that is, coupling between integration levels is associated with the emergence of at least three features: fractal geometry of space–time, in which growth occurs, conditional probability of events, which reduces sensitivity to the initial conditions, and entropy. The latter one determines both a capability of the supramolecular system for transfer of biologically relevant information and evolution of intercellular interactions. There is a limit for self-organization of cells into structures of higher order defined by the Fibonacci constant. A relationship between sigmoidal dynamics and the Feigenbaum diagram suggests that both growth and self-organization occur with parameters within the Mandelbrot set. The set of non-interacting, infiltrating cancer cells becomes topologically dense. It has the highest entropy. The global spatial fractal dimension approaches the integer value. Hence, the coefficient of cellular expansion is a novel quantitative measure of biological tumour aggressiveness. It is based on complexity of intercellular interactions. Neither biological complexity can be reduced to physical one, nor be fully mathematized. Computer simulations may help to elucidate details of tumourigenesis. The mathematical models should be expressed in the algebraic form of fractal sheaves and fractional equations.

---

P. Waliszewski (✉)
Department of Urology, Ameos Clinics, Landrat-Beushausen-Strasse 26, 31061 Alfeld, Germany
e-mail: complexityresearch@yahoo.com



I. Balaz and A. Adamatzky (eds.), *Cancer, Complexity, Computation*,
Emergence, Complexity and Computation 46,
https://doi.org/10.1007/978-3-031-04379-6_2

## 1 Introduction

Multicellular spatial transformations underlie both morphogenesis and tumourigenesis. Both proliferation and migration of cells may proceed in the similar timescale. Morphogenesis comprises teleological organization of cells into specialized tissues and organs as well as their complementary integration into one organism. Tumourigenesis results in formation of tissue with both distorted spatial organization of cells and non-teleological function. A variety of molecular aberrations modifies intra- and intercellular feed-back interactions so much that transformed cells loose polarity, cease local cooperation, evade apoptosis, and grow autonomously. They escape the neuro-immuno-hormonal surveillance owing to both damage at different levels of intraorganismic integration and immunological malfunction. Tumour progression ends up in metabolic failure in different organs and death of multicellular organism.

In response to environmental factors or physical forces, tissues can disintegrate or counterbalance the perturbation. In that latter case, molecular, cellular, or neuro-immuno-hormonal mechanisms will be activated. Organism reacts with alterations in circulation, inflammation, degeneration, or tumourigenesis [1]. Tumourigenesis is a complex, non-complementary, temporo-spatial, multistep tissue phenomenon. It seems to be initiated by alterations in cellular dynamic network of genes and their regulatory protein elements that occur in a specific kind of cells called stem cells. Those cells differentiate towards normal phenotype and renew a tissue of origin. In response to chronic perturbations, stem cells may undergo a multistep transformation towards cancer stem cells; a source of malignant phenotypes present within a malignant tumour [2, 3]. Neither cancer cells can appear as a result of de-differentiation in short living matured cells [4], nor that process is just a phase transition seen in physical systems [5]. While phenomena that resemble phase transition, such as the GTPase-controlled dynamic chain reaction of G-actin-ATP polymerization do occur in cells, both the compact and teleological organization of macromolecules excludes such simple scenario [6]. In addition, some steps of phenotype transformation are irreversible owing to both accumulation of molecular defects in cellular dynamic network and their transfer to the next cell generation. In parallel to transformation of stem cells into cancer stem cells the primary complementarity of the multilevel interactions within the organism undergoes alterations owing to aging or environmental factors, such as radiation, smoking, alcohol, carcinogens, or oncoviruses. Those changes occur at different integration levels, what results in both malfunction of neuro-immuno-hormonal system and natural selection of cell phenotypes. In consequence, malignant cells progress and form metastases eventually. Transformed cells form initially geometric patterns as normal cells do. Their growth is correlated, and temporo-spatial intercellular associations persist over some period. Pattern formation is the essence of emergence and self-organization, that is, a spontaneous formation of globally coherent pattern out of local interactions. On one hand, self-organization of cells into tissue structures of the higher order, such as glands is limited by the Fibonacci constant (see Appendix 9.5). On the other hand, diverse patterns formed by cancer cells evolve until they attain both maximal complexity and maximal entropy. At that stage, malignant cells cease intercellular interactions and metastasize spontaneously following those two principles that seem to determine the natural course of disease [7].

## 2 From Quantum Events to Phenotype Transformation

All macromolecules, either proteins, RNAs or DNA, undergo spontaneous deterministic chaotic oscillations at the quantum level caused by physical forces whose average value in time is zero [8]. For example, aberrant activities of various proteins including DNA polymerases, DNA methyltransferases, DNA demethylases, ATP-dependent chromatin remodelers, or histones were observed in response to the fluctuations of energy. All those alterations occur in parallel and may contribute to the phenotype transformation towards malignancy. Those thermodynamic fluctuations occur in response to a variety of energy impulses, such as heat, acoustic or electromagnetic waves, radiation with high-energy particles alpha, ultraviolet radiation, X-rays etc. It is well-known now that DNA mutations are initiated as quantum jumps [9, 10]. A hydrogen bond joins base pairs in DNA. There exists a double well potential along a hydrogen bond separated by a potential energy barrier. The double well potential is asymmetric with one well deeper than the other, so the proton normally rests in the deeper well. For a mutation to occur, the proton must have tunneled into the shallower of the two potential wells undergoing a tautomeric shift, that is, a move from one position in pyridine or pyrimidine to another one. If DNA replication takes place in that critical, instable state, the base pairing rule for DNA may be jeopardized causing a mutation [11]. In the presence of heavy metal ions, radicals, or chemical compounds, fluctuations stabilize and will be fixed in DNA structure as various gene or chromosome aberrations [12]. DNA sequence can also change owing to cell infection with some viruses with oncogenic potential, such as human papilloma virus (HPV), Epstein-Barr virus (EBV), hepatitis B virus (HBV), hepatitis C virus (HCV), human T-lymphotrophic virus (HTLV) or human immunodeficiency virus (HIV) [13]. Some of gene mutations may already be present in germline cells and are of hereditary nature.

DNA mutations occur with frequency $10^{-6}$/cell divisions (reviewed in [14]). In addition, errors in DNA synthesis during replication or repair occur at random and in parallel in different somatic stem cells [15]. Those errors lead to DNA damage and occur with a frequency of $10^4$–$10^6$/cell/day [16]. The frequency of gene mutations can even be increased in the synergistic manner if phenotype gets growth advantage over the non-mutated cells during the process of natural selection [17]. Fortunately, only about 2% of all DNA defects occur in exons, that is, in the gene parts that are transcribed to messenger RNA to proteins. Exon sequences are rich in Alu elements, CpGs islands, and (G+C) content. They possess varied multifractality and histogram entropy as measured by the chaos-game representations of gene structure [18, 19]. Since majority of DNA defects occur in the non-coding regions of DNA with low or no multifractality, those defects are neutral or lethal for normal-appearing stem cells. They do not participate in appearance of malignant phenotype.

Proteins with altered structure are usually eliminated from cells unless the areas modified by mutations are covered by the rest of molecular structure; the case best exemplified by the dimerized steroid receptors and their hydrophobic domains [20]. If alterations appear in specific locations of protein macromolecules [21], or if a novel macromolecular structure confers benefit favored by natural selection of cells, or if their presence initiates some processes, such as lysine acetylation, leading to activation of transcription factors [22], cell phenotype may change irreversibly. In parallel, aberrant DNA methylation activates or inactivates various genes. Importance of DNA methylation for development of normal tissues can be best seen during embryogenesis or aging. That process operates on unchanged DNA sequence of CpGs that are heavily methylated in embryonic stem cells (MESC). It is controlled by some transcription factors, such as Oct4, Sox2, Klf4, and c-myc [23]. Hypomethylation of the MESC sequences may occur together with hypermethylation at stem cell PolyComb Group Target genes (PCGTs) three years in advance of the first morphological neoplastic changes. The MESC hypomethylation progresses from the stage of primary carcinoma up to the metastatic stage [24]. If the process of DNA hypermethylation occurs in genes with altered sequence or in their promoters, cell shape, cell adhesion, or rate of cell division may change. For example, methylation of $\varepsilon$-cadherin promoter decreases expression of that adhesion molecule on surface of cancer cell and enables both a change of cell shape and its migration [25].

Appearance of malignant phenotype relies not only upon some critical DNA alterations. Subcellular organelles, such as mitochondria, lysosomes or endoplasmic reticulum also contribute to the appearance of the novel phenotype. It is impossible to describe all changes that occur in those organelles in details. Briefly, calcium-dependent stimulation of three matrix-located key enzymes, that catalyze reactions upstream of or within the Krebs cycle, regulates the entire cell metabolism. The calcium uptake by mitochondria takes advantage of their close positioning to the calcium-releasing channels of endoplasmic reticulum. There is a system of specialized proteins that sustain the calcium exchange between those organelles and activate apoptosis in normal cells if mitochondrial calcium load is excessive. That precise, fine regulation will be replaced during tumourigenesis by the hyperactivation state associated with appearance of some additional biochemical pathways. Those pathways regulate synthesis of reactive oxygen species, modify cellular metabolism, influence energy production, stimulate cell proliferation and promote tumourigenesis [26–28]. Endoplasmic reticulum is involved in protein, lipid, and steroid synthesis. It serves as a transportation system engaged in protein folding conforming to a nontrivial minimal surface, that is, a surface that locally minimizes its area according to surrounding's pressure [29]. Endoplasmic reticulum possesses a fine-tuned transmembrane luminal surveillance system composed of a circuit of three proteins called unfolded protein response. That system activates in response to a buildup of misfolded proteins known as endoplasmic reticulum stress. Since that system requires ATP as the energy source, mitochondria undergo changes to increase both their number around endoplasmic reticulum and contacts with that structure. Some transcriptional and translational measures as well as apoptotic signaling will be used to restore the balance. The excessive gene activation, increased protein secretion, hypoxia, oxidative stress, growth signals, inadequate amino acid supplies, glucose deprivation, and lactic acidosis that occur during tumourigenesis also end up in both endoplasmic reticulum stress and hyperactivation of the system of unfolded protein response. The hyperactivated system sends then signals to both the autophagy system and cytoskeleton. That latter action facilitates cytoskeleton remodeling and the epithelial-to-mesenchymal phenotypic transition; a prerequisite condition for metastasis formation [30]. Malignant cells have different ratio of F-actin to G-actin, possess altered cytoskeleton or lose it, what results in disappearance of the "lumen-basement membrane" polarity. Interestingly, ATP synthesis occurs in transformed cells through the low-efficient, faster anaerobic glycolysis. Molecular events that drive aberrant proliferation of malignant cells upregulate glycolysis and downregulate pyruvate mitochondrial flux without impairing mitochondrial function (Warburg effect) (reviewed in [31]). In consequence, cancer cells have both up to 17-fold less ATP and lower electric transmembrane potential as compared to their normal counterparts. Also, they distribute energy among a lower number of degrees of freedom, what implies lower stability of growing tumour tissue in comparison with the normal counterpart [32]. However, cancer cells continue to exploit oxidative phosphorylation to produce intermediates for enhanced nucleotide synthesis using other substrates, alternative to pyruvate, such as glutamate and fatty acids.

While all the above-mentioned changes that occur at the microscopic scale also enable a continuous flow of information, energy, or matter typical of normal cells, malignant phenotype, characterized at the macroscopic scale by autonomous and immortalized growth, is non-complementary with the other tissues of multicellular organism. Owing to interactions with microenvironment, that also undergoes some critical changes (reviewed in [33]), cancer cells diversify rapidly into subpopulations with different dynamics of growth. Tumour tissue, that emerges, loses the teleological character of its normal counterpart. In particular, the ratio between the surface area and volume of cells has an enormous impact on their behavior, functions, and heat exchange with environment. While phenotypically normal cells can achieve a large ratio of surface area to volume with an elaborately convoluted surface, like the microvilli lining the small intestine or internal branching of bronchial tree, characterized well by the global spatial fractal dimensions, disappearance of some specialized adhesion molecules either in external cell membrane of cancer cells or in their mesenchymal microenvironment results in shape modification [34]. For example, metastatic cancer cells lose their fractal borders. They become small, spheroidal objects. DNA synthesis dominates in those cells. Both the ratio of cytoplasm to nucleus and the ratio of surface to volume decrease. The latter one approaches the limit typical of the sphere, that is, 3/r, where r stands for the sphere radius. Since a low surface area to volume ratio is a strong inhibiting force for thermodynamic processes that minimize free energy, those processes slow down. Those cells synthesize fewer functional proteins, and, therefore, are not able to perform many biological activities typical of their normal counterparts.

## 3 The Quantitative Measures of Biological Tumour Agressiveness

Cancer cells form a colony, a kind of autonomous cellular dynamic system. Those cells develop first a network of tunneling nanotube connections within the colony spanning hundreds of microns. Those nanotube connections enable intercellular transport of ions or subcellular organelles, such as mitochondria and facilitate the intracellular co-operation [35]. Cancer cells interact also with normal cells in their environment, such as endothelial cells and fibroblasts. Those interactions initiate both angiogenesis, that is, formation of blood vessels and metastasis formation [36]. Angiogenesis within a colony of cells with malignant phenotype enables much more effective circulation of oxygen and substrates in that tissue system, what increases its dynamics of growth. That colony can also increase in size. Cancer cells undergo epithelial-to-mesenchymal phenotypic transition and spread over the body and colonize the distant organs as metastases. Genetic analysis revealed that first metastases possess a similar profile of gene mutations as primary carcinomas. However, secondary metastases, that originate from the primary ones, contain cells from different clones with the mixed profiles of gene defects [37]. The spatio-temporal evolution of colonies of cancer cells as well as their invasion seems to underlie the Allee principle. The Allee principle relates population size with its growth rate [38]. Metabolic alterations initiated in the local lesion at the level of subcellular organelles modify eventually intracellular metabolism, intercellular interactions, and influence multicellular organism composed of 3–4 $\times$ $10^{13}$ cells. All those events determine the course of tumourigenesis as a disease.

Malignant tumour is a local tissue structure composed, in the case of carcinomas, not only of epithelial cells with malignant phenotype, but also fibroblasts, muscle cells, and blood vessels. Evolution of geometric patterns formed by cancer cells proceeds from highly ordered structures with both low complexity and low entropy, such as glands to structure with no organization with both high complexity and entropy, such as cell infiltrates. Since a digitalized image is a snapshot of some mathematical function of two spatial variables that determine localization of cancer cells, tumour structure can be measured objectively, unequivocally, and in a synthetic manner by a set of complexity, diversity and homogeneity measures. Those measures define the quantitative system for stratification of prostate cancer patients into classes of equivalence called also classes of complexity [39–42].

The application of the quantitative measures revealed some fundamental facts. First, natural evolution of the histological tumour patterns is associated with the increase in both entropy and complexity [39]. That process is described by the fundamental relationship between the global spatial information fractal dimension $D_1$ and entropy. Malignant cells change the epithelial phenotype into the mesenchymal one using a specialized genetic circuit composed of two interconnected chimeric modules and metastasize [43]. It is important to notice that metastasis formation occurs at the maximal values of both entropy and global as well as local fractal dimensions [7].

Second, there is a quantitative limit for self-organization of cells, either normal or malignant, into tissue structures of the higher order such as glands. That limit is defined by the Fibonacci constant [7] (see Appendix 9.5). Intercellular interactions can be characterized quantitatively by a novel measure called the coefficient of cellular expansion. That coefficient is a function of the global spatial fractal dimension $D_0$ that characterizes a dynamic process occurring within the surrounding Euclidean space with the integer dimension [44]. More specifically, it is defined as a ratio of the fractal dimension to the Euclidean dimension minus one. That function allows a classification of dynamic systems into three categories. If the fractal dimension equals the Euclidean dimension, then the system is a non-interacting one, such a neutral gas closed in a container. If the fractal dimension is lower than the Euclidean dimension, then the system is an interacting one. If the fractal dimension is greater than the Euclidean dimension, then the system has a fractal memory, that is, stores information about the former events.

If the global fractal dimension is different from the Euclidean integer dimension, connectivity between the interacting cells, that is, interconnectedness which denotes the existence of complex, dynamic relationships in a population of cells leading to the spatial and temporal emergence of global features in the system that would never appear in a single cell existing out of the system, increases. Then, the limit for the ratio of the integer Euclidean dimension of the space in which proliferation, growth and self-organization of cells occur to the global spatial capacity fractal dimension $D_0$ equals the Fibonacci constant $\Phi$. Accordingly, tumour aggressiveness can now be better described in terms of complexity of intercellular interactions rather than by an imprecisely defined similarity of tumour structure to the normal one. It can be measured quantitatively by the coefficient of cellular expansion with the frame of reference defined by the Fibonacci constant [7] (see Appendix 9.5). It is worth to notice that the above quantitative relationship indicates that both tumour formation and tumour expansion depend not only on the potential of interacting cancer cells but also on the support of available mesenchymal microenvironment [33, 45]. That conclusion supports also the course of anharmonic Morse-like potential associated with sigmoidal dynamics of tumour growth in early tumourigenesis, before angiogenesis takes place, and the interplay between drift and diffusion [46, 47] (see Appendices 9.1–9.3). Moreover, interactions within malignant tumour can be represented by a set of undirected graphs, with emerging holes characterized by the additional topological measures of homology, such as the local clustering coefficient or the Betti numbers, and analyzed by methods of algebraic topology. From that perspective, either normal or tumour tissue is represented topologically by a fractal sheaf [42].

## 4 The Weibull Distribution of Cancer Incidence

Morphogenesis or tumourigenesis represent the most dynamic natural phenomena, in which physical complexity meets biological complexity. Those phenomena comprise various microscopic processes including regular ones, that is, processes with dynamics that is predictable for any time-point, (e.g., sinusoidal glycolytic oscillations) and chaotic deterministic processes, that is, processes with dynamics that is unpredictable in time, but is characterized by certain regularities, (e.g., the existence of strange attractor for a biochemical reaction). The fundamental theorem of calculus, the Stokes theorem on manifolds states that a total change on the boundary of a dynamic system equals the sum of little changes on inside the system [48]. By analogy, the sum of local dynamic microscale processes in a cell or tissue, in which random events with probability distribution co-exist with the chaotic deterministic ones are integrated into a global, predictable, macroscopic dynamics.

The phenotype change depends on the co-existing alterations of both DNA sequence and methylation status as well as their influence on DNA multifractality, on activity of cellular network of genes and their regulatory elements, on the cell microenvironment, and on size and structure of cell population [14, 33]. The Vogelstein model assumes that colon tumourigenesis depends exclusively on the appearance of specific molecular defects in some limited number of critical, gate-keeper genes in response to random external events. According to that model, tumourigenesis is a deterministic linear evolutionary phenomenon [4]. Most people would have to develop colon cancer with the exponential distribution of incidence in a function of time (age). Yet, the incidence of colon or prostate cancer in a function of time is described by the Weibull distribution with the shape parameter larger than one [49, 50]. The incidence decreases after reaching its maximum at some age and some people will never develop cancer. The Weibull distribution of cancer incidence supports the view that tumourigenesis is a quasi-deterministic phenomenon determined early in life by an accumulation of both deterministic and non-deterministic molecular defects in proliferating eucaryotic cells [51] and influenced by both the local (microenvironment) [33] and global (neuro-immuno-hormonal system) networks [2, 3]. That accumulation may be facilitated by quantum effects leading to the conjugated appearance of changes in DNA sequence [11]. Natural selection of clones plays also important role [4]. Some of those defects are neutral for mutation spreading in a population of cancer cells. The neutral evolution occurs in about 30% of human cancers, such as colon or stomach one [52]. The other mutations are advantageous or unfavorable for proliferation of cells with defects. The spatial distribution of cells with gene mutations also plays some role. The structured spatial distribution of mutated cells offers the highest dynamics of cell proliferation versus the intermediate one for mass-action distribution or the lowest one for the hierarchical distribution [53].

tumourigenesis in human bladder provides a model that shows how a difference in a profile of molecular defects determines activation or inactivation of some important metabolic pathways and leads to a difference in patterns of macroscopic growth. For transitional cell carcinoma, two distinct genetic pathways have been identified (see for details [54]). The first one starts with aberrations of the fibroblast growth factor receptor (FGFR-3) gene that is responsible for recurrent low-aggressive superficial bladder cancer. The second one is associated with aberrations of p53 gene that leads to an ultimately highly aggressive muscle-invasive cancers. The origin of the aberrations in those pathways is thought to be both multifactorial and multifocal. On one hand, genetic, structural, and numerical chromosomal alterations accumulate and contribute to the biological progression of the disease. On the other hand, methylation of promoter regions of genes inactivates gene promoters, and leads to silencing of important genes. It is known that methyl groups can influence protein-DNA interactions without altering the DNA sequence or base pairing in the urothelium, even before neoplastic lesions can be seen. The frequency of those aberrations increases with disease progression. The polymorphic variations in the DNA sequence appear to be associated with the inherited transitional cell carcinomas. Finally, both molecular defects in genes controlling the programed cell death and DNA repair are particularly important for immortalization of transformed cells in any type of cancer. Mutations, genetic drift, and selection act on billions of cancer cells and their microenvironment leading to tumours' emergent behavior and tree-like cancer evolution with branching patterns of genomic alterations [55]. Cancer is a dynamic evolutionary entity with significant heterogeneity across primary and secondary tumours or patients with tumours of the same histological type. Subclonal populations of cancer cells with a variety of molecular defects interact each other in the antagonizing or synergizing manner. Those interactions play a role in emergence of resistance to therapy or influence patient's survival [51, 55].

The Weibull distribution of cancer incidence suggests that transformed cells of young patients grow and self-organize within a molecular network that possesses different integration than the old patients. Indeed, there is a clinically relevant difference in dynamics of cancer disease between a population of younger and older patients. That difference can be well-seen particularly in prostate cancer, where disease has usually greater dynamics and cancer is more aggressive in younger patients [49–51]. On one hand, the potential of individual cells for growth decreases with age. On the other hand, mutations do accumulate in all cells during aging. In addition, interactions with microenvironment influence phenotypic features of cancer cells [33]. The unchanged microenvironment facilitates both differentiation and self-organization of even highly malignant cells as was observed in experiments with metastatic prostate cancer cells LNCaP grown in chicken embryo (data unpublished). Genomic instability, telomere alterations, and microenvironment changes are aging-related processes. Even if some aging cells develop malignant phenotype after the host reaches certain critical age, they do not have enough potential to form an autonomous colony, and to metastasize. In fact, cells with normal-appearing phenotype in the senescent organism slow down dividing. Tissues undergo the process of atrophy characterized among other by reduction in both size of cells and their potential for both intercellular interactions and co-ordination. All those phenomena may explain the appearance of the Weibull distribution in cancer incidence.

## 5 Cell as a Complex Supramolecular Dynamic System

An insight into perhaps less well-known features of eucaryotic cells, cellular organelles, and macromolecules is necessary to comprehend both the origin and evolution of neoplastic diseases in the specific context of both biological complexity and complementarity. It should also be kept in mind that it took billions of years before supramolecular cellular systems were able to reach their optimum for both molecular self-organization and integration.

According to the classical biological theory formulated in 1839 by Mathias Jacob Schleiden and Theodor Schwann [56, 57], cells are the fundamental units that determine structure, organization and function of organisms. The autonomous character of cells is determined by two elements. First, semipermeable bilayer lipoprotein membranes separate cells from their environment. Matter is transported through those membranes by passive, that is, simple or facilitated diffusion, without the energy input, along the concentration gradient or by active diffusion using chemical energy, against the concentration gradient. Specialized proteins, such as ion channels, cell junctions, adhesion molecules, hormone receptors, and enzymes are anchored to those membranes. Using receptor proteins and their capability for reversible non-covalent binding, cells not only control intra- and intercellular transport of matter, but also respond to a variety of signals, regulate shape, and migrate [58]. Some of those receptors can also detect mechanical or electromagnetic forces. Then, the complexes ligand-receptor or receptors activated by physical energy transduce information along signalling pathways, what ends up first in a response of genes, then the regulatory protein network, and finally cells. Second, each cell contains hereditary genetic information necessary to initiate and to regulate all metabolic processes including process of cell division. The classical biological model of a cell assumes tacitly a one-directional flow of information from nuclear DNA through mRNAs to the functional protein level. That model does not, however, indicate multidirectional processes responsible for the temporo-spatial co-ordination of growth in multicellular organism, nor explains how cells regulate their size and shape [4]. Cells are objects that interact each other, co-operate, self-replicate, grow, differentiate, mature, adopt, migrate, self-organize into tissue structures, and self-eliminate (apoptosis). Metabolic networks of coupled exergonic, that is, spontaneous and endergonic, that is, non-spontaneous molecular reactions determine in a teleological manner capability of cells for carrying out chemical, electrical or mechanical work.

From the perspective of science of complexity, a living cell is a complex (i.e. emergent network of multiple cross-interacting elements sensitive to initial conditions, with multiple equilibria and morphological patterns), multidimensional, multiscale, non-linear (i.e. lack of the proportional relationship between input and the outcome), quasi-deterministic (i.e. the co-existence of deterministic and non-deterministic events), thermodynamically opened, non-ergodic, complementary supramolecular dynamic system. That system is in non-equilibrium and possesses both a multidirectional flow of information and a distributed rather than centralized control [4, 59]. That supramolecular system has numerous non-linear intra- or intercellular interactions spanning different levels of organization, both spatial and temporal scales, processes and phenomena [4, 8]. Non-linear interactions are the essence of both physical and biological complexity [60, 61]. Those interactions also determine the emergence of spatial or temporal fractal geometry seen in biochemical processes or tissues [62].

Eucaryotic cells emerged most likely owing to both evolution of intracellular intermolecular couplings and natural selection of phenotypes. Those objects are composed of various macromolecules organized in a specific, teleological manner. Molecular interactions do not occur in those systems at random. Interactions are not only non-linear, but also frequently synergistic, that is, the result is qualitatively unique and quantitatively larger than a simple sum of disconnected events. Feedback loops of non-linear interactions are responsible for growth or decay at the exponential rate. The coordination of interactions in time and in space occurs precisely in a decentralized manner with sensitivity to changes in intra- and extracellular conditions. Connectivity between elements of the complex system is the other pillar of complexity. It changes Euclidean geometry of time–space that will be now redefined by topology of networks with the global and local spatial or temporal fractal dimensions, degree of connectivity of elements and their spatio-temporal location. Connectivity enables self-organization of cells into structures of higher order such as glands responsible for both secretion and absorption. Despite of the hierarchical anatomy, all elements in the organism have interconnected each other.

It must be emphasized that eukaryotic cells are not perfect supramolecular systems. For example, they use the dimerized steroid receptors instead of the monomers to induce gene expression in nuclei. That dimerization has been forced by mutations encoding the hydrophobic domains of steroid receptors and represents no evolutionary benefit for cells [58]. Also, cells accumulated many viral DNA sequences that play no important role in metabolic activity [13]. Despite those marginal changes, cells and their macromolecules are sufficiently optimized for the effective self-organization into tissue structures and well-adapted to changes of environmental conditions. That optimization took about 4 billion years. For example, RNA molecules were not only able to select RNA pool starting from the initial random sequence chimeric oligonucleotides formed by untemplated polymerization and self-replicate [20], but also form ribozymes. Those are specialized RNA molecules with catalyst activity present in ribosomes engaged in both cleavage and ligation of RNA and DNA, as well as RNA splicing, tRNA synthesis, or peptide bond formation during protein synthesis [63, 64]. Enzymatic protein macromolecules present everywhere in cells catalyze complex biochemical reactions in a reproducible manner with high accuracy and efficacy at the minimal use of ATP energy [65].

## 6 The Unique Interactions in Supramolecular Cellular System: Quantum Entanglement, Tunneling, Coherence and Chirality

Specific features of macromolecules cause that eucaryotic cells are different from any known physical systems. Also, their transformation to malignant phenotype is more complex than just a simple phase transition in homogeneous physical systems, in which a single control parameter, such as temperature, volume or pressure can shift the system near the equilibrium to the new phase. This is because normal cells are characterized by highly complementary organization and fine-tuned functions. Tuning is usually done by specialized enzymatic systems. If the existing complementarity is broken, some altered macromolecules may create spontaneously novel molecular couplings or signalling pathways that continue to drive transformation to malignant phenotype.

Intracellular biochemical reactions are of teleological nature [66, 67]. Those reactions occur usually at the low amounts of reagents. Some genes, such as cyclins can be expressed as a few transcripts per cell. Nonetheless, the appropriate mRNAs encode enzymes that regulate important biological functions in both a selective and powerful manner during the short period of cell division. Furthermore, intracellular reactions occur in a limited volume. Substrates cannot gain the required activation energy due to the Brownian motions to overcome the energy threshold. Unlike in thermodynamics, where both passive diffusion along a gradient of concentration and mechanical collisions of molecules with sufficient energy leads to chemical reactions with rates dependent on temperature, concentrations, particle size, pressure and the presence of catalysts, enzymes assure a precise, fine-tuned transformation of substrates. Macromolecules of enzymes or receptors are usually attached to membranes as vector units, that is, a kind of nanochips, in which a product of one reaction serves as a substrate for the next one. Signalling pathways, such as G-protein-adenylate cyclase-cAMP-phosphorylase, or a process of mitochondrial ATP synthesis exemplify the reaction vector units [68–70].

Both the stereochemical capability of enzymes for three-dimensional redistributions of intramolecular van der Waals forces in a range of 0.1 nm [71–73] and chirality of molecules, that is, a geometric asymmetry that excludes superimposition of molecular structure on its mirror image by any combination of rotations and translations [74] underlay intracellular biochemical reactions. That spatial stereochemical recognition is so precise that cells can bind actively some substrates, such as glucose, an energy source, and inhibit absorption of its isomer, D-mannose. Some experimental data suggest that enzymes use both quantum entanglement [75] and proton tunneling [76] in their interactions with DNA or RNA. The phenomenon of entanglement occurs if at least a pair of particles interact in a way such that the quantum state of each particle of the pair cannot be described independently of the state of the other one, including the situation when the particles are separated by a large distance. That phenomenon implies novel, non-trivial understanding of biological reality including the Bell's concepts of superdeterminism and non-locality [77–79]. For example, collective quantum effects might play a role in control of cell size or in the conjugated appearance of cancer mutations.

Tunneling is a phenomenon in which a wavefunction representing the entire physical system propagates through a potential barrier. That phenomenon occurs usually in objects smaller than 1–3 nm [80]. Quantum interactions are not well visible among molecular cellular events at first glance. However, protein ion channels in cellular membranes use energy of chemical bonds in ATP molecules for tunneling and active transport of kalium, sodium, and calcium ions. Also, a system of cytochromes in mitochondria called the electron transport chain utilizes the electron tunneling phenomenon in the complex process of ATP synthesis that spans different scales of organization from the quantum level up to the organelle one [81]. It seems that macromolecular systems with quantum states perform quantum computations detecting all quantum states at once [82] and choosing the most optimal one as in the Grover algorithm [83].

On one hand, quantum effects emerge in molecules not larger than 2 nm, such as DNA or within small areas of protein macromolecules. Their surrounding vibrates, is noisy, and rich in water molecules. In that environment, both quantum decoherence and decay of quantum information occur quickly owing to the vibrational motions of atoms present in larger protein macromolecules of about 10 nm in size [84]. Perhaps the co-existence of the relatively small molecular areas, in which quantum effects may emerge, with the larger macromolecular structure that facilitates decoherence, enables a control over the emergence of the coherent states. Those states can then appear for a particular purpose, such as electron or ion transfer and disappear quickly. In that way, cells, cellular organelles, such as mitochondria, or cytoskeleton, that is, a network of interlinking protein filaments composed of microfilaments (actin polymers 7 nm in size), intermediate filaments (a variety of protein polymers 10 nm in size) and microtubules (tubulin polymers 25 nm in size) can easier adapt oneself to environmental signals by energy release, shape optimization, or cell division [85, 86]. Those events are associated with a change in cell structure mimicking a phase transition. That change depends, however, on a chain of complex biochemical reactions including a rapid synthesis of the specialized proteins. For example, protein LAF-1, electrostatically disordered protein, drives intracellular phase separation of P granules, RNA and protein assemblies, into liquid droplets, and is necessary for RNA-protein interactions [87]. On the other hand, organisms developed a complex network of the long-range neuronal, hormonal, and immunological mechanisms of the receptor-ligand type that co-ordinates effectively multiple activities in multicellular organism. Apparently, processes widespread in physical systems, such as diffusion or phase transition are not sufficient to ensure the effective transfer of both chemical signals and information in order to integrate the entire system. For example, simple diffusion is effective on short intracellular distances only. In some real biological settings, diffusion of an ion across a 100-cm-long neuron with a typical diffusion coefficient $D = 10^{(-5)}$ $cm^2/s$ would take about 10 years according to the relationship between the squared length of diffusion and the diffusion coefficient multiplied by two.

## 7 Complementarity

Despite of the impression evoked by anatomical studies, multicellular organism does not really operate according to the hierarchical organization of tissues and organs. The principle of complementarity describes better what happens during self-organization of cells. There exists a complex network of precisely regulated couplings that connect different elements, ranges, time scales, and levels of organization in the organism.

Those couplings ensure the optimal flow of information, energy and matter between cells, tissues, and organs as well as with environment starting from the quantum level. It took billions of years before multicellular structures reached that level of both complementarity and optimization. The whole produces unique combined effects. Majority of those effects, such as vision, speech, walking, hunting, thinking, playing music, mood, consciousness is co-determined by the interactions between the whole and its environment as well as by the biological context. Neither they can be explained with reference to both macromolecular and cellular features of the lower level of integration (strong emergence), nor be simulated in silico. An embryo that evolves from a single fertilized maternal cell incorporates different extracellular elements into its complex metabolizing cellular network. Cells migrate long distances, far greater than their linear sizes. They differentiate depending on location, hormonal influences, and functional operations. A gradual, discrete, and harmonious growth of the entire multicellular structure follows. Finally, central nervous system develops as a co-ordination center. However, its operations depend on both impulses and substrates provided by the peripheral organs, such as muscles, lungs, kidneys, or heart as well as electric impulses coming from the cellular level of integration. That whole is a non-ergodic object, that is, a dissipative dynamic system that does not fill up the available time–space evenly like diffusion or Brownian movement do. Furthermore, rules of the whole may influence the lower levels of integration. For example, a depressive mood may end up in sluggishness, obesity, and loss of muscle tissue as well as susceptibility to diseases. A coherent religious mood engages areas of brain responsible for social cognition, brings psychical comfort, increases pain tolerance, alerts immunological system and enhances resistance to infection or cancer [88].

Complementarity can be seen already at the level of quantum particles. For thermonuclear reaction to take place, a proton must be able to combine with a neutron to produce a deuteron, a gradually burnt fuel. Simultaneously, the same proton must be unable to combine with another proton to prevent the explosive course of the reaction. It is complementarity of the nuclear forces maintained in the narrow range that ensures the proper course of the reaction; either the weaker or stronger nuclear forces would inhibit proton-neutron interaction or facilitate the destructive proton-proton interaction [89].

Complementarity does not appear in the quasi-deterministic dynamic cellular network owing to randomness. Complementarity emerges in a pervasive manner along a process of self-organization of multicellular tissue structure. Multiple complex feedbacks develop to stabilize the growing organism. Complementarity is therefore the intrinsic feature of both complexity and connectivity in the self-organizing system. Moreover, the emerging time–space, in which all microscopic processes occur, gets fractal topology. On June 10, 1854, the important day in the history of mathematics, Bernhard Riemann gave a habilitation lecture at the Georg-August University of Göttingen, Germany. According to Riemann, geometry of time–space is much more than just the static arena for physical events [90]. Dynamics of the underlying phenomena influences the geometry of time–space and vice versa. By studying geometry, one can get information about both the underlying dynamics and complexity of the system. In the case of the interacting cellular systems, time–space emerges *in statu nascendi* in consequence of the multiple, nonlinear interactions within the non-ergodic system. During that process, time and space not only couple each other, but also get fractal structure [41, 46, 47, 62, 90] (see Appendix 9.3).

Stuart Kaufmann proposed a universal abstract scenario of self-organization for some autonomous objects, such as cells. Those objects utilize several autocatalytic cycles and search actively for energy sources. They can not only decrease or increase barriers of potential energy, but also couple both spontaneous and non-spontaneous thermodynamic reactions. In that way, cells do work by streaming energy in a teleological manner for both self-replication and self-organization. They also use that energy to construct some constraints for growth [91].

Cellular respiration is a good example of the precisely self-regulated, complex biological process that holds the principle of complementarity. Regulation of that process starts at the quantum level as a kind of the game between elemental particles, electrons and protons. Those fermions exchange energy and activate large protein complexes to transfer their energy gradually through the electron transfer chain up to the final products, ATP and oxygen radicals [68–70]. From the perspective of thermodynamics, synthesis of ATP from glucose is a complex process comprising both exergonic and endergonic steps with the total available Gibbs free energy of $\Delta G = -686$ kcal/mol. Would be that reaction an exclusively spontaneous process, energy of glucose bonds would be converted immediately into heat and light. Neither chemical energy of ATP could be collected for self-organization, self-replication or cell migration, nor any useful tissue activity could occur. Owing to coupling with the non-spontaneous reactions both ATP synthesis and ATP accumulation within cells are possible with efficiency of 40% at heat production of 60%. The process is initiated by cytoplasmic glycolysis, that is of oscillatory nature [92]. It continues as the intramitochondrial Krebs cycle. Cytochrome macromolecules are anchored to the internal mitochondrial membrane in the ordered manner. Electron transport occurs along the electrochemical potential gradient. Energy of electrons will be accepted by oxygen molecules to produce oxygen radicals. At that quantum fractal process, the ATP proton pump maintains the proton gradient across the mitochondrial membranes to ensure the counterclockwise direction of movement of the $\alpha$-subunit of the enzyme; an element crucial for ATP synthesis and its release [68, 69]. Mathematical modelling indicates that radical oxygen species are not just by-products of mitochondrial oxidative phosphorylation. That is a teleological process with fractal dynamics, in which radical oxygen species play a role of a molecular clamp. That clamp connects electron transfer at the quantum level with transformations of both oxygen and organic molecules engaged in the Krebs cycle [70]). All macromolecules involved in cellular respiration compose a complementary system. They possess a capability to oscillate between the non-equilibrium and equilibrium state as well as to recognize the infinitesimal, yet sufficient stereochemical changes in the three-dimensional structure of cytochromes, what results in accepting electrons, pumping protons and opening the internal ion channels. The entire process occurs at relatively narrow window of temperature, pH, and ion concentrations [93].

Eucaryotic cells as supramolecular systems are composed of a variety of interconnected subsystems with different functions, such as nucleus, nucleoli, mitochondria, ribosomes, lysosomes, membranes etc. Owing to mutual connections between those subsystems, any change in one of them will cause some changes in the other ones [94]. This may end up in a change of phenotype, that is a set of all features that a given cell type has. One should realize that phenotype is a dynamic category. It is phenotype, not genotype, that undergoes natural selection during the interactions of cells with environment and selects the most favourable patterns of gene expression. Since dynamic cellular network reacts to different environmental signals and transfers them down to the gene level, some novel traits may develop and be reinforced by natural selection; the statement best exemplified by metaplastic transformation of epithelial cells exposed to the unfavorable conditions existing in surgical conduits, fistulas, or chronic inflammation [95].

In general, depending on both cell type and their metabolic activity, expression comprises 10,000–80,000 genes [91]. The number of possible gene states in that network represents then a number $2^{80,000}$. However, genes are conjugated in a kind of teleological circuits that are activated or inactivated in response to environmental signals in a regular or chaotic modus. If the modus of gene activation/inactivation is regular, slow, and periodic, that is, if genes remain active or inactive for a longer period, then the number of such circuits in relationship to gene number in the entire network scales according to the polynomial function and equals square root of the number of genes, that is, about 282. Those circuits determine a limited number of phenotypes that not only compose important tissue structures, such as intestinal crypts, but also co-operate each other within those structures in a complementary manner. If genes are activated and inactivated in the chaotic deterministic regime, circuit number scales exponentially with the size of the network. In that extreme regime, each gene is a Boolean function of all genes in the network. Hence, the estimated number of circuits is represented by the gigantic number, the square root out of $2^{80,000}$ all possible gene states [91]. The latter scenario seems to be typical of cancer cells with multiple gene alterations in comparison with their normal counterparts. The number of those molecular defects changes especially in more advanced tumour stages, in patients undergoing chemo- or radiotherapy [96–99]. Those alterations are in part responsible for resistance of cancer cells to anticancer drugs that develops during therapy. Molecular changes during sequential therapy of metastatic prostate cancer with abiraterone followed by enzalutamide or vice versa cause that both progression-free survival and PSA reduction vary significantly [100].

It should be noticed that the principle "one gene-one phenotypic trait" is not a universal one. The relationship between the number of genes involved in the emergence of a phenotypic traits and the number of those traits is non-bijective. That means that one gene may co-influence the emergence of more than one trait, and one trait may be dependent on activation of many genes. G. Mendel has just assumed that phenotype is determined by genotype unequivocally and, therefore, subordinated to the latter one. Mendel needed that assumption to generalize results of the genetic experiments and formulate the Law of Mendelian Inheritance [101]. His assumption became a keystone of the deductive strategy of molecular reductionism in cellular biology. Phenomenology permits, however, an alternative epistemological model of a relationship between genotype and phenotype grafted in a more universal Husserl's philosophy [102]. Both categories, genotype and phenotype, represent here complementary aspects of the same common, complex entity. They are coupled through dynamic cellular network with multidirectional flow of information between its elements [4]. Probability of events in that entity depends on both up- and down-regulation of genes. It is also dependent on gene mutations as well as time. That probability underlies the law of generalized diffusion, in which probability possesses algebraic shape of Boltzmann-type equilibrium [103]. In fact, the extraordinary equilibrium and complementarity of molecules, factors, or physical forces can be seen in all complex cellular biological phenomena. Any imbalance leads here to both functional alterations and structural deformations.

In the above-discussed context, cancer cells appear in the highly organized, complementary, multicellular organism owing to the local intracellular accumulation of diverse molecular defects. Those defects lead to the deterministic chaotic rather than regular way of activation or inactivation of genes predicted by Stuart Kaufmann. Profiles of gene expression in each tumour or at each stage of cancer disease as well as molecular defects can be different. Those findings were confirmed in experimental studies [18, 19, 51, 96–99]. The appearance of some unique metabolites activates both novel molecular couplings and metabolic pathways [54, 96, 97]. This alters gradually cellular network and ends up in the weak emergence of malignant phenotype. Although cancer cells interact each other and may also interact to some degree with normal cells, they do not self-organize into the complementary tissue. Owing to excretion of lytic enzymes, cancer cells cause destruction of their microenvironment [104]. Metastasis formation appears to be the ultimate purpose of cells with malignant phenotype. It is associated with increment of both entropy and complexity in tissue system [7, 39]. According to the Bekenstein Bound, the amount of information present in the finite fractals representing the idealized distribution of cancer cell nuclei in adenocarcinomas is limited, and changes according to the ratio 1:1.94:2.64. A similar relationship can be seen in changes of entropy in the function of the global fractal dimensions (the coefficient of correlation 0.904). Entropy determines the natural course of cancer disease [105].

## 8 Fractal-Probabilistic Dualism and Fractal Time–space

The first theoretical considerations as to the origin of frequency distributions in biology suggesting probabilistic course of cellular phenomena were published in the beginning of the XX century [106]. It is well accepted now that intracellular expression of genes, mRNAs or proteins is of stochastic nature. In other words, there exists a probability function, which describes distribution of intensity of gene expression in a single cell [103, 107, 108]. Majority of intracellular microscale processes are conditionally interconnected in both fractal space and time. In contrast to some physical systems, such as industrial chemical reactions or thermodynamic processes, intra- and intercellular interactions are not random. Those interactions are either deterministic chaotic, that is, they occur at the interface of randomness and order, or regular one [8]. Those deterministic chaotic phenomena are not seen at first glance. Growth of cells and tissues at the macroscale is usually predictable. In the physical thermodynamic phenomena mentioned above, random microscale interactions end up in the emergence of a predictable macroscale dynamics with the Maxwell–Boltzmann probability distribution of energy or particle speed. That distribution of probability appears in a system at equilibrium with maximal entropy and fixed energy. Similarly, if the cell number is constant in a given interval of time, each cell produces chemical energy of ATP, and transfers some of that energy to the other cells, than the Maxwell–Boltzmann probability distribution function describes also that process. In general, there is some special flow of information between the microscopic and macroscopic level of organization that excludes randomness at the latter one. In particular, the above-mentioned quantum molecular transformations exclude randomness in majority of intra- and intercellular phenomena replacing it by some equilibrium of regular and deterministic chaotic events in fractal geometric space. Hence, the emerging mathematical description of that kind of phenomena at the interface of micro- and macroscale is fractional calculus, not just classical differential calculus [109]. The principle of probability distribution implies that whatever events occur at the microscale molecular level, intercellular interactions as well as global dynamics remain relatively stable and evolve in a similar manner. For example, each malignancy engages different molecular events for both promotion of tumour growth and tissue invasion. Those differences in molecular mechanisms can be seen in malignancies originating from the same tissue, such as breast or prostate and undergoing both a temporal and spatial evolution or in malignancies originating from closely related tissues, such as bladder and ureter epithelium [110–113]. Therefore, molecular biological models of tumourigenesis are not universal. However, malignant cells share many morphological similarities. In the extremal case, morphology of non-differentiated cancer cells is identical independently of tumour origin.

Molecular cellular processes are both non-linear and bi-stable or multistable. Those processes possess discrete states, may reveal oscillations of both substrates and products or can adjust their dynamics according to the memory of the input strength. In other words, the same input may release two different, molecularly stable outputs. That kind of dynamics is described as hysteresis phenomenon. In the mathematical sense, hysteresis phenomenon appears if a multi-branch operator transforms the extreme values of the input into branch transitions. Hysteresis was identified in a broad class of cellular phenomena, such as protein folding [114], DNA-protein interactions [115], signaling with bistable stages [116]. For example, some enzymes, such as caspase, catalyze the same reaction into different products depending on physical parameters of the intracellular environment [117]. Concentrations of cyclins in cells undergoing division follow a hysteresis curve. In consequence, there is required a much higher concentration of cyclins during the shift from the G2 phase to mitosis than during mitosis itself [118, 119]. Also, binding of oxygen to hemoglobin and its dissociation into tissues is an example of the hysteresis phenomenon. Those allosteric reactions are also synergistic, that is, binding of oxygen molecules to hemoglobin is facilitated the more oxygen atoms bind to one hem molecule [120].

According to Corning's hypothesis, synergistic effects drive cooperative relationships within supramolecular cellular system providing functional advantages in relation to survival and reproduction that have been favoured by natural selection. Indeed, saturation curves for hemoglobin and for myoglobin are sigmoidal and hyperbolic, respectively. That means that at low oxygen pressure, for example, in peripheral tissues, hemoglobin has a large capacity for binding with oxygen while myoglobin is already saturated in 100%. This situation exemplifies teleological character of both macromolecules. Hemoglobin serves as the oxygen carrier. Myoglobin is the oxygen reservoir for muscle cells. It is worth to notice that supramolecular synergistic systems may react in a non-linear way to perturbations, such as mutations, so that the outcome may be greater than the sum of the individual component alterations [121]. The last, but not least, hysteresis stands behind a feedback relationship. That kind of relationship underlies any normal or pathologic regulatory process with exponential dynamics. It enables adaptation of cellular systems in changing environment including self-organization of cancer cells.

Sigmoidal dynamics deserves more detailed consideration owing to its universal character. It is typical for many molecular cellular phenomena including the relationship between a signal and a response of molecular cellular network as well as gene expression induced or inhibited by a variety of transcription factors. In general, that dynamics emerges if some inhibiting force appears in the interacting system of elements as a regulatory constraint of the process [46, 47, 122]. For example, that kind of response can be observed during enzymatic reactions, such as DNA synthesis or during some cellular processes, such as monocytes activation during inflammation. In those processes, sigmoidal dynamics emerges because of inhibition by products [123]. That kind of dynamics can also be seen during proliferation of cancer cells in culture in vitro or in small animals (reviewed in [46, 47, 122]), but not in larger organisms, such as humans. In that latter case, cancer cells proliferate or decay exponentially. Cells growing rapidly in culture or in small organism have both limited resources and space, what leads to inhibition of proliferation. Their growth is exponential in the first phase, then comes to inhibition, and cell population reaches plateau. Since larger organisms possess more resources, and much larger fractal space available for expansion, sigmoidal dynamics does not emerge. The sigmoidal dynamics can be described mathematically by the logistic function or by the Gompertz function [46, 47, 122, 124]. Interestingly, if the continuous variable of time will be replaced by the discrete steps, both functions can be transformed to the same Feigenbaum diagram [47]. That diagram reveals the co-existence of both deterministic chaos and order during the temporal evolution of cellular system [46]. Both functions reflect the equilibrium of regular states and chaotic states with the same Feigenbaum constant. Both that equilibrium and volume of the available complementary Euclidean space determine temporal and spatial expansion of a process with sigmoidal dynamics, such as tumour growth [44, 46]. That latter finding indicates the importance of microenvironment for tumourigenesis. The Gompertz function is also a solution of the Markovian model of cellular growth and is related to the entropy function [125] (see Appendix 9.4). In consequence, mathematical description of the process of growth requires the Fokker–Planck equation. That equation reveals that growth of multicellular structure, such as cancer depends on two main forces, that is, drift and diffusion. That relationship can be best seen in the Morse-like anharmonic potential associated with the sigmoidal dynamics function. The potential function is a solution of the special Schrödinger-like differential operator [47] (see Appendix 9.1). It reveals that drift plays a dominant role in the first stage of growth. Diffusion drives growth after the potential function reaches its minimum. At that stage, interactions of cells with microenvironment play a decisive role for successful establishing of cancer colony. In general, the entire class of models for local growth with reset originate from the Fokker–Planck equation. Those models indicate that different probability distribution functions lead to different growth rates. For example, the Gompertzian probability distribution is associated with the exponential growth rate and the Weibull one leads to the power law growth rate that determines fractal dynamics in space–time with the maximum entropy state at stationarity [126].

The last, but not least, the Feigenbaum diagram of those sigmoidal dynamics represents a real domain of the Mandelbrot set. The Mandelbrot set forms in turn an index into the Julia set. Each point of the Mandelbrot set defines some specific Julia set by matching its constant c value. Since values of c chosen from within the Mandelbrot set are connected while those from the outside of the Mandelbrot set are disconnected, the corresponding Julia set is either connected or disconnected. It has been proved that maximal entropy is a curvature measure of the Julia set [127]. Since self-organization reaches its limit determined by the Fibonacci constant, biological complex phenomena require dynamics with parameters at the edge of the Mandelbrot set or close to it. Those chaotic deterministic processes with fractal dimensions are critical for morphogenesis or tumourigenesis. The latter one loses, however, fractal structure while maximizing entropy during natural tumour evolution.

Features of time and space do not play a major role in biology. However, it is interesting that both time and space, in which a nonlinear biological process with sigmoidal dynamics, such as growth occurs are coupled each other and get fractal dimensions [46, 47, 122, 128]. For example, Gompertzian dynamics couples time and space through the linear function of their logarithms. Moreover, the spatial fractal dimension is a function of both scalar time and the temporal fractal dimension [47] (see Appendix 9.3). In particular, the relationship between sigmoidal dynamics and the Feigenbaum diagram exemplifies how the interplay between regular and chaotic processes influence the macroscopic integration of the system. Any alteration of that coupling occurs through modification of intercellular interactions. A change of shape of the entire complex tissue system, flow of information or dynamics of intra- and intercellular processes end up in the modification of gene expression and vice versa [129]. Those alterations are reflected by a change of the mean values of the global spatial fractal dimensions.

# 9 Appendix

Coupling of molecular events in the microscale quasi-deterministic cellular network has several consequences. First, fractal dynamics, which is normally sensitive to the initial conditions does not dominate at the macroscale level. Tissue growth is very stable, predictable, and independent of the initial conditions. This is conditional probability at the microscale level, which is responsible for the emergence of a single, simplistic, macroscopic dynamics of growth in that fractal space–time (Eqs. (1)–(12)). Conditional probability works as a kind of a filter that eliminates the influence of the chaotic processes at the microscale level on the macroscale level dynamics leading to the Fokker–Planck equation (Eq. 9). The Markov process examplifies the simpliest model of couplings between the preceding events and the succeeding ones. The Gompertz function appears as a part of the solution in the relationship with distribution of the conditional probability (Eq. 12). Second, both time and space, in which such a supramolecular cellular dynamic system exists, possesses fractal structure; a fact exemplified by Eqs. (19)–(23). Those equations indicate that coupling of both time and space occurs through the appropriate dynamics of growth described by the contraction-mapping. Third, entropy depends on the probability distribution. Molecular coupling stabilizes the growing supramolecular system. The entropy of the system of interacting molecules decreases as compared to the entropy of the same set of non-interacting molecules (Eq. 27). Thus, the amount of missing information about the system decreases. The system is more predictable after than before coupling. Since entropy function remains related to dynamics of growth, it is described by the function of time, and increases along time (Eq. 31). This is the value of the entropy function which determines a capability of the supramolecular system for coding or transfer of biologically relevant information (Eq. 33). Self-organization of interacting supramolecular systems ends up in formation of structures (patterns of growth) of higher order. That process is limited by the Fibonacci constant (Eqs. (34)–(46)).

Space is defined here by a system of the geometrical co-ordinates. Those co-ordinates build up a volume, in which the nonlinear dynamic process occurs. Time is a parameter, which takes the sense of the evolutional co-ordinate.

## 9.1 *The Chapman–Kolmogorov Approach*

Division of eucaryotic cells occurs during a well-defined cell cycle. The cell cycle is a series of molecular biological events that take place in the cell and lead to its replication. The cell grows, accumulating nutrients needed for mitosis and duplicating its DNA during the interphase. The cell splits itself into two distinct cells during the mitosis. Expression of two key classes of regulatory proteins, i.e., cyclins and cyclin-dependent kinases, determine a cell's progress through the cell cycle. Cell cycle checkpoints are used by cells to control the progress of the cell cycle. Checkpoints allow verification of critical processes or repair of DNA damage at specific points, such as the $G_1/S$ point. Cells cannot proceed to the next phase until checkpoint requirements have been met. Cells that are fully differentiated enter a state of quiescence called $G_0$ phase, in which they cease division process for long periods of time. Non-differentiated, actively proliferating cells can also enter $G_0$ phase under certain circumstances (reviewed in [130]). Differentiation is a non-Markovian process with continuous time and long-range memory.

Without a loss of generality, proliferation of a special class of cells, such as cancer stem cells in early tumourigenesis can be described as the Markov process containing both a continuous and a discrete part. In general case, a joint probability in such process can be expressed in terms of transition probabilities as in (1):

$$P(x, t + \Delta t; z, t|x_0, t_0) = P(x, t + \Delta t|z, t)P(z, t|x_0, t_0) \tag{1}$$

Since

$$P(x) = \int P(x, y)dy \tag{2}$$

i.e., continuous summing a joint probability *P(x)* over all values of the variables *x* eliminates that variable, then using this principle and integrating Eq. (2) one gets the Chapman–Kolmogorov equation (3):

$$P(x, t + \Delta t) = \int_O P(x, t + \Delta t|z, t)P(z, t|x_0, t_0)dz \tag{3}$$

By definition of time derivative and using the normalization condition (4):

$$\int_O P(z, t + \Delta t|x, t) = 1 \tag{4}$$

one gets (5):

$$\begin{aligned} \frac{\partial P(x,t)}{\partial t} &= \lim_{\Delta t \to 0} \frac{1}{\Delta t}(P(x, t + \Delta t) - P(x, t)) = \\ \lim_{\Delta t \to 0} \frac{1}{\Delta t} &\int_O P(x, t + \Delta t|z, t)P(z, t) - P(z, t + \Delta t)P(x, t)dz \end{aligned} \tag{5}$$

in which area of integration *O* corresponds to at least two processes, i.e., a continuous one in the infinitezimal surroundings of *x* and a discrete one outside that surroundings. Expanding the integrand into a Taylor series one gets Eq. (6) with a component reflecting the discrete part of the stochastic process

$$\frac{\partial P(x,t)}{\partial t} = -\sum_j \frac{\partial}{\partial x_j}\left[U_j(x,t)P(x,t)\right] + \frac{1}{2}\sum_{jk} \frac{\partial^2}{\partial x_j \partial x_k}\left[D_{jk}(x,t)P(x,t)\right]$$
$$+ \int_O [V(x|z,t)P(z,t) - V(z|x,t)P(x,t)dz] \tag{6}$$

in which $U_j(x, t)$ stands for potential, known also as a drift vector; a measure of the internal interactions within dynamic cellular system and $D_{jk}(x)$ represents a diffusion coefficient known also as a diffusion matrix; a measure of the external interactions of dynamic cellular system. If drift equals zero, then Eq. (6) becomes a well-known diffusion equation.

Potential $U(x, t)$ determines evolution of dynamic cellular system towards a stationary point in space and in time. In such point, all forces acting in dynamic system are in equilibrium. It is both diffusion and interactions with microenvironment that pushes dynamic system to leave the minimum of potential. Otherwise, growth of the colony of cancer cells would quit in that minimum. This maintains growth of the colony until angiogenesis starts and provides impulses for novel dynamics [128].

## 9.2 The Markovian Model of Tumour Growth

Cell division in a population of rapidly proliferating cells, which do not differentiate, occurs in a continuous manner with short-range memory, i.e., the conditional probability is determined by the most recent state, and does not depend on the initial state $(x_O, t_O)$. It can be described as a stochastic Markov process of probability transitions. Then, Gompertzian dynamics, but not Verhulst dynamics, emerges in that simpliest model of coupling between the succeeding stage and its predecessor [47, 122].

Let us consider a small cellular colony with less than $10^6$ cancer cells growing within a normal tissue environment. First, let those cells possess broad autonomy. Let metabolic exchange through the gap junctions with normal surrounding cells and with each other be very weak or does not exist. Second, there is no blood vessels in the colony. Feeding of cells occurs by diffusion. Third, cancer cells continue to proliferate spontaneously owing to many molecular defects. Fourth, there is a minimal reaction of the external tissue systems, such as the neuroimmunohormonal one or the internal mechanisms, such as apoptosis. Finally, cancer cells belong to a single clone. Cells do not undergo differentiation or do not express multiple transitional phenotypes. There is a difference between the time-scales of molecular signalling, i.e., femtoseconds to milliseconds, cellular growth, i.e., hours and cellular proliferation, i.e., days. Single cells in the colony integrate molecular signals much faster than the colony expands in space–time. There is no memory of the state at previous timepoints in that tissue object. Then, it is possible to describe a growth trajectory under those assumptions as a Markov chain of transitions for each timepoint by (7):

$$P(x^0, \ldots, x^n) = P(x^0, \ldots,^{n-1})P\langle x^n|x^{n-1}\rangle = P(x^0)\prod_{j=1}^{n} P\langle x^j|x^{j-1}\rangle \quad (7)$$

in which $P(x^0, \ldots, x^n)$ is a probability that the growing cellular colony is at the positions $x^0, \ldots, x^n$ at the timepoints $0, \ldots n$; $P < x^n|x^{n-1} >$ is a conditional probability that between timepoints $n - 1$ and $n$ the growth succeeds from the position $x^{n-1}$ to $x^n$.

Since tissue growth occurs simulataneously in space and in time, it is particularly interesting to introduce the probability $P$ as a function of geometrical spatial variable $x$ and scalar time $t$. A speed of both processes is usually not large. So, the spatial expansion of cellular system $\Delta x = x - k$ in the time step $\Delta t$ will also not be large. A change of the probability $P$ in the infinitesimal interval of time can be described by differential Eq. (8), in which such change results from a difference between the probabilities of the jump from the position $k$ to $x$ and the probabilities of return owing to verification of critical processes or repair of DNA damage at the checkpoints:

$$\frac{\partial P(x,t)}{\partial t} = \sum_k P(k,t)P\langle x|k\rangle - \sum_k P(x,t)P\langle k,x\rangle \quad (8)$$

This leads to (9), which has a form of the Fokker–Planck equation; an integral part of (6):

$$\frac{\partial P(x,t)}{\partial t} = -\sum_j \frac{\partial}{\partial x_j}\left[U_j(x,t)P(x,t)\right] + \frac{1}{2}\sum_{j,k} \frac{\partial^2}{\partial x_j \partial x_k}\left[D_{jk}(x,t)P(x,t)\right] \quad (9)$$

The stationary differential Markov process is specified by the probability distribution $P(M(t), t)$ given by (10) [125]:

$$P(M - M_0, t) = \int_{-\infty}^{\infty} \frac{dk}{2\pi} e^{-ik(M-M_0)} e^{-\beta t|k|^\alpha} \quad (10)$$

in which $t$ stands for scalar time, $k$ is the Fourier variable, $\alpha$, $\beta > 0$ are real, constant factors, and $dM(t)$ stands for the fluctuations.

It is worth to notice that there is a key relationship between such the conditional probability density $P(v, t)$ and the Gompertz function. Indeed, the conditional probability density $P(v, t)$ can be expressed in the form of the Fourier transform taken with respect to the variable $(v - v_0 e^{-\lambda t})$ containing the Gompertz function $f(t)$ (11):

$$P(v,t|v_0) = \int_{-\infty}^{\infty} \frac{dk}{2\pi}\left(e^{\left(-ik\int_0^{M(t)} e^{-\lambda(t-\tau)dM(\tau)}\right)}\right) f(t)^{|k|^\alpha} \quad (11)$$

in which *v(t)* is the dynamical variable, i.e., the velocity of the division process and *f(t)* stands for the Gompertz function given by (12):

$$f(t) = e^{a(1-e^{-bt})} \tag{12}$$

## 9.3 Fractal Time–Space

Cells grow both in space and in time [46, 47, 128]. Let us assume that both variables, the spatial $x$ and the temporal $t$, are coupled each other in a linear manner as in (13) into a single, complex spatio-temporal variable:

$$\theta = \mu x + t \tag{13}$$

Then, the appropriate equation relating the function of probability distribution *P(x, t)* and the potential function *U(x, t)* is given by (14):

$$-\frac{1}{D}\frac{\partial^2 P(\theta)}{\partial\theta^2} + \frac{D}{4}P(\theta) + U(\theta)P(\theta) = 0 \tag{14}$$

Indeed, let us calculate the appropriate derivatives of the Eq. (14), i.e., *d/dt*, *d/dx*, and $d^2/dx^2$. Let put them into the well-known equation of diffusion. We should get (15) with a single spatio-temporal variable:

$$\left(\frac{\partial}{\partial\theta} - D\mu^2\frac{\partial^2}{\partial\theta^2} - U(\theta)\right)\Pi(\theta) = 0 \tag{15}$$

For $\mu = 1/D$ and for

$$0 < \Pi(\theta) = Pe^{-\frac{\theta}{2}} < 1 \tag{16}$$

we can develop (15) into (17):

$$-\frac{1}{D}\frac{\partial}{\partial\theta}\left(\frac{\partial}{\partial\theta}\Pi(\theta)\right) + \frac{\partial}{\partial\theta}\Pi(\theta) + U(\theta)\Pi(\theta) = 0 \tag{17}$$

and finally, by calculating the appropriate complex derivatives arrive to (14). The Gompertz function and the anharmonic Morse-like potential *U(θ)* are related each other through the one-dimensional differential operator (Eq. 18), in which $a$ represents a depth of the potential well, $b$ is a range parameter [47, 128].

$$\left(-\frac{1}{b}\frac{\partial^2}{\partial t^2} + \left(ae^{-bt} - \frac{1}{2}\right)^2 - \frac{1}{4}\right) f(t) = 0 \tag{18}$$

This linear coupling of variables can be defined as a function with both spatial and temporal fractal dimension. It is known from experimental data that the number of cells (or their volume) changes in time $t$ according to the Gompertzian function $f(t)$ (12). A volume of the spheroid $V$ is given by (19):

$$f(t) = V = nV_k \tag{19}$$

in which $V_k$ is a mean volume of a single cell, n stands for a number of cells in the spheroid.

From (12) to (19), and from the fact that the Gompertzian function is a fractal, (e.g., it can be fitted with the function $f(t) = at^b$ with very high accuracy, a coefficient of nonlinear regression $R >> 0.95$ for $n^3$ $100$ pairs of co-ordinates), in which $a$ stands for a scaling coefficient, $b_t$ is a temporal fractal dimension, (i.e., any real number), $t$ is scalar time, we get (20):

$$V = V_k F(t_0) e^{a(1-e^{-bt})} = V_k f(t_0) a t^{b_t} = V_0 a t^{b_t} \tag{20}$$

The volume $V$ of the spheroid can also be expressed as a function of scalar geometrical variable $x$ (i.e., a radius of a family of the concentric spheres covering the entire spheroid) by (21):

$$V = a_1 x^{b_s} \tag{21}$$

in which $a_1$ stands for a scaling coefficient, $b_s$ is a spatial fractal dimension after scalar time $t_1$, $x$ is a scalar, geometrical variable, which locates an effect in space.

If the initial value of the temporal fractal dimension $b_{t0}$ for cellular population expanding in space is different from the fractal dimension $b_t$ during the other stages of the process ($t = t_n$), then, from (20) to (21), we get (22):

$$V = a_1 x^{b_s} = V_0 a t^{b_t} = a_0 x^{b_{s0}} a t^{b_t} \tag{22}$$

in which $a$, $a_0$, and $a_1$ stand for the scaling coefficients, $b_t$ is the temporal fractal dimension, $b_{s0}$ and $b_s$ are the spatial fractal dimensions after time $t_0$ and $t$, respectively, $x$ is a geometrical variable. Hence, we get (23) that relates space and time and defines the geometrical variable $x$ as a function of the scalar time $t$.

$$\ln x = \frac{1}{b_s - b_{s_0}} \ln \frac{a_0 a}{a_1} + \frac{b_t}{b_s - b_{s_0}} \ln t \tag{23}$$

in which $t$ stands for scalar time, $x$ is geometrical variable, $b_s$ is the spatial fractal dimension, $b_t$ is the temporal fractal dimension [46].

## 9.4 Entropy and Sigmoidal Dynamics

It is worth to notice that the proposed Markov model implies at least three important consequences. First, entropy, (i.e. missing information) $H_M$ of such the Markov chain of the coupled molecular reactions defined by (1) or (7) is always lower than entropy of the set of random and independent biochemical reactions $H_R$. Indeed, entropy is defined as the expected value of missing information $H_p$:

$$H_P = H(X) = -\sum_{j=1}^{N} p_j \log p_j \tag{24}$$

in which $p = (p_1, p_2, \ldots p_j)$, $j \in N$, is a probability density function over a generic variable $X$, and if $p_j = 0$, then $H_p = 0$, log is a natural logarithm, providing a unit of measure. Hence, the conditional entropy $H(X_k|Y_{k-1})$ of the $X_k$ reaction stands for which conditional information is determined when the state $Y_{k-1} = i$, is given by (25):

$$H(X_k|Y_{k-1} = i) = -\sum_j p_{ij} \log p_{ij} \tag{25}$$

The conditional entropy of the Markov chain $H_C$ is given by (26):

$$H_C = H(X_k|X_{k-1}) = -\sum_i p_i \sum_j p_{ij} \log p_{ij} \tag{26}$$

Finally, we get (27) for the $n$ first steps of the Markov chain $X_1, X_2, \ldots, X_n$ from (24) to (26), the principle of additivity of independent random events, and from the analog principle for the conditional probabilities (compare [131]):

$$\begin{aligned} H_M &= H(X_1) + \sum_{k=2}^{k=n} H\langle X_k \mid X_{k-1}\rangle = -\sum p_j \log p_j + (n-1)H_C < nH_P \\ &= -\sum_j p_j \log p_j = H_R \end{aligned} \tag{27}$$

Second, Gompertzian dynamics of growth can be normalized, i.e., growth dynamics of various tissue systems can be described by a single normalized Gompertz function $f_N(t)$ (28). In fact, this normalized Gompertz function is both a dynamics function $f_N(t)$ and a probability function $p_N(t)$ (see for details [47]).

$$f_N(t) = e^{-e^{(-bt)}} = p_N(t) \tag{28}$$

Consider a coupling of probability function $p_N(t)$ and antiprobability function $-log\ p_N(t)$, in which $r = b$ (29):

$$\frac{dp_N(t)}{dt} = -rp_N(t) \log p_N(t) \tag{29}$$

This equation defines a relationship between entropy $H(t)$ and the normalized Gompertzian dynamics of growth $p_N(t)$ (30) [132]:

$$p_N(t) = \int \frac{\partial p_N(t)}{\partial t} dt = -r \int p_N(t) \log p_N(t) dt = rH(t) \tag{30}$$

Finally, from (29) to (30) we get (31):

$$H(t)_{Gompertz} = \frac{1}{b} e^{-e^{(-bt)}} \tag{31}$$

According to Shannon theorem, of all the continuous distribution densities for which the standard deviation exists and is fixed, the Gaussian, (i.e. normal) distribution has the maximum value of entropy $H$ (32) [133]:

$$H_{Gauss} = -\int_{-\infty}^{\infty} \frac{e^{-\frac{t^2}{2\sigma^2}}}{\sqrt{2\pi\sigma^2}} \log \frac{e^{-\frac{t^2}{2\sigma^2}}}{\sqrt{2\pi\sigma^2}} dt = \frac{1}{2} \log(2\pi e \sigma^2) \tag{32}$$

In the case of growing supramolecular cellular system, entropy or missing information $H(t)$ is a function of time related with dynamic function of growth in fractal space–time. For $b = 1$ both the normalized Gompertz function (28) and the entropy function (31) overlap each other. However, for most cellular systems $b << 1$.

Third, there is a relationship between the number of elements in the Markov chain and entropy. If $M_p(n)$ stands for several Markov chains of the length $n$ with the total probability $p$, $0 < p < 1$, there exists the same limit for each probability $p$ that equals entropy $H$ (33) [134].

$$\lim_{n\to\infty} \frac{\log M_p(n)}{n} = H \tag{33}$$

If a total number of states of the supramolecular cellular system equals 2 m, then the number of molecular reactions interconnected in the Markov chains of the length $n$ is 2 nm. It is clear from (33) that only 2 nH Markov chains with probability $1 - \varepsilon$, $\varepsilon > 0$ will be involved in transfer of biologically relevant information.

## 9.5 The Fibonacci Constant and Self-Organization of Cells

Let us consider density $\rho$ of some geometric objects in the classic, Euclidean space with integer dimension w. We define it by the following equation

$$\rho = N/V = N_1 r^{-w} \tag{34}$$

in which $N_1$ is the number of objects in the given volume $V$ of a sphere, $r$ is a radius of the unit sphere, $w$ is geometric dimension of the space.

Let us define the corresponding density $\rho_f$ of the same objects in space with the spatial fractal dimension $b$.

$$\rho_b = N_b/V_b = N_b r^{-b} \tag{35}$$

After logarithmic transformation,

$$\ln N_1/\rho = w \ln r \tag{36}$$

and,

$$\ln N_b/\rho_b = b \ln r \tag{37}$$

Hence,

$$w/b = (\ln N_1/\rho)/(\ln N_b/\rho_b) \tag{38}$$

and,

$$N_b/\rho_b = (N_1/\rho)^{b/w} \tag{39}$$

The quotient $N_1/\rho$ is a measure of object expansion into space volume, (e.g., cells). Let us mark this quotient by $V$, and the corresponding expression for fractal space by $V_b$.

Then,

$$V_b = V^{b/w} \tag{40}$$

We define the relative volume by the equation

$$\kappa = V_b/V = V^{\sigma} \tag{41}$$

in which

$$\sigma(b) = (b/w) - 1 \tag{42}$$

is a relative differential coefficient of object expansion into space and a function of the fractal dimension. It possesses the following features: a. If the dynamic process or objects occupy the entire Euclidean space, then $b = w$, $\sigma = 0$, and $\kappa = 1$. In other words, if the classic space is completely occupied by geometric objects, there is no free region available for expansion as in the case of neutral gas, and the relative differential coefficient of object expansion into space equals zero; b. If the process or objects possesses fractal dimension, then $b < w$, $\sigma < 0$, and $\kappa < 1$. Thus, the classic Euclidean space is not occupied completely, and there is some free subspace; a fact reflected by a value of the coefficient $\sigma(b)$ lower than zero, and value of relative volume lower than one [44].

Let us consider a golden rectangle of the length $b$ and the width $w$.

If $b < w$, then it holds Eq. 1:

$$\frac{b}{w} = \frac{w-b}{b} = \left(\frac{w}{b}\right) - 1 \tag{43}$$

Hence, we get Eq. 44:

$$\left(\frac{w}{b}\right)^2 - \left(\frac{w}{b}\right) - 1 = 0 \tag{44}$$

where

$$\left(\frac{w}{b}\right) = \left(\frac{1}{\frac{b}{w}}\right) = \frac{1+\sqrt{5}}{2} = 1.618 = \Phi \tag{45}$$

Hence

$$\left(\frac{b}{w}\right) = \frac{1}{1.618} = 0.618 = \frac{1}{\Phi} \tag{46}$$

where $\Phi$ stands for the Fibonacci constant; the irrational number and the limit of self-organization in dynamic systems with fractal dimension [7]. Then, the global capacity fractal dimension has a value of 1.236 and the corresponding coefficient of cellular expansion –0.382.

## References


1. Virchow, R.: Die Cellular pathologie in ihrer Begründung auf physiologische und pathologische Gewebelehre. Verlag A. Hirschwald, Berlin (1858)
2. Reya, T., Morrison, S.J., Clarke, M.F., Weissman, I.L.: Stem cells, cancer, and cancer stem cells. Nature **414**(6859), 105–111 (2001)
3. Gonzales-Villarreal, C.A., Quiros-Reyes, A.G., Islas, I.F., Garza-Trevino, E.N.: Colorectal cancer stem cells in the progression to liver metastasis. Front. Oncol. (2020). https://doi.org/10.3389/fonc.2020.01511
4. Waliszewski, P.: Complexity, dynamic cellular network, and tumorigenesis. Pol. J. Pathol. **48**(4), 235–241 (1997)
5. Davies, P.C., Demetrius, L., Tuszynski, J.A.: Cancer as a dynamical phase transition. Theor. Biol. Med. Model. **8**(30), 1–16 (2011)
6. Niedergang, F.: Phagocytosis. Encycl. Cell Biol. **2**, 751–757 (2016)
7. Waliszewski, P.: The Fibonacci constant and limits of tissue self-organization; local complexity measures in evaluation of the risk of metastasis formation. Banach Center Publ **124**, 143–157 (2021). In: Banaszak G., Krason P., Milewski J., Waliszewski P. (eds.) Arithmetic Methods in Mathematical Physics and Biology II. Bedlewo, August 3–11, (2018) https://doi.org/10.4064/bc124-12
8. Waliszewski, P., Molski, M., Konarski, J.: On the holistic approach in cellular and cancer biology: nonlinearity, complexity, and quasi-determinism of the dynamic cellular network. J. Surg. Oncol. **68**(2), 70–78 (1998)
9. Huxley, J.: The dawn of quantum biology **474**, 272–274 (2011)
10. McFadden, J., Al-Khalili, J.: A quantum mechanical model of adaptive mutation. Biosystems **50**, 203–211 (1999)
11. Cooper, W.G.: Roles of evolution, quantum mechanics and point mutations in origins of cancer. Cancer Biochem. Biophys. **13**(3), 147–170 (1993)
12. Kryachko, E.S.: The origin of spontaneous point mutations in DNA via Loewding mechanism of proton transition in DNA base pairs. Cure with covalent base parity. Int. J. Quantum Chem. (2002). https://doi.org/10.1002/qua.975
13. Zapatka, M., Borozan, I., Brewer, D.S., Iskar, M., Grundhoff, A., Alawi, M., Desai, N., Sültmann, H., Moch, H., Cooper, C.S., Eils, R., Ferretti, V., Lichter, P.: The landscape of viral associations in human cancers. Nat. Genet. **52**, 320–330 (2020)
14. Loewe, L., Hill, W.G.: The population genetics of mutations: good, bad, and indifferent. Philos. Trans. R Soc. B **365**, 1153–1167 (2010)
15. Tomasetti, C., Lu, L., Vogelstein, B.: Stem cell division, somatic mutation, cancer etiology, and cancer prevention. Science **355**, 1330–1334 (2017)
16. Robert, L., Ollion, J., Robert, J., Song, X., Matic, I., Elez, M.: Mutation dynamics and fitness effects followed in single cells. Science **359**, 1283–1286 (2018)
17. Cairns, J., Overbay, J., Miller, S.: The origin of mutants. Nature **335**, 142–145 (1988)
18. Moreno, P.A., Velez, P.E., Martinez, E., Garreta, L.E., Diaz, N., Amador, S., Tischer, I., Gutierrez, J.M., Naik, A.K., Tobar, F., Garcia, F.: The human genome: a multifractal analysis. BMC Genomics **12**(506), 1–17 (2011)
19. Oliver, J.L., Bernaola-Galvan, P., Guerrero-Garcia, J., Roman-Roldan, R.: Entropic profiles of DNA sequences through chaos-game-derived images. J. Theor. Biol. **160**, 457–470 (1993)
20. Hochberg, G.K.A., Liu, Y., Marklund, E.D., Metzger, B.P.H., Laganowsky, A., Thornton, J.W.: A hydrophobic ratchet entrenches molecular complexes. Nature **588**, 503–508 (2020)
21. Plon, S.E., Eccles, D.M., Easton, D., Foulkes, W.D., Genuardi, M., Greenblatt, M.S., Hogervorst, F.B.L., Hoogerbrugge, N., Spurdle, A.B., Tavtigian, S.V.: Sequence variant classification and reporting: recommendations for improving the interpretation of cancer susceptibility genetic test results. Hum. Mutat. **29**(11), 1282–1291 (2008)
22. Ononye, O.E., Sausen, Ch.W., Balakrishnan, L., Bochman, M.L.: Lysine acetylation regulates the activity of nuclear pif1. J. Biol. Chem. **295**(46), 15482–15497 (2020)

23. Lu, Y., Brommer, B., Tian, X., Krishnan, A., Meer, M., Wang, Ch., Vera, D., Zeng, Q., Yu, D., Bonkowski, M., Yang, J.-H., Zhou, S., Hoffmann, E., Karg, M., Schultz, M., Kane, A., Davidson, N., Korobkina, E., Chwalek, K., Rajman, L., Church, G., Hochedlinger, K., Gladyshev, V., Horvath, S., Levine, M., Gregory-Ksander, M.S., Ksander, B.R., He, Z., Sinclair, D.A.: Reprogramming to recover youthful epigenetic information and restore vision. Nature **588**(7836), 124–129 (2020)
24. Zhuang, J., Jones, A., Lee, S.H., Ng, E., Fiegl, H., Zikan, M., Cibula, D., Sargent, A., Salvesen, H.B., Jacobs, I.J., Kitchener, H.C., Teschendorff, A.E., Widschwendter, M.: The dynamics and prognostic potential of DNA methylation changes at stem cell gene loci in women's cancer. PLoS Genet. **8**(2), e1002517, 1–12 (2012)
25. De Craene, B., Berx, G.: Regulatory networks defining EMT during cancer initiation and progression. Nat. Rev. Cancer **13**, 97–110 (2013)
26. Verschoor, M.L., Ungard, R., Harbottle, A., Jakupciak, J.P., Parr, R.L., Singh, G.: Mitochondria and cancer: past, present, and future. Biomed. Res. Int. **2013**, 612369 (2013)
27. Vyas, S., Zaganjor, E., Haigis, M.C.: Mitochondria and cancer. Cell **166**(3), 555–566 (2016)
28. Davidson, S.M., van der Heiden, M.G.: Critical functions of the lysosome in cancer biology. Ann. Rev. Pharm. Toxicol. **57**, 481–507 (2016)
29. Terasaki, M., Shemesh, T., Kasthuri, N., Klemm, R.W., Schalek, R., Hayworth, K.J., Hand, A.R., Yankova, M., Huber, G.: Stacked endoplasmic reticulum sheets are connected by helicoidal membrane motifs. Cell **154**(2), 285–296 (2013)
30. Oakes, S.A.: Endoplasmic stress signaling in cancer cells. Am. J. Pathol. **190**, 934–946 (2020)
31. Zong, W.X., Rabinowitz, J.D., White, E.: Mitochondria and cancer. Mol. Cell. **61**(5), 667–676 (2016)
32. Waliszewski, P., Konarski, J.: Neuronal differentiation and synapse formation occur in space and time with fractal dimension. Synapse **43**(4), 252–258 (2002)
33. Fane, M., Weeraratna, A.T.: How the aging microenvironment influences tumor progression. Nat. Rev. Cancer **20**(2), 89–106 (2020)
34. Janiszewska, M., Candido Primi, M., Izard, T.: Cell adhesion in cancer: beyond the migration of single cells. J. Biol. Chem. **295**, 2495–2505 (2020)
35. Vignais, M.L., Nakhle, J., Griessinger, E.: Tunneling nanotubes (TNTs): intratumoral cell-to-cell communication. In: Encyclopedia of Cancer, Boffetta P., Hainaut P. (eds.) Hallmarks of Cancer, pp. 513–522. Academic (2019). ISBN 978-0-12-812485-7
36. Lugano, R., Ramachandran, M., Dimberg, A.: Tumor angiogenesis: causes, consequences, challenges, and opportunities. Cell Mol. Life Sci. **77**, 1745–1770 (2020)
37. Reeves, M.Q., Kandyba, E., Harris, S., Del Rosario, R., Balmain, A.: Multicolour lineage tracing reveals clonal dynamics of squamous carcinoma evolution from initiation to metastasis. Nat. Cell Biol. **20**(6), 699–709 (2018)
38. Allee, W.C., Bowen, E.: Studies in animal aggregation: mass protection against colloidal silver among goldfishes. J. Exp. Zool. **61**(2), 185–207 (1932)
39. Waliszewski, P.: The quantitative criteria based on the fractal dimensions, entropy and lacunarity for the spatial distribution of cancer cell nuclei enable identification of low or high aggressive prostate carcinomas. Front. Physiol. Fract. Physiol. **7**(34), 1–16 (2016)
40. Tanase, M., Waliszewski, P.: On complexity and homogenity measures in predicting biological aggressiveness of prostate cancer; implication of the cellular automata model of tumor growth. J. Surg. Oncol. **112**(8), 791–801 (2015)
41. Waliszewski, P.: Computer-aided image analysis and fractal synthesis in the quantitative evaluation of tumor aggressiveness in prostate carcinomas. Front. Oncol. Genitourin. Oncol. **6**, 110 (2016)
42. Waliszewski, P., Banaszak, G.: On fractal and topological measures in human colon carcinomas. In: Proceedings 2021 23rd International Conference on Control Systems and Computer Science, Bucharest, 26–28 May 2021. https://doi.org/10.1109/CSCS52396.2021.00042
43. Lu, M., Jolly, M.K., Levine, H., Onuchic, J.N., Ben-Jacob, E.: MicroRNA-based regulation of epithelial-hybrid-mesenchymal fate determination. PNAS **110**(45), 18144–18149 (2013)

44. Waliszewski, P., Konarski, J., Molski, M.: On the modification of fractal self-space during tumor progression. Fractals **8**(2), 195–203 (2000)
45. Chandolia, B., Bajpai, M.: Epithelial Mesenchymal Interactions. Lambert Academic Publishing, Saarbrücken (2016)
46. Waliszewski, P., Konarski, J.: On time-space of nonlinear phenomena with Gompertzian dynamics. Biosystems **80**, 91–97 (2005)
47. Waliszewski, P.: A principle of fractal-stochastic dualism and Gompertzian dynamics of growth and self-organization. Biosystems **82**(1), 61–73 (2005)
48. Stewart, J.: Calculus - Early Transcendentals, 7th edn. Brooks/Cole Cengage Learning, p. 1122 (2021)
49. Siegel, R.L., Miller, K.D., Jemal, A.: Cancer statistics 2019. CA Cancer J. Clin. **69**(1), 7–34 (2019)
50. Cancer Research UK Prostate cancer incidence statistics (2019). https://www.cancerresearchuk.org/health-professional/cancer-statistics/statistics-by-cancer-type/prostate-cancer/incidence#heading-One
51. Erickson, A., Hayes, A., Rajakumar, T., Verrill, C., Bryan, R.J., Hamdy, F.C., Wege, D.C., Woodcock, D.J., Miller, I.G., Lamb, A.D.: A systemic review of prostate cancer hetrogeneity: understanding the clonal ancestry of multifocal disease. Eur. Urol. Oncol. **4**(3), 358–369 (2021)
52. Williams, M.J., Werner, B., Barnes, Ch.P., Graham, T.A., Sottoriva, A.: Identification of neutral tumor evolution across cancer types. Nat. Genet. **48**, 238–244 (2016). https://doi.org/10.1038/ng.3489
53. Wodarz, D., Komarova, N.L.: Mutant evolution in spatially structured and fragmented expanding populations. Genetics **216**(1), 191–203 (2020)
54. Szymanska, K., Bosman, F.T., Hainaut, P.: Bladder cancer: pathology, genetics, diagnosis and treatment. In: Boffetta P., Hainaut P. (eds.) Encyclopedia of Cancer, 3rd edn., pp. 122–133. Elsevier, Amsterdam (2019)
55. Gerlinger, M., Rowan, A.J., Horswell, S., Math, M., Larkin, J., Endesfelder, D., Gronroos, E., Martinez, P., Matthews, N., Stewart, A., Tarpey, P., Varela, I., Phillimore, B., Begum, S., McDonald, N.Q., Butler, A., Jones, D., Raine, K., Latimer, C., Santos, C.R., Nohadani, M., Eklund, A.C., Spencer-Dene, B., Clark, G., Pickering, L., Stamp, G., Gore, M., Szallasi, Z., Downward, J., Futreal, P.A., Swanton, C.: Intratumor heterogeneity and branched evolution revealed by multiregion sequencing. N. Engl. J. Med. **366**, 883–892 (2012)
56. Schwann, Th.: Mikroskopische Untersuchungen über die Übereinstimmung in der Struktur und dem Wachstum der Thiere und Pflanzen. Sander, Berlin (1839) http://www.deutschestextarchiv.de/book/show/schwann_mikroskopische_1839
57. Schwann, Th., Schleyden, M.J.: Microscopical Researches into the Accordance in the Structure and Growth of Animals and Plants. Printed for the Sydenham Society, London (1847). http://vlp.mpiwg-berlin.mpg.de/library/data/lit28715
58. Bongrand, P.: Ligand-receptor interactions. Rep. Prog. Phys. **62**, 921–968 (1999)
59. Stein, D.L. (ed.): Lectures in the Science of Complexity. Adison-Wesley Pub. Co., Advanced Book Program, Redwood, CA (1989)
60. West, B.J., Geneston, E.L., Grigolini, P.: Maximizing information exchange between complex networks. Phys. Rep. **468**, 1–99 (2008)
61. Kirilyuk, A.: New mathematics of complexity and its biomedical applications. In: Banaszak, G., Milewski, J., Waliszewski, P. (eds.) Arithmetic Methods in Mathematical Physics and Biology, vol. 109, pp. 57–81. The Banach Center Publications, Warsaw (2016). https://doi.org/10.4064/bc109-0-5
62. Waliszewski, P., Konarski, J.: Tissue as a self-organizing system with fractal dynamics. Adv. Space Res. **28**(4), 545–548 (2001)
63. Kim, S.Ch., Zhou, L., Zhang, W., O'Flaherty, D.K., Rondo-Broveto, V., Szostak, J.W.: A model for the emergence of RNA from a prebiotically plausible mixture of ribonucleotides, arabinonucleotides, and 2′-deoxynucleotides. J. Am. Chem. Soc. **142**(5), 2317–2326 (2020)

64. Micura, R., Höbartner, C.: Fundamental studies of functional nucleic acids: aptamers, riboswitches, ribozymes and DNAzymes. Chem. Soc. Rev. **49**, 7331–7353 (2020)
65. Zhang, Y., Narlikar, G.J., Kutateladze, T.G.: Enzymatic reactions inside biological condensates. J. Mol. Biol. 166624 (2020)
66. Meyer, S.C.: Darwin's doubt: the explosive origin of animal life and the case for intelligent design. HarperOne (2014). ISBN-10 0062071483
67. Berlinski, D.: The deniable Darwin and other essays. Discovery Inst 2009 (2009). ISBN-10 0979014131
68. Tilokani, L., Nagashima, S., Paupe, V., Prudent, J.: Mitochondrial dynamics: overview of molecular mechanisms. Essays Biochem. **62**, 341–360 (2018)
69. Rossi, A., Pizzo, P., Filadi, R.: Calcium, mitochondria and cell metabolism: a functional triangle in bioenergetics. Biochim. Biophys. Acta Mol. Cell Res. **1866**(7), 1068–1078 (2019). https://doi.org/10.1016/j.bbamcr.2018.10.016
70. Waliszewski, P., Skwarek, R.: Deterministic chaos and mitochondrial synthesis of reactive oxygen species. In: Proceedings 2017 21st International Conference on Control Systems and Computer Science, Bucharest, 29–31 May 2017, pp. 356–363. https://doi.org/10.1109/CSCS.2017.55
71. Langmuir, I.: The constitution and fundamental properties of solids and liquids. J. Am. Chem. Soc. **39**, 1848 (1917)
72. Hermann, J., DiStasio, R.A., Tkatchenko, A.: First principles models for van der Waals interactions in molecules and materials: concepts, theory, and applications. Chem. Rev. **117**, 4714–4758 (2017)
73. Woods, L.M., Dalvit, D.A.R., Tkatchenko, A., Rodriguez-Lopez, P., Rodriguez, A.W., Podgornik, R.: Materials perspective on Casimir and van der Waals interactions. Rev. Mod. Phys. **88**(4), 045003 1–48 (2016)
74. Lord Kelvin, P.R.S.: The Molecular Tactics of a Crystal, p. 27. The Clarendon Press, Oxford (1894)
75. Kurian, P., Dunston, G., Lindesay, J.: How quantum entanglement in DNA synchronizes double-strand breakage by type II restriction endonucleases. J. Theor. Biol. **391**, 102–112 (2016)
76. Hay, S., Scrutton, N.S.: Good vibrations in enzyme-catalysed reactions. Nat. Chem. **4**, 161–168 (2012)
77. Bell, J.S.: On the Einstein-Rosen-Podolsky paradox. Physics **1**(3), 195–200 (1964)
78. Chen, E.K.: Bell's theorem, quantum probabilities, and superdeterminism. In: Knox E., Wilson A. (eds.) The Routledge Companion to the Philosophy of Physics (2020). arXiv:2006:08609v2 [quant-phys]
79. O'Callaghan, J.: "Schrödinger's Bacterium" Could Be a Quantum Biology Milestone. Scientific American, 29 October 2018 (2018)
80. Trixler, F.: Quantum tunnelling to the origin and evolution of life. Curr. Org. Chem. **17**(16), 1758–1770 (2013)
81. Chance, B.: The energy-linked reaction of calcium with mitochondria. J. Biol. Chem. **240**(6), 2729–2748 (1965)
82. Engel, G., Calhoun, T.R., Read, E.L., Ahn, T.K., Mancal, Th., Cheng, Y.Ch., Blankenship, R.E., Fleming, G.R.: Evidence for wave maker energy transfer through quantum coherence in photosynthetic systems. Nature **446**(7137), 782–786 (2007)
83. Grover, L.K.: From Schrödinger's equation to quantum search algorithm. Am. J. Phys. **69**(7), 769–777 (2001)
84. Hoyer, S., Sarovar, M., Whaley, K.B.: Limits of quantum speedup in photosynthetic light harvesting. NJP **12**, 065041 (2010)
85. Fletcher, D.A., Mullins, R.D.: Cell mechanics and the cytoskeleton. Nature **463**(7280), 485–492 (2010)
86. Goldstein, R.E., van de Meent, J.W.: A physical perspective on cytoplasm in streaming. Interface Focus **5**(4), 20150030 (2015)

87. Elbaum-Garfinkle, S., Kim, Y., Szczepaniak, K., Chih-Hiung Chen, C., Eckmann, Ch.R., Myong, S., Brangwynne, C.P.: The diordered P granule protein LAF-1 drives phase separation into droplets with tunable viscosity and dynamics. Proc. Natl. Acad. Sci. USA **112**, 7189–7194 (2015)
88. Koenig, H.G.: Religion, spirituality, and health: a review and update. Adv. Mind Body Med. **29**(3), 19–26 (2015)
89. Davis, P.C.W.: Time variation of the coupling constants. J. Phys. A: Math. Gen. **5**, 1296–1304 (1972)
90. Weber, H.M., Dedekind, R.: Bernhard Riemann's gesammelte mathematische Werke und wissenschaftlicher Nachlass. Cambridge University Press, Cambridge (2013)
91. Kaufmann, S.: Investigations, pp. 159–209. Oxford University Press, Oxford (2000)
92. Termonia, Y., Ross, J.: Oscillations and control features in glycolysis: analysis of resonance effect. PNAS **76**(6), 3563–3566 (1981)
93. Mair, T., Warnke, Ch., Tsuji, K., Müller, S.C.: Control of glycolytic oscillations by temperature Biophys. J. **88**(1), 639–646 (2005)
94. Nelson, D.L., Cox, M.M.: Lehninger Principles of Biochemistry, 8th edn. Freeman WH and Co. (2021)
95. Yang, W.J., Cho, K.S., Rha, K.H., Lee, H.Y., Chung, B.H., Hong, S.J., Yang, S.C., Choi, Y.D.: Long-term effects of ileal conduit urinary diversion on upper urinary tract in bladder cancer. Urology **68**(2), 324–327 (2006)
96. Okon, K., Dyduch, G., Bialas, M.B., Milian-Ciesielska, K., Szpor, J., Leszczynska, I., Tyrak, K., Szopinski, T., Chlosta, P.: Image analysis discloses differences in nuclear parameters between ERG+ and ERG- prostatic carcinomas. Pol. J. Pathol. **71**(1), 20–29 (2020)
97. Vlajnic, T., Bubendorf, L.: Molecular pathology of prostate cancer: a practical approach. Pathology **53**(1), 36–43 (2021)
98. Lozano, R., Castro, E., Aragón, I.M., Cendon, Y., Cattrini, C., Lopes-Casas, P.P., Olmos, D.: Genetic aberrations in DNA repair pathways: a cornerstone of precision oncology in prostate cancer. Br. J. Cancer **124**, 552–563 (2021)
99. Josefsson, A., Larsson, K., Freyhult, E., Damber, J.E., Welen, K.: Gene expression alterations during development of castration-resistant prostate cancer are detected in circulating tumor cells. Cancer **12**(1), 39 (2019)
100. Casinello, J., Dominguez-Lubillo, T., Gomez-Barrera, M., Hernando, T., Parra, R., Asensio, I., Casado, M.A., Moreno, P.: Optimal treatment sequencing of abiraterone acetate plus prednisone and enzalutamide in patients with castration-resistant metastatic prostate cancer: a systematic review and metaanalysis. Cancer Treat Rev. **93**, 102152 (2021)
101. Mendel, G.: Experiments in plant hybridization (1865)
102. Husserl, E.: Ideen zu einer reinen Phänomenologie und phänomenologischen Philosophie. Erstes Buch: Allgemeine Einführung in die reine Phänomenologie. Max Niemeyer Verlag, Halle (Saale) (1913)
103. Held, T., Nourmohammad, A., Lässig, M.: Adaptive evolution of molecular phenotypes. J. Stat. Mech.: Theory Exp. P09029 (2014)
104. Quintero-Fabián, S., Arreola, R., Becerril-Villanueva, E., Torres-Romero, J.C., Arana-Argáez, V., Lara-Riegos, J., Ramírez-Camacho, M.A., Alvarez-Sánchez, M.E.: Role of matrix metalloproteinases in angiogenesis and cancer. Front. Oncol. **9**, 1370 (2019)
105. Waliszewski, P.: The circular fractal model of adenocarcinomas and tumor aggressiveness. Banach Center Publ. **109**, 183–196 (2016). https://doi.org/10.4064/bc109-0-12
106. Baas-Becking, L.G.M., Drion, E.F.: On the origin of frequency distributions in biology. Acta. Biotheor. **1**, 133–150 (1936)
107. Kuznetsov, V.A., Knott, G.D., Bonner, R.E.: General statistics of stochastic gene expression in eucaryotic cells. Genetics **161**(3), 1321–1332 (2002)
108. Cao, Z., Grima, R.: Analytical distribution for detailed models of stochastic gene expression in eucaryotic cells. Proc. Natl. Acad. Sci. USA **117**(9), 4682–4692 (2020)
109. Magin, R.L.: Fractional calculus models of complex dynamics in biological tissues. Comput. Math. Appl. **59**(5), 1586–1593 (2010)

110. Campoy, E.M., Branham, M.T., Mayorga, L.S., Rogue, M.: Intratumor heterogeneity index in breast carcinomas based on DNA methylation profile. BMC Cancer **19**, 328 (2019)
111. Swanton, Ch.: Intratumor heterogeneity: evolution through space and time. Cancer Res. **72**(19), 4875–4882 (2012)
112. Wu, X.R.: Urothelial tumorigenesis: a tale of divergent pathways. Nat. Rev. Cancer **5**, 713–725 (2005)
113. Castillo-Martin, M., Domingo-Domenech, J., Karni-Schmidt, O., Matos, T.: Molecular pathways of urothelial development and bladder tumorigenesis. Urol. Oncol. **28**(4), 401–408 (2010)
114. Andrews, B.T., Capraro, D.T., Sulkowska, J.I., Onuchic, J.N., Jennings, P.A.: Hysteresis as a marker for complex, overlapping landscapes in proteins. J. Phys. Chem. Lett. **4**, 180–188 (2013)
115. Chatterjee, A., Kaznessis, Y.N., Hu, W.S.: Tweaking biological switches through a better understanding of bistability behavior. Curr. Opin. Biotechnol. **19**, 475–481 (2008)
116. Angeli, D., Ferrell, J.E., Jr., Sontag, E.D.: Detection of multistability, bifurcations, and hysteresis in a large class of biological positive-feedback systems. Proc. Natl. Acad. Sci. USA **101**, 1822–1827 (2004)
117. Eissing, T., Conzelmann, H., Gilles, E.D., Allgoewer, F., Bullinger, E., Scheurich, P.: Bistability analyses of a caspase activation model for receptor-induced apoptosis. J. Biol. Chem. **279**, 36892–36897 (2004)
118. Pomerening, J.R., Sontag, E.D., Ferrell, J.E.: Building a cell cycle oscillator: hysteresis and bistability in the activation of Cdc2. Nat. Cell Biol. **5**(4), 346–351 (2003)
119. Solomon, M.J.: Hysteresis meets the cell cycle. Proc. Natl. Acad. Sci. USA **100**, 771–772 (2003)
120. Vesper, M.D., de Groot, B.L.: Collective dynamics underlying allosteric transitions in hemoglobin. PLoS Comput. Biol. **9**(9), e1003232 (2013)
121. Peter, A.: Corning, The Synergism Hypothesis: A Theory of Progressive Evolution. McGraw Hill, New York (1983)
122. Waliszewski, P., Konarski, J.: A mystery of the Gompertz curve. Losa, G.A., Merlini, D., Nonnenmacher, Th.F., Weibel, R.E. (eds.): Fractals in Biology and Medicine, vol. IV, pp. 277–286. Birkhäuser, Basel (2005)
123. Mytych, J., Romerowicz-Misielak, M., Koziorowski, M.: Long-term culture with lipopolysaccharide induces dose-dependent cytostatic and cytotoxic effects in THP-1 monocytes. Toxicol. In Vitro **42**, 1–9 (2017)
124. Baas Becking, L.G.M.: On the analysis of sigmoid curves. Acta. Biotheor. **8**, 42–59 (1946)
125. West, B., Bologna, M., Grigolini, P.: Physics of Fractal Operators. Springer (2003)
126. Biro, T.S., Telcs, A., Neda, Z.: Entropic distance for nonlinear master equation. Universe **4**, 10 (2018). https://doi.org/10.3390/universe4010010
127. DeMarco, L., Lindsey, K.: Convex shapes and harmonic caps. Arnold Math. J. **3**, 97–117 (2017)
128. Waliszewski, P.: A principle of fractal-stochastic dualism, couplings, complementarity and growth. Control Eng. Appl. Inform. **11**(4), 45–52 (2009)
129. Waliszewski, P., Molski, M., Konarski, J.: On the relationship between fractal geometry of space and time in which a system of interacting cells exists and dynamics of gene expression. Acta Biochim. Pol. **48**(1), 209–220 (2001)
130. Elledge, S.J.: Cell cycle checkpoints: preventing an identity crisis. Science **274**, 1664–1672 (1996)
131. Kullback, S.: Information Theory and Statistics. Dover Publications, New York (1997)
133. Waliszewski, P., Konarski, J.: The complex couplings and Gompertzian dynamics. In: Novak, M.M. (ed.) Complexus Mundi Emergent Patterns in Nature, pp. 343–344. World Scientific Publishing, Singapore (2006)
133. Shannon, C.E.: A mathematical theory of communication. Bell Syst. Tech. J. **27**, 629–630 (1948)
134. Borowkow, A.A.: Kurs teorii wierojatnostiej (Russian). Nauka, Moskwa (1972)

# Quantitative In Vivo Imaging to Enable Tumour Forecasting and Treatment Optimization

**Guillermo Lorenzo, David A. Hormuth II, Angela M. Jarrett, Ernesto A. B. F. Lima, Shashank Subramanian, George Biros, J. Tinsley Oden, Thomas J. R. Hughes, and Thomas E. Yankeelov**

**Abstract** Current clinical decision-making in oncology relies on averages of large patient populations to both assess tumour status and treatment outcomes. However, cancers exhibit an inherent evolving heterogeneity that requires an individual approach based on rigorous and precise predictions of cancer growth and treatment response. To this end, we advocate the use of quantitative in vivo imaging data to calibrate mathematical models for the personalized forecasting of tumour development. In this chapter, we summarize the main data types available from both common and emerging in vivo medical imaging technologies, and how these data can be used to obtain patient-specific parameters for common mathematical models of cancer. We then outline computational methods designed to solve these models, thereby enabling their use for producing personalized tumour forecasts in silico, which, ultimately, can be used to not only predict response, but also optimize treatment. Finally, we discuss the main barriers to making the above paradigm a clinical reality.

G. Lorenzo (✉) · E. A. B. F. Lima · S. Subramanian · G. Biros · J. T. Oden · T. J. R. Hughes
Oden Institute for Computational Engineering and Sciences, The University of Texas at Austin, 201 E. 24th Street, Austin, TX 78712-1229, USA
e-mail: guillermo.lorenzo@utexas.edu; guillermo.lorenzo@unipv.it

E. A. B. F. Lima
e-mail: ernesto.lima@utexas.edu

S. Subramanian
e-mail: shashanksubramanian@utexas.edu

G. Biros
e-mail: biros@oden.utexas.edu

J. T. Oden
e-mail: oden@oden.utexas.edu

T. J. R. Hughes
e-mail: hughes@oden.utexas.edu

G. Lorenzo
Department of Civil Engineering and Architecture, University of Pavia, via Ferrata 3, 27100 Pavia, Italy



I. Balaz and A. Adamatzky (eds.), *Cancer, Complexity, Computation*,
Emergence, Complexity and Computation 46,
https://doi.org/10.1007/978-3-031-04379-6_3

**Keywords** Cancer · Computational oncology · Magnetic resonance imaging · Finite element analysis · Isogeometric analysis · Finite differences · Model selection · Sensitivity analysis · Inverse problems · Patient-specific models · Optimal control theory

## 1 Introduction to Tumour Forecasting

Cancers are highly *heterogeneous* diseases supported by diverse biological mechanisms occurring, interacting, and evolving at multiple spatial and temporal scales [74]. These phenomena span from the phenotypic and genotypic cellular diversity within the tumour to the regional variations of the tumour microenvironment (e.g., vasculature and extracellular matrix), which can result in epigenetic changes in cancer cells or gradients in nutrient availability. Hence, the heterogeneous nature of cancer makes each patient's case unique. However, established, standard-of-care methods determine diagnosis, stage, treatment regimen, and response to treatment according to historical population averages. This paradigm only enables the observation of cancer evolution and the outcome of treatment at fixed time points, offers a limited individualization of disease management, and largely ignores the intrinsic heterogeneity of cancers, which may result in treatment failure [45, 117]. Thus, a new clinical paradigm that effectively integrates the spatiotemporal dynamics of tumour growth and treatment response to identify effective clinical strategies for each patient is desperately needed. We posit that mathematical modeling informed by clinically-relevant data can provide the framework to address this challenge [90, 122, 123].

*Computational oncology* is a rapidly growing field that attempts to leverage mathematical models of the key biological mechanisms that characterize cancer to predict

D. A. Hormuth II · A. M. Jarrett
Oden Institute for Computational Engineering and Sciences and Livestrong Cancer Institutes, The University of Texas at Austin, 201 E. 24th Street, Austin, TX 78712-1229, USA
e-mail: david.hormuth@utexas.edu

A. M. Jarrett
e-mail: ajarrett@utexas.edu

E. A. B. F. Lima
Texas Advanced Computing Center, The University of Texas at Austin, J.J. Pickle Research Campus, Building 205, 10100 Burnet Rd (R8700), Austin, TX 78758, USA

T. E. Yankeelov
Departments of Biomedical Engineering, Diagnostic Medicine, and Oncology, Oden Institute for Computational Engineering and Sciences, Livestrong Cancer Institutes, The University of Texas at Austin, 201 E. 24th Street, Austin, TX 78712-1229, USA
e-mail: thomas.yankeelov@utexas.edu

Department of Imaging Physics, MD Anderson Cancer Center, 1515 Holcombe Blvd, Houston, TX 77030, USA how a patient's tumour will grow and respond to treatment [90, 122, 123]. Computer simulations of these models provide personalized *tumour forecasts*, designed to ultimately assist oncologists in clinical decision-making. For example, tumour forecasts may predict disease progression, thereby providing much needed guidance on the optimal intervention strategy early in the course of therapy. Indeed, we hypothesize that treatment optimization can be achieved through the development and rigorous validation of practical mathematical models and efficient computational methods that can provide accurate personalized predictions of cancer development and treatment response.

A fundamental challenge in computational oncology is accomplishing the *patient-specific parameterization* of the biological mechanisms involved in cancer models (e.g., tumour cell mobility, proliferation and death rates, or therapy efficacy). In general, these parameters are extremely difficult to measure in vivo in human tumours. However, medical imaging may provide a viable source of data for this purpose. Clinical oncology currently focuses on *anatomical imaging* for the diagnosis, treatment, monitoring, and assessment of therapeutic response of solid tumours [111] (e.g., measuring tumour size, identifying invasion into adjacent structures, and detecting metastasis). Unfortunately, anatomical imaging frequently fails to capture the heterogeneous underlying biology within tumours. Alternatively, *quantitative imaging* techniques enable the measurement of clinically-relevant biological features of tumours, such as tumour cell density, blood volume fraction, and perfusion [42]. Thus, these quantitative imaging techniques can be used to assess the spatiotemporal evolution of a cancer's heterogeneous architecture, morphology, growth dynamics, and response to therapy, thereby providing the necessary data to parameterize predictive models of tumour growth and treatment [41, 122].

In this chapter, we will discuss how quantitative imaging can be used to enable tumour forecasting and optimization of therapeutic response. We will begin by identifying relevant quantitative imaging data types and how they are incorporated into existing image-based models of cancer growth and treatment. We will also provide insights into the technical aspects of numerical implementation, model calibration, and model selection. Then, we will introduce a promising framework to optimize patient treatment plans. We will conclude with a discussion of the barriers to successfully translating image-based computational tumour forecasting into patient care.

## 2 Relevant Data Types from Medical Imaging

While measuring tumour size throughout therapy is central in oncological response assessments [111], the dynamics of tumour size changes are often temporally downstream of intratumoral biological and physiological responses to therapy. Magnetic resonance imaging (MRI) and positron emission tomography (PET) provide non-destructive and non-invasive 3D quantitative measurements of biological properties within and around the tumour. Hence, the acquisition of these imaging data at several timepoints is well-suited to initialize and parameterize mathematical models of tumour growth and treatment response. In this section, we will briefly introduce the relevant MRI and PET measurements that have been commonly used in computational oncology (see Fig. 1 for representative images of these techniques). For a detailed review of advanced MRI and PET techniques in oncology, the reader is respectively referred to [28, 42].

## 2.1 Diffusion Weighted Magnetic Resonance Imaging

Diffusion weighted (DW-) MRI is an established technique that has been applied in oncology as a noninvasive assessment of cellularity changes during treatment [84]. DW-MRI is sensitive to the diffusion of water molecules within tissue. In a DW-MRI experiment, water molecules are first tagged based on their spatial location. Then, after a short delay of typically 20–60 ms, a second spatial-encoded tag is applied. During this delay, water molecules move throughout the tissue due to diffusion. If the water molecules do not travel far, the first spatial-encoded tag can be largely removed by the second spatial-encoded tag and there is no loss (or gain) in signal intensity. However, if the water molecules move throughout the domain, there is a net-difference between the two spatial-encoded tags resulting in a decrease in signal intensity. Thus, the signal intensities within each voxel in the resulting image are "weighted" based on water diffusion. In practice, several diffusion weighted experiments are performed with different settings (e.g., varied diffusion-sensitizing gradient amplitudes of the magnetic field) to spatially quantify the apparent diffusion coefficient ($ADC$) of water. However, water diffusion in tissue is heavily restricted by cells, macromolecules, and extracellular structures. Hence, these physical barriers reduce the measured $ADC$. This phenomenon has been observed in several studies showing an inverse correlation between $ADC$ and cellularity [4, 8, 55]. Following these reports' results, $ADC$ can be used to estimate cellularity using:

$$N(\mathbf{x}, t) = \theta \left( \frac{ADC_w - ADC(\mathbf{x}, t)}{ADC_w - ADC_{\min}} \right) \tag{1}$$

where $\theta$ represents the maximum tumour cell carrying capacity for an imaging voxel (determined by the voxel dimensions and assumptions in cell geometry and packing density), $ADC_w$ is the $ADC$ of free water at 37°C (i.e., $2.5 \cdot 10^{-3}$ mm$^2$/s; [116]), $ADC(\mathbf{x}, t)$ is the $ADC$ value at a given 3D position $\mathbf{x}$ and time $t$, and $ADC_{\min}$ is the minimum $ADC$ value observed within the tumour. Figure 1 displays a representative $ADC$ map from breast and brain cancer. While there are significant correlations between cellularity and the measured $ADC$, cellularity is not the sole factor in dynamic changes in $ADC$. Changes in cell size, cell permeability, and tissue tortuosity may alter the measured $ADC$ [84]. Other diffusion-based imaging approaches can also report on cell size [55] and diffusion anisotropy [105]. The reader is referred to [58] for a technical review of DW-MRI and its applications in oncology.

## 2.2 Dynamic Contrast-Enhanced Magnetic Resonance Imaging

Dynamic contrast-enhanced (DCE-) MRI consists of the rapid acquisition of a series of heavily $T_1$-weighted images before, during, and after the injection of a $T_1$ altering-contrast agent (typically a Gadolinium chelate) to probe vascular properties in tissue [121]. Using a pre-contrast $T_1$ map, any post-contrast $T_1$ changes can be related to the concentration of the contrast agent. Thus, each image voxel yields a signal intensity time course that can be related to the concentration of the contrast agent within that voxel. The subtraction images obtained from pre- and post-contrast enhanced images are often used to identify tumour regions, which usually show areas of rapid and intense enhancement due to their higher and more permeable vascularity than the neighboring healthy tissue. The dynamics of signal intensity are commonly analyzed with a two-compartment pharmacokinetic model describing the extravasation of the contrast agent from the plasma space to the tissue space [121]. The solution to this model is given by

$$C_t(\mathbf{x},t) = K^{trans}(\mathbf{x}) \int_0^t C_p(u) e^{-\frac{K^{trans}(\mathbf{x})}{v_e(\mathbf{x})}(t-u)} du + v_p(\mathbf{x}) C_p(t), \tag{2}$$

where $C_t(\mathbf{x},t)$ is the concentration of the contrast agent in tissue at position $\mathbf{x}$ and time $t$, $C_p(t)$ is the concentration of the contrast agent in the plasma space at time $t$, $K^{trans}(\mathbf{x})$ is the volume transfer constant from the plasma to tissue space, $v_e(\mathbf{x})$ is the extravascular-extracellular volume fraction, and $v_p(\mathbf{x})$ is the plasma volume fraction. Importantly, $C_t$, $K^{trans}$, $v_e$, and $v_p$ are all voxel-specific and are related to structural (cell density) and physiological (vessel permeability and perfusion) properties. $C_p(t)$ can be measured directly for individual subjects from a large artery within the image field of view or can be replaced with a population-based estimate [65]. Figure 1 shows $K^{trans}$, $v_e$, and $v_p$ maps from a preclinical and a clinical study.

## 2.3 Molecular Imaging with Positron Emission Tomography

PET relies on the injection of a radiopharmaceutical (or PET tracer) to generate image contrast. As there is no endogenous signal, PET has excellent sensitivity to detect and localize the distribution of radiopharmaceuticals throughout the body. Several radiopharmaceuticals have been developed to probe tumour properties, such as glucose metabolism (*via* $^{18}$F-fludeoxyglucose or $^{18}$FDG), hypoxia (*via* $^{18}$F-fluoromisonidazole or $^{18}$F-MISO), cellular proliferation (*via* $^{18}$F-Flurodeoxythymidine [119]), and receptor status (e.g., $^{64}$Cu-diethylenetriaminepentaacetic acid Trastuzumab for HER2+ positive cancers [79]). We will primarily focus on $^{18}$FDG and $^{18}$F-MISO as they are well-established in oncology, but the analysis techniques are similar for other PET tracers.

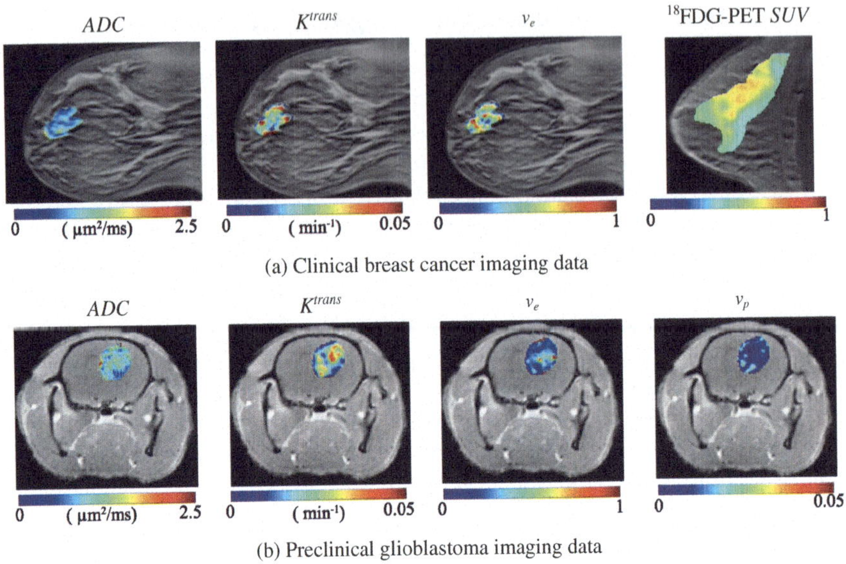


**Fig. 1** Representative quantitative imaging measurements from clinical and preclinical settings. **a** Parameter maps extracted from DW-MRI, DCE-MRI, and $^{18}$FDG-PET through the central slice of a breast tumour (all MR data are from the same patient, PET data are from a different individual). DW-MRI provides estimates of $ADC$, while DCE-MRI yields estimates of $K^{trans}$ and $v_e$. The $^{18}$FDG-PET $SUV$ map shows increased glucose uptake within the breast tumour relative to surrounding tissue. **b** Parameter maps acquired in a preclinical murine model of glioblastoma from DW-MRI ($ADC$) and DCE-MRI ($K^{trans}$, $v_e$, and $v_p$)

In an $^{18}$FDG-PET study, a single image is acquired following the injection of the glucose analogue $^{18}$FDG, which cells uptake in a similar fashion to glucose. However, once internalized, $^{18}$FDG is phosphorylated and trapped intracellularly. The resulting image intensities are proportional to the concentration of $^{18}$FDG within each voxel. PET tracer uptake can be quantified using the standardized uptake value ($SUV$), which is the ratio of the concentration of $^{18}$FDG radioactivity in tumour tissue to the total injected dose and divided by the patient's body weight. In oncology studies, contrast between tissues is typically generated due to variations in glucose uptake due to an overexpression of glucose transporters and hexokinase activity in tumour cells relative to healthy cells [14]. This difference in $^{18}$FDG uptake is also shown in Fig. 1.

Likewise, in $^{18}$F-MISO PET a single image is also acquired following the injection of $^{18}$F-MISO, which is a radiopharmaceutical that produces images sensitive to oxygen concentration in tissue [88]. After $^{18}$F-MISO is internalized by cells, it is reduced to produce a radical anion. In normoxic or oxygen-rich environments, oxygen accepts the electron from the radical anion enabling $^{18}$F-MISO to leave the cell. Conversely, in hypoxic or oxygen-poor environments, the radical anion of $^{18}$F-MISO binds to other intracellular macromolecules trapping it within the cell. Thus, the concentration of $^{18}$F-MISO and the produced PET signal within a voxel are inversely proportional to the oxygen concentration. $^{18}$F-MISO uptake is quantified using the standardized uptake value ($SUV$) or the oxygen enhancement ratio ($OER$), which is the ratio of signal intensity in tumour relative to blood.

## 3 Image-Based Mathematical Models of Cancer

Medical imaging provides an excellent way to develop, calibrate, and validate personalized mathematical models of cancer evolution and treatment response [122] for three main reasons. First, medical imaging enables the in vivo measurement of relevant biological properties in tumour and healthy tissues, which would otherwise be impractical or impossible to measure in individual patients. Second, medical imaging data can be obtained frequently throughout the clinical management of the patient's tumour, which enables model calibration. Third, medical imaging data are acquired on a regular voxel grid, which facilitates their computational processing. In this section, we discuss common image-based models of tumour growth and treatment response that leverage the quantitative imaging data types introduced in Sect. 2. Figure 2 shows simulation outputs of many of the models discussed in this section.

### *3.1 Baseline Tumour Growth Models*

In mathematical oncology [90], the *logistic growth* model is one of the simplest and most common approaches to describe changes in tumour volume [11] or cell number [6] over time. It is a flexible model that can be adapted to in vitro and in vivo data alike. The formulation of the logistic growth model over a certain tissue region of interest follows the partial differential equation (PDE):

$$\frac{\partial N(\mathbf{x}, t)}{\partial t} = k(\mathbf{x}) N(\mathbf{x}, t) \left(1 - \frac{N(\mathbf{x}, t)}{\theta}\right), \tag{3}$$

where $N(\mathbf{x}, t)$ is the *tumour cell density* at position $\mathbf{x}$ and time $t$, $k(\mathbf{x})$ is a spatially-varying net proliferation rate, and $\theta$ is the carrying capacity. The image-informed applications of this model have typically been posed voxelwise, such that $N(\mathbf{x}, t)$ is redefined as the number of cancer cells within the voxel in position $\mathbf{x}$ at time $t$. Atuegwu et al. [6] used this approach to predict tumour growth in breast cancer patients receiving neoadjuvant chemotherapy. First, they used Eq. (1) to estimate $N(\mathbf{x}, t)$ from $ADC$ maps obtained *via* DW-MRI. The estimates of $N$ at baseline (pre-treatment) and after one cycle of therapy were used to determine $k(\mathbf{x})$. Then, Atuegwu et al. used their model equipped with the resulting $k(\mathbf{x})$ to predict $N$ at the conclusion of therapy. They observed a strong correlation between the predictions and data estimates of $N$ over the entire tumour (Pearson correlation coefficient, PCC,

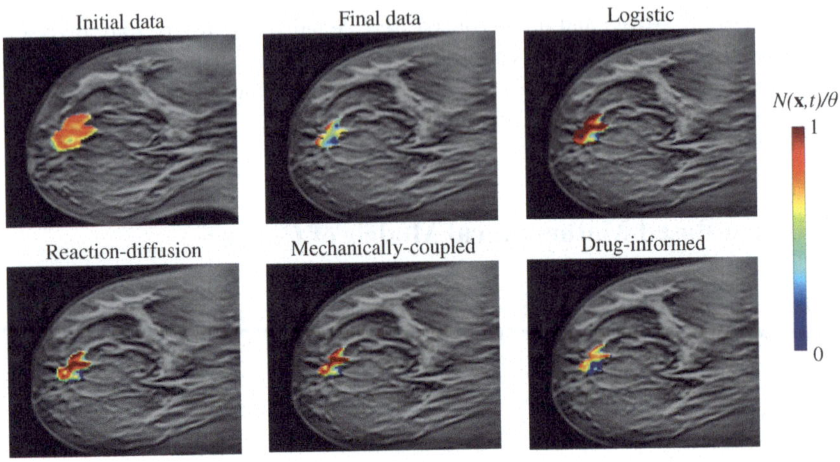


(a) Clinical breast cancer setting

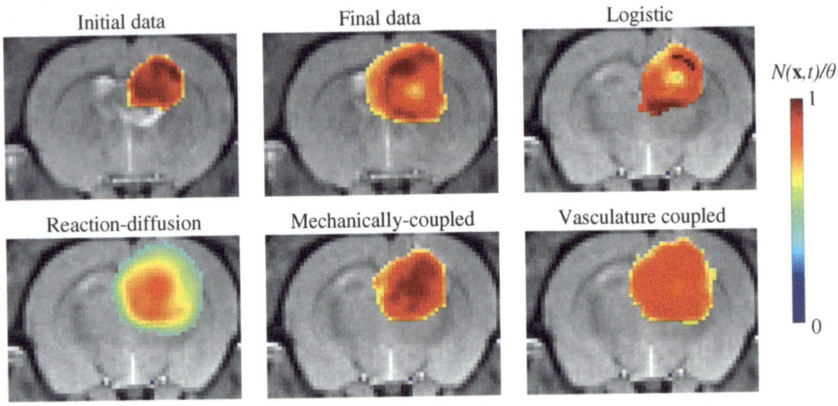


(b) Preclinical glioblastoma setting

**Fig. 2** Differences between image-based models in clinical and preclinical settings. **a** Example from a clinical breast cancer model, where measured initial, measured final, and model forecasts of the final distributions of tumour cell density are shown. The logistic model fails to capture the expansion of the tumour into nearby tissue, while the remaining models incorporating a diffusion term perform better. **b** Example from a preclinical model of glioblastoma, where measured initial, measured final, and model predicted final distributions of tumour cell density are shown. The logistic model also fails to predict the expansion, but does predict an intratumoral low cell density area

of 0.95) and for individual voxels (PCC = 0.70). Figure 2 shows an example of this approach for a clinical breast cancer model and a preclinical glioblastoma model.

However, the logistic growth model fails to capture the potential movement of cells that may occur over time. To overcome this limitation, the logistic growth model can be extended to a *reaction-diffusion* model given by the PDE

$$\frac{\partial N(\mathbf{x},t)}{\partial t} = \nabla \cdot (D\nabla N(\mathbf{x},t)) + k(\mathbf{x})N(\mathbf{x},t)\left(1 - \frac{N(\mathbf{x},t)}{\theta}\right), \tag{4}$$

where the first term on the right is a diffusion term describing the movement of tumour cells with a constant diffusion coefficient $D$, while the second term on the right is a reaction term represented by the logistic growth of cancer cells.

Equation (4) is well established in computational oncology [35, 37, 50, 89, 91, 106, 115, 118]. The work of Swanson et al. [7] in high grade gliomas showed one of its first image-informed applications, using anatomical MRI data to provide segmentation of tumour boundaries and fixed cell-density counts in enhancing and non-enhancing disease. The spatiotemporal changes in tumour boundaries were used to estimate a constant tumour-specific proliferation rate $k$ and tissue-specific $D$, which were then used to predict patient survival. This approach has had promising results in relating growth kinetics to patient outcomes [7, 81]. However, it does not describe the intratumoral heterogeneity of cell density. Hormuth et al. [39] addressed this limitation by estimating $N$ from $ADC$ maps obtained *via* DW-MRI using Eq. (1) in a murine model of glioma. In this study, animals were imaged up to seven times over ten days. The first three imaging datasets were used to initialize $N$ as well as to calibrate $k(\mathbf{x})$ and $D$. In a separate calibration, a spatially-constant $k$ (i.e., tumour specific) was also calibrated along with $D$. The calibrated model was then used to predict $N(\mathbf{x}, t)$ at the remaining imaging visits. While both calibration scenarios overestimated future tumour growth, the predictions with a spatially-varying $k(\mathbf{x})$ rendered lower tumour volume errors, higher Dice correlation coefficients, and higher concordance correlation coefficients (CCC; all $p < 0.05$). These results highlight the importance of accounting for the intratumoral heterogeneous dynamics to obtain accurate tumour forecasts and the promising potential of quantitative imaging to provide the required data for this purpose. Other studies have used alternative quantitative imaging measures to inform the reaction-diffusion model in Eq. (4); examples include incorporating anisotropic diffusion *via* diffusion tensor imaging (DTI, a form of DW-MRI [105]) [53, 106], using cell density measurements *via* contrast-enhanced computed tomography [118], and estimating cell phenotypes *via* DCE-MRI [91].

Alternatively, *phase-field models* are another common paradigm to describe tumour growth [1, 66, 70, 71, 120]. The *phase field* $\phi(\mathbf{x}, t)$ identifies healthy tissue (e.g., $\phi = 0$) from tumour tissue (e.g., $\phi = 1$), showing a smooth and thin transition between either region. Phase-field models usually focus on the dynamics of tumour morphology through the evolution of the healthy-tumour interface, which is implicitly defined by a phase-field isosurface. These models rely on more complex physics than those described above, usually requiring the definition of an energy functional that drives tumour growth [31, 66, 67]. Phase-field modeling has been notably applied in brain tumours [1, 66], prostate cancer [70, 71], and tumour angiogenesis [113, 120]. However, there is a paucity of studies using quantitative imaging data to inform phase-field models. This is partly due to their more complex dynamics, which usually requires a higher number of parameters, larger and richer patient-specific datasets, more advanced numerical methods, and more computational resources.

We identified two illustrative works that use imaging measurements to initialize, calibrate, and/or validate phase-field models of cancer. Lima et al. [66, 67] have been investigating model selection to identify the best formulation of murine glioma growth according to longitudinal anatomical MRI tumour measurements, including an array of phase-field models also accounting for the local tumour-induced mechanical stress field (see Sect. 3.2). Their work shows that phase-field models are plausible formulations of tumour growth and in [66] they emerge as the best models indeed. Additionally, Agosti et al. [1] developed a phase-field model of glioblastoma multiforme that uses quantitative DTI data to define anisotropic tumour cell motility and nutrient diffusion. Their work focuses on the prediction of tumour recurrence after surgical resection and subsequent radiotherapy. By accounting for post-surgery changes in tissue architecture, they obtained a Jaccard index of 0.71 post-radiotherapy.

## *3.2 Mechanically-Coupled Models*

Local mechanical tissue properties and tumour-induced mechanical stresses are known to affect cancer growth dynamics [34, 47, 80]. For example, Helmlinger et al. [34] observed that tumour spheroid growth in vitro was increasingly inhibited as the substrate matrix stiffness was augmented. Uncontrolled tumour growth can also severely deform healthy tissue structures, thereby adversely impacting patient health and quality of life. Therefore, several mathematical models of cancer couple tumour growth dynamics with local mechanical equilibrium [18, 35, 40, 50, 69, 71, 115].

A common approach [115] is to dampen the diffusion coefficient in Eq. (4) with a function of local tissue stress:

$$D(\mathbf{x}, t) = D_0 e^{-\gamma_v \sigma_{vm}(\mathbf{x},t)}, \tag{5}$$

where $D(\mathbf{x}, t)$ is now a spatially and temporally varying diffusion coefficient, $D_0$ is the tumour cell diffusion coefficient in the absence of stress, $\gamma_v$ is an empirical coupling constant, and $\sigma_{vm}(\mathbf{x}, t)$ is the von Mises stress. Here, $\sigma_{vm}(\mathbf{x}, t)$ is used to summarize the local mechanical stress field, which is calculated assuming quasistatic linear elastic equilibrium with tissue-specific mechanical properties:

$$\nabla \cdot \left(\lambda \left(\nabla \cdot \mathbf{u}\right) \mathbf{I} + \mu \left(\nabla \mathbf{u} + \nabla \mathbf{u}^T\right)\right) - \gamma_N \nabla N = \mathbf{0}, \tag{6}$$

where $\lambda$ and $\mu$ are the Lamé coefficients (related to the tissue's Young modulus $E$ and Poisson's ratio $\nu$), $\mathbf{u}$ is the displacement field due to tumour cell growth, and $\gamma_N$ is another empirical coupling constant. In Eq. (5), the first term on the left-hand side represents the linear elastic tissue response to the local tumour-induced forces described by the second term on the left. Weis et al. [115] used tumour cell number estimates from DW-MRI data using Eq. (1) to initialize and calibrate a mechanically-coupled reaction-diffusion model of breast cancer growth during neoadjuvant chemotherapy consisting of Eqs. (4)–(6). Their work shows that the mechanically-coupled model rendered more accurate predictions of $N$ (PCC=0.85) than the baseline reaction-diffusion model (PCC= −0.29). Several subsequent studies have also used Eqs. (5)–(6) to couple mechanics to breast and brain tumour dynamics [40, 50], as shown in Fig. 2. Lima et al. further considered a mechanical inhibition of tumour proliferation following a similar formulation to Eq. (5) [66, 67]. Moreover, the prostate cancer model of Lorenzo et al. [71] extended Eq. (5) to combine the measure of mechanical tissue distortion *via* $\sigma_{vm}$ with hydrostatic stress, which is not captured by $\sigma_{vm}$ and contributes to a more precise description of intratumoral stress.

Other mathematical models couple local mechanics to tumour growth dynamics through a reaction-advection-diffusion equation [35, 69] in which the tumour cell drift velocity is related to the displacement field, thereby explicitly simulating the displacement of cells due to mechanical deformation. Additionally, Wong et al. [118] leveraged a hyperelastic biomechanical model. Interestingly, in this study tumour dynamics was described using a reaction-diffusion model in which the proliferation rate $k(\mathbf{x})$ was calibrated with $^{18}$FDG-PET $SUV$ data, as follows:

$$k(\mathbf{x}) = \frac{\alpha SUV(\mathbf{x}) - \beta c(\mathbf{x}, t)}{c(\mathbf{x}, t)\,(1 - c(\mathbf{x}, t))}, \tag{7}$$

where $SUV(\mathbf{x})$ is the standardized uptake value at position $\mathbf{x}$, $\alpha$ and $\beta$ are unknown constants to be calibrated, and $c(\mathbf{x}, t)$ is the cell volume fraction estimated from computed tomography at position $\mathbf{x}$ and time $t$.

### 3.3 Vasculature-Coupled Models

Co-opting of local vasculature and recruitment of new blood vessels *via* angiogenesis is a critical component of cancer development that is needed to support growth past 2–3 mm$^3$ in size [30, 46]. Thus, understanding the evolving distribution and function of the tumour-supporting vasculature is crucial to accurately model tumour growth and treatment response. There is an extensive literature on mathematical models of tumour angiogenesis [113]. However, very few describe this phenomenon at the imaging/tissue scale [37, 109] or personalize it for individual tumours.

Hormuth et al. [37] developed a murine model of glioma growth coupled with angiogenesis that was initialized and calibrated using tumour cell number estimates obtained from DW-MRI *via* Eq. (1) and estimates of the blood volume fraction extracted from DCE-MRI. The spatiotemporal evolution of tumour cells and vasculature was described using two coupled reaction-diffusion equations. In this model, vasculature influenced the direction of tumour growth and was coupled to the carrying capacity. Similarly, tumour cells also influenced the direction of vasculature evolution. The animals were imaged up to seven times over a period of ten days. Model parameters were calibrated using the first three imaging datasets, and then used in a forward evaluation of the model to predict tumour growth at the remaining imaging time points. The authors observed that their model resulted in less than 10.3% error in tumour volume predictions and less than 9.4% error at the voxel-level for all prediction time points. Figure 2 shows an example of this approach in a pre-clinical model of glioblastoma.

Roque et al. [91] developed a vasculature-informed preclinical reaction-diffusion model of breast cancer accounting for normoxic, hypoxic, and necrotic cancer sub-populations along with nutrient dynamics, which regulates normoxic cell proliferation as well as the normoxic-hypoxic and hypoxic-necrotic transfer rates. While not explicitly evolving the tumour-supporting vascular network, the authors used vasculature-derived parameters obtained from DCE-MRI (e.g., blood flow, mean transit time, and maximum enhancement) to initialize all model variables and calibrate key parameters. While the study results suggested that further model development is needed to capture individual differences in tumour growth, this work is a unique effort to identify tumour subpopulations using quantitative imaging data.

## 3.4 Radiotherapy

Radiotherapy is a common treatment for many cancers [77]. However, intratumoral heterogeneity may result in significant variations in treatment response, which may ultimately lead to poor therapeutic outcomes [9, 30]. Image-based modeling could prove valuable to predict the response to radiotherapy and hence optimize treatment protocols for individual patients. To this end, several studies have investigated incorporating imaging measures from PET [89] and MRI [33, 36, 38, 67] into reaction-diffusion based models to characterize patient response to radiotherapy. The usual approach to model radiotherapy effects is by instantaneously killing a fraction of tumour cells at treatment times [33, 36, 38, 67, 89]. This strategy may be further combined with a transient or permanent reduction in tumour cell proliferation [36, 38, 67]. These radiotherapy effects are usually modeled as a function of the prescribed dose, which may also account for local tumour cell and vascular densities.

Rockne et al. [89] adapted the glioblastoma model by Swanson et al. [7] to explicitly incorporate cell death due to radiotherapy based on $^{18}$F-MISO PET data. Oxygen concentration and the degree of hypoxia in tumours are known to significantly impact response to radiotherapy [112]. Thus, Rockne et al. used $^{18}$F-MISO PET to assess the level of hypoxia by calculating the $OER$, which is then used along with the usual linear quadratic model of radiotherapy response [22] to calculate cell survival, $S$, as follows:

$$S = \exp\left(-\alpha\,(OER(\mathbf{x}))\left(d + \frac{d^2}{\alpha/\beta\,(OER(\mathbf{x}))}\right)\right), \tag{8}$$

where $d$ is the prescribed radiation dose while $\alpha\,(OER(\mathbf{x}))$ and $\alpha/\beta\,(OER(\mathbf{x}))$ are radiosensitivity parameters as a function of the $OER$ at position $\mathbf{x}$. Rockne et al. observed that predictions by a model featuring Eq. (8) outperformed those obtained with a model with uniform radiosensitivity (1.1% *vs* 14.6% error in tumour volume, respectively).

### 3.5 Chemotherapy

Chemotherapy is another common treatment for most cancers [77]. Unlike the localized nature of radiation therapy, chemotherapy relies on drugs that are administered systemically throughout the body. While chemotherapy has traditionally leveraged cytotoxic drugs (i.e., promoting cell death), recent approaches also use drugs targeting specific cancer cell markers to decrease proliferation or triggering particular immune responses. Similar to radiation therapy, challenges for modeling chemotherapies stem from quantifying how much drug is distributed in the tumour (and healthy tissues) and patient-specific treatment efficacy. Multiple mathematical models have been proposed to describe the effect of chemotherapy on tumour growth [107, 110, 124], but only a few are informed by quantitative imaging data.

In particular, the contribution by Jarrett et al. [50] extended the mechanically-coupled reaction diffusion model consisting of Eqs. (4)–(6) by including the dynamic effect of chemotherapy in the tumour growth equation:

$$\frac{\partial N(\mathbf{x},t)}{\partial t} = \nabla \cdot (D(\mathbf{x},t)\nabla N(\mathbf{x},t)) + k(\mathbf{x})N(\mathbf{x},t)\left(1 - \frac{N(\mathbf{x},t)}{\theta}\right) - \alpha C_d(\mathbf{x},t)N(\mathbf{x},t) \quad (9)$$

where $\alpha$ is the patient-specific drug efficacy and $C_d(\mathbf{x},t)$ is the drug concentration in the tissue. $C_d(\mathbf{x},t)$ was approximated patient-wise by means of the two-compartment model commonly used to analyze the contrast agent pharmacokinetics in DCE-MRI data (see Sect. 2.2). This approach has two central limitations: it assumes that the drug and the contrast agent have similar dynamics, and that all chemotherapies explicitly induce tumour cell death. However, Jarrett et al. showed that their drug-informed model predictions outperformed those of the mechanically-coupled reaction-diffusion model without the drug term when compared to patient-specific estimates of tumour cell density extracted from DW-MRI *via* Eq. (1) at the end of chemotherapy; in particular, the CCC improved from 0.85 to 0.99 ($p < 0.01$).

## 4 Computational Methods to Solve Image-Based Cancer Models

Mechanistic models of cancer usually consist of coupled, nonlinear PDEs. Using the appropriate numerical strategies, these cancer models can be solved and rendered as a computer simulation of the spatiotemporal development of a patient's tumour; i.e., a tumour growth forecast. In this section we will provide an elementary description of the Finite Difference Method (FDM) [64], Finite Element Analysis (FEA) [43] and

Isogeometric Analysis (IGA) [21]. All these numerical methods have been widely used to solve PDEs in science and engineering.

## 4.1 The Finite Difference Method

The FDM relies on a direct approximation of the derivatives involved in the PDEs of the model by means of Taylor series expansions [64]. To apply the FDM, we define a global time interval for the simulation $[0, T]$ and a geometric domain $\Omega$ consisting of a 3D box that includes the tumour-harboring organ. Let us discretize $[0, T]$ with a constant time step $\Delta t$, leading to a partition in time subintervals $[t_n, t_{n+1}]$, such that $t_{n+1} - t_n = \Delta t$, $t_0 = 0$, $t_{n_t} = T$, and $n = 0, \dots, n_t - 1$. We discretize $\Omega$ with a uniform 3D cartesian grid composed of $n_p = n_x n_y n_z$ nodes numbered $A = 1, \dots, n_p$, where $n_x$, $n_y$ and $n_z$ are the number of nodes in each spatial direction. Let $\mathbf{g} = (i, j, k)$ further denote the grid coordinates of each node, such that $i = 0, \dots, n_x - 1$, $j = 0, \dots, n_y - 1$, and $k = 0, \dots, n_z - 1$. Then, the spatial coordinates of each node $A$ can be written as $\mathbf{x}_A = \mathbf{x}_0 + \mathbf{h} \circ \mathbf{g}$, where $\mathbf{h} = (h_x, h_y, h_z)$ is a vector holding the grid spacing in each spatial direction. Figure 3 illustrates an FDM grid in 2D.

The standard FDM uses first-order approximations of the time and spatial derivatives in the PDE at the grid nodes and at a certain time instant $\tilde{t} \in [t_n, t_{n+1}]$. For example, the time derivative in Eq. (4) would be approximated by

$$\frac{\partial N(\mathbf{x}, t)}{\partial t} \approx \frac{N(\mathbf{x}_A, t_{n+1}) - N(\mathbf{x}_A, t_n)}{\Delta t} = \frac{N^{n+1}_{i,j,k} - N^{n}_{i,j,k}}{\Delta t} \tag{10}$$

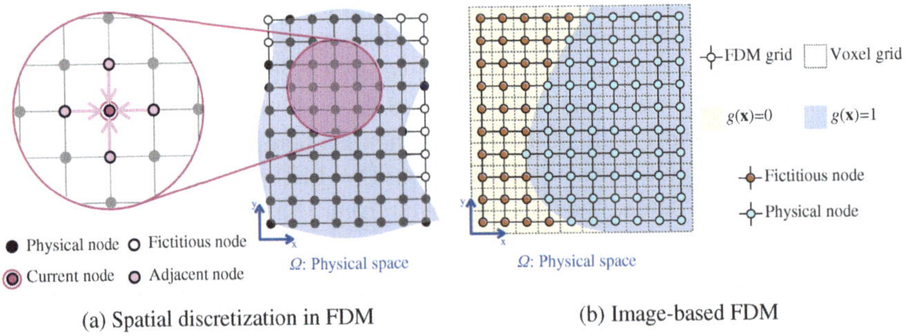


**Fig. 3** The Finite Difference Method (FDM). **a** In FDM, the model equations are solved on a rectangular grid of physical nodes, which approximates the physical space $\Omega$ representing the problem's geometry. Ancillary fictitious nodes may also be required to construct the FDM grid. Spatial derivatives are approximated with linear combinations of the model solution at times $t_n$ and $t_{n+1}$ on each node and adjacent neighbors in each grid direction. **b** The FDM can be formulated using imaging voxel data by defining one node per voxel. The organ segmentation can be used to build a map $g(\mathbf{x})$ to define $\Omega$ and hence identify physical ($g(\mathbf{x}) = 1$) and fictitious nodes ($g(\mathbf{x}) = 0$)

on each node at $t = t_{n+1}$. Higher-order derivatives are recursively approximated with first-order approximations of the subsequent lower-order derivatives. Ultimately, the FDM method reduces the PDE on every node to an algebraic equation involving a combination of values of the PDE solution on the current and adjacent nodes in each spatial direction (see Fig. 3) at instants $t_n$ and $t_{n+1}$. Then, the general strategy is to recursively use the known nodal values of the PDE solution at $t_n$, $\{N(\mathbf{x}_A, t_n)\}_{A=1,\ldots,n_p}$, to calculate the nodal values at $t_{n+1}$, $\{N(\mathbf{x}_A, t_{n+1})\}_{A=1,\ldots,n_p}$.

Depending on the choice of $\tilde{t}$, there are three common FDM approaches in practice: the explicit Euler method ($\tilde{t} = t_n$), the implicit Euler method ($\tilde{t} = t_{n+1}$), and the Crank–Nicolson method ($\tilde{t} = t_n + \Delta t/2$) [64]. The explicit method enables the direct calculation of the PDE solution on the grid nodes at $t_{n+1}$, $\{N(\mathbf{x}_A, t_{n+1})\}_{A=1,\ldots,n_p}$, from the nodal values of the solution at $t_n$. This approach involves a minimal computational cost, which has been exploited for the recursive model resolutions involved in the patient-specific calibration of image-based models of brain and breast cancer [37, 50]. For example, the application of the explicit method to Eq. (4) yields

$$\frac{N^{n+1}_{i,j,k} - N^n_{i,j,k}}{\Delta t} = D\left(\frac{N^n_{i-1,j,k} - 2N^n_{i,j,k} + N^n_{i+1,j,k}}{h_x^2} + \frac{N^n_{i,j-1,k} - 2N^n_{i,j,k} + N^n_{i,j+1,k}}{h_y^2} + \frac{N^n_{i,j,k-1} - 2N^n_{i,j,k} + N^n_{i,j,k+1}}{h_z^2}\right) + k_{i,j,k} N^n_{i,j,k}\left(1 - \frac{N^n_{i,j,k}}{\theta}\right) \quad (11)$$

at every grid node, where we have denoted $k(\mathbf{x}_A) = k_{i,j,k}$. Note that in Eq. (11), we can directly compute $N^{n+1}_{i,j,k}$ from a linear combination of nodal values at $t_n$. However, the explicit method usually requires small time steps to ensure numerical stability. The implicit and Crank–Nicolson methods lead to a system of $n_p \times n_p$ algebraic equations whose resolution provides $\{N(\mathbf{x}_A, t_{n+1})\}_{A=1,\ldots,n_p}$. These FDM schemes are computationally more intensive, but show better numerical stability and enable the use of larger time steps. Application of these methods to nonlinear PDEs like Eq. (4) results in a nonlinear algebraic system, which can be solved with Newton's method by iteratively solving the corresponding linearized system [64]. Alternatively, an implicit-explicit method can use an implicit approach for the diffusion operator and an explicit scheme for the nonlinear logistic term [64, 92], which leads to a linear algebraic system. Currently, multiple sparse-matrix algorithms enable a computationally efficient resolution of most linear systems emanating from the application of the implicit, Crank–Nicolson, and implicit-explicit methods [64].

Boundary conditions (BCs) in FDM are applied to the grid nodes lying on the boundary of $\Omega$ (i.e., $\partial\Omega$). The usual approach is to fix the value of the PDE solution on them (Dirichlet BCs) or to approximate a differential boundary condition with an FDM scheme (Neumman and Robin BCs) [64]. However, organ borders have complex geometries that rarely coincide with the FDM cartesian grid. This leads to the partition of $\Omega$ into the physical domain, corresponding to the tumour-harboring organ, and a fictitious domain, as shown in Fig. 3. FDM codes label grid nodes as physical or fictitious and only solve the PDE on the former. Additionally, FDM codes need to identify the physical nodes closer to organ borders to apply BCs.

FDM may be appealing for image-based cancer models because of its simplicity, rapid implementation, and that the cartesian grid can naturally fit the voxel datasets obtained with the imaging technologies described in Sect. 2, as shown in Fig. 3. However, the FDM neglects the approximation of the organ geometry and simply relies on placing sufficient grid nodes to capture the organ's border. This impedes an accurate implementation of BCs and may also compromise the resolution of geometry-sensitive problems (e.g., mechanics). FEA and IGA overcome these limitations, also providing superior numerical results that are supported by a strong and rigorously demonstrated mathematical basis [21, 43].

## 4.2 Finite Element Analysis and Isogeometric Analysis

### 4.2.1 General Framework

The central constituents of FEA and IGA are (i) the weak or variational formulation of the strong form of the model, and (ii) a robust approximation of this variational formulation using finite-dimensional function spaces with powerful approximation properties [21, 43]. To define and illustrate these ideas, let us start by considering the stationary heat equation over a certain physical domain $\Omega$:

$$\nabla \cdot (\kappa \nabla u(\mathbf{x})) + f(\mathbf{x}) = 0, \tag{12}$$

where $\kappa$ is the constant heat conductivity, $u(\mathbf{x})$ is the spatial map of temperatures over $\Omega$, and $f(\mathbf{x})$ is a heat source. We further consider homogeneous Dirichlet BCs (i.e., $u(\mathbf{x}) = 0$ on $\partial\Omega$), which together with Eq. (12) constitute the *strong form* of the problem. To derive the *weak form* or variational formulation of this PDE model, we define the *trial function space* $\mathcal{U}$, where the PDE solution resides, and the *weighting function space* $\mathcal{V}$. To this end, we choose $\mathcal{U}, \mathcal{V} \subset \mathcal{H}^1$, which is the Sobolev space of square-integrable functions with square-integrable first derivatives. Standard FEA and IGA follow a Bubnov–Galerkin approach. For our heat problem, this translates in $\mathcal{U} = \mathcal{V}$ with functions $u \in \mathcal{U}$ and $w \in \mathcal{V}$ verifying $u(\mathbf{x}) = w(\mathbf{x}) = 0$ on $\partial\Omega$. The interested reader is referred to [21, 43] for a rigorous construction of $\mathcal{U}$ and $\mathcal{V}$. We obtain the weak form of our heat problem as follows: we multiply all terms in Eq. (12) by an arbitrary $w(\mathbf{x}) \in \mathcal{V}$, integrate the PDE in space over $\Omega$, and integrate the diffusive term by parts using the divergence theorem recalling that $w(\mathbf{x}) = 0$ on $\partial\Omega$, which cancels the boundary integral. As a result, the weak form is

$$\int_\Omega \nabla w(\mathbf{x}) \cdot (\kappa \nabla u(\mathbf{x}))\, d\mathbf{x} - \int_\Omega w(\mathbf{x}) f(\mathbf{x}) d\mathbf{x} = 0, \tag{13}$$

which accounts for both the PDE and the BCs [21, 43]. Let us define the finite-dimensional subspaces $\mathcal{U}^h \subset \mathcal{U}$ and $\mathcal{V}^h \subset \mathcal{V}$ to approximate the infinite-dimensional spaces $\mathcal{U}$ and $\mathcal{V}$, respectively. We choose a set of basis functions $\{B_A(\mathbf{x})\}_{A=1,\dots,n_p}$

spanning $\mathcal{U}^h$ and $\mathcal{V}^h$, where $n_p = \dim(\mathcal{U}^h) = \dim(\mathcal{V}^h)$. This enables us to *discretize* the weak form in space. Now, our aim is to find $u^h(\mathbf{x}) = \sum_{B=1}^{n_p} u_B B_B(\mathbf{x})$ in $\mathcal{U}^h$ satisfying Eq. (13) for any $w^h(\mathbf{x}) = \sum_{A=1}^{n_p} w_A B_A(\mathbf{x})$ in $\mathcal{V}^h = \mathcal{U}^h$. In these expressions, the coefficients $u_B$ and $w_A$ are real constants. Using the definition of $w^h$ and recalling that all $w_A$ are arbitrary, we can simplify Eq. (13) to the Galerkin form

$$\int_\Omega \nabla B_A(\mathbf{x}) \cdot \left(\kappa \nabla u^h(\mathbf{x})\right) d\mathbf{x} - \int_\Omega B_A(\mathbf{x}) f^h(\mathbf{x}) d\mathbf{x} = 0, \tag{14}$$

for all $A = 1, \ldots, n_p$ and where $f^h(\mathbf{x}) = \sum_{A=1}^{n_p} f_A B_A(\mathbf{x})$. Then, by introducing $u^h(\mathbf{x}) = \sum_{B=1}^{n_p} u_B B_B(\mathbf{x})$ in Eq. (14) and rearranging terms, we obtain

$$\sum_{B=1}^{n_p} u_B \int_\Omega \nabla B_A(\mathbf{x}) \cdot (\kappa \nabla B_B(\mathbf{x}))\, d\mathbf{x} = \int_\Omega B_A(\mathbf{x}) f^h(\mathbf{x}) d\mathbf{x}, \tag{15}$$

for all $A = 1, \ldots, n_p$. Equation (15) corresponds to a linear algebraic system $\mathbf{KU} = \mathbf{F}$, where $\mathbf{F} = \{F_A\}$ and $\mathbf{K} = \{K_{AB}\}$ are given by

$$F_A = \int_\Omega B_A(\mathbf{x}) f^h(\mathbf{x}) d\mathbf{x} \quad \text{and} \quad K_{AB} = \int_\Omega \nabla B_A(\mathbf{x}) \cdot (\kappa \nabla B_B(\mathbf{x}))\, d\mathbf{x}. \tag{16}$$

The solution $\mathbf{U} = \{u_B\}$ provides the coefficients to determine the FEA or IGA approximation $u^h(\mathbf{x}) = \sum_{B=1}^{n_p} u_B B_B(\mathbf{x})$ to our original model in Eq.(12). In this process, the construction of the finite spaces $\mathcal{U}^h$ and $\mathcal{V}^h$ along with the basis $\{B_A(\mathbf{x})\}_{A=1,\ldots,n_p}$ are key steps that ultimately control the convergence and accuracy of the numerical scheme, and that exhibit methodological differences between FEA and IGA.

Standard FEA uses piecewise Lagrangian polynomial bases to approximate $u^h(\mathbf{x})$ [43]. The piecewise architecture of FEA bases enables the partition of $\Omega$ in a mesh of $n_e$ subregions termed *elements*, as shown in Fig. 4. FEA bases are also *isoparametric* [43], which is a crucial property enabling the use of the same basis functions to describe the geometry $\Omega$ of our problem by means of a function $G(\mathbf{x}) = \sum_{A=1}^{n_p} \mathbf{x}_A B_A(\mathbf{x})$, where $\mathbf{x}_A$ are the physical coordinates of a known set of points over the elements termed *global nodes*.

FEA bases $\{B_A(\mathbf{x})\}_{A=1,\ldots,n_p}$ are built from a canonical *local basis* defined on a *parent element* $\hat{\Omega}$, which is common for all the elements in the mesh. The local basis $\{b_a(\boldsymbol{\xi})\}_{a=1,\ldots,n_b}$ is composed of $n_b$ functions constructed on the local coordinate system $\boldsymbol{\xi}$ of the parent element (see Fig. 4). Each local basis function $b_a$ is associated to a unique point in the parent element termed *local node*, with local coordinates $\boldsymbol{\xi}_a$ (see Fig. 4). For each element $e$ in the mesh, we can build a *geometric mapping* from the parent element given by $\mathbf{x}^e(\boldsymbol{\xi}) = \sum_{a=1}^{n_b} \mathbf{x}_a^e b_a^e(\boldsymbol{\xi})$, where $\mathbf{x}_a^e$ are the physical coordinates of the local nodes of element $e$ in $\Omega$. Hence, we can repeatedly map the parent element and the local basis to each of the elements in the mesh (see Fig. 4), thereby obtaining the definition of the local basis over each element $e$, i.e., $\{b_a^e(\mathbf{x}^e)\}_{a=1,\ldots,n_b}$. In this process, we also build a *connectivity array* of the form $A = c(e, a)$ to identify the global nodes $\mathbf{x}_A$ and global basis functions $B_A(\mathbf{x})$ associated to each element's local

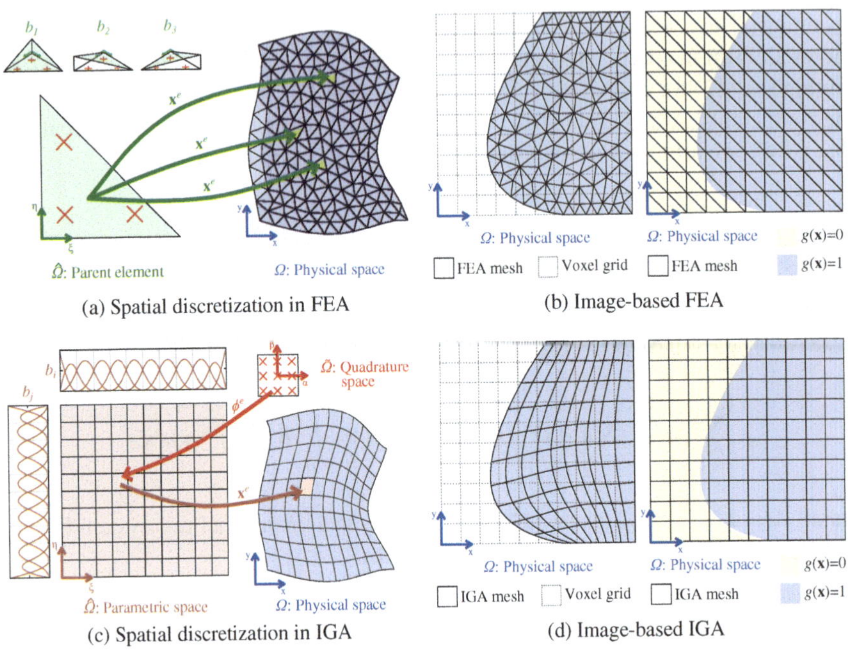


**Fig. 4** Finite Element Analysis (FEA) and Isogeometric Analysis (IGA). **a** In FEA, we approximate the physical domain by repeatedly mapping a common parent element $\hat{\Omega}$ over $\Omega$ using a geometric map $\mathbf{x}^e$ to generate each element in the mesh. The parent element supports a local function basis $\{b_1, b_2, b_3\}$, which, once mapped to $\Omega$ from each element, contributes to the definition of a global function basis. This is used to approximate the model equations in variational form and is usually integrated using quadrature rules also defined on the parent element (red crosses). **b** Unstructured FEA meshes can be built to match the segmentation of an organ extracted from medical images (left). Alternatively, immersed-boundary approaches define a FEA mesh matching the voxel grid and a map $g(\mathbf{x})$ to identify the physical domain in which the model will be solved (right). **c** In IGA, the physical domain is approximated with a topologically equivalent parametric space $\hat{\Omega}$, that is globally mapped onto $\Omega$. The parametric space results from the tensor product of univariate piecewise spline basis $\{b_i\}_{i=1,\dots,n_i}$ and $\{b_j\}_{j=1,\dots,n_j}$. The resulting multivariate spline basis is used to approximate the model equations in variational form. These can be integrated using quadrature rules (red crosses) defined over the quadrature space $\tilde{\Omega}$, which is mapped to $\Omega$ *via* composition of ‘q23s $\phi^e$ and $\mathbf{x}^e$ for each element. **d** IGA meshes can be built to match the segmentation of an organ extracted form medical images (left). Alternatively, immersed-boundary strategies define an IGA mesh aligning with the voxel grid and a map $g(\mathbf{x})$ to identify the physical domain in which the model will be solved (right)

nodes $\mathbf{x}_a^e$ and local functions $b_a^e(\mathbf{x}^e)$ [43]. Note that the combination of all element geometric mappings ultimately renders the geometric function $G(\mathbf{x})$ representing all $\Omega$.

The geometric map $\mathbf{x}^e$ is *invertible*, such that we can use its inverse to map each element $e$ in the mesh back to the parent element. This *pull-back* enables us to integrate any basis function over the common parent element and always use the same quadrature rule. This is a key idea to efficiently calculate vector $\mathbf{F}$ and matrix $\mathbf{K}$, using processes called *formation* and *assembly*. This strategy consists of looping over the elements of the mesh, such that for each element $e$ we (i) use the connectivity array to identify the local nodes and basis functions, (ii) pull them back to the parent element, (iii) calculate the integrals participating in Eq. (16) using Gaussian quadrature, and (iv) assemble the resulting local values $f_a^e$ and $K_{ab}^e$ by summing them into their corresponding global counterparts $F_A$ and $K_{AB}$ as indicated by the connectivity array. Note that $F_A$ and $K_{AB}$ will receive a contribution from each of the elements sharing node $A$. In step (ii) we can also pull back spatially varying functions over each element, such as $f(\mathbf{x})$ or even $u^h(\mathbf{x})$ (e.g., in nonlinear problems). Thus, the processes of formation and assembly capitalize on the piecewise definition of FEA bases over the elements to efficiently calculate the integrals in Eq. (16) specifically wherever the basis functions are defined, instead of performing an inefficient integration over the whole physical domain $\Omega$ [43].

IGA is considered a generalization of FEA because it relies on the same core ideas. However, IGA employs more sophisticated polynomial functions coming from computer graphics because its root idea is to use the functions *exactly* describing a computer-generated geometric model of $\Omega$ (e.g., an engineering design, an organ segmentation) to numerically solve the PDE problem posed on such geometry [21]. Conversely, in FEA we first choose the basis to approximate the solution $u^h(\mathbf{x})$ and then we use it to describe $\Omega$, which usually results in an approximation of $\Omega$ as well. Thus, IGA bases are *geometrically-exact* and isoparametric. IGA bases also show higher global continuity, which ultimately yields superior accuracy [21].

Non-Uniform Rational B-splines (NURBS) define the most usual function space in IGA [21, 25]. Univariate NURBS bases are globally defined by a *knot vector*, which is a set of non-decreasing coordinates termed *knots* enabling the definition of all NURBS basis functions over a segment *parametric space* $\hat{\Omega}$ (see Fig. 4). Multivariate NURBS bases are defined by the tensor product of univariate NURBS bases. Likewise, the tensor product of the corresponding knot vectors results in the definition of the complete IGA mesh formed by $n_e$ elements in a multivariate *parametric space* $\hat{\Omega}$, which is topologically equivalent to the physical space $\Omega$. The isogeometric elements are formed by the knot lines in each parametric direction (see Fig. 4). Each of the resulting $n_p$ basis functions $B_A(\boldsymbol{\xi})$ defined in $\hat{\Omega}$ is associated to a *control point*, with parametric coordinates $\boldsymbol{\xi}_A$ and physical coordinates $\mathbf{x}_A$. This results in the definition of an invertible *global geometric mapping* $\mathbf{x}^g(\boldsymbol{\xi}) = \sum_{A=1}^{n_p} \mathbf{x}_A B_A(\boldsymbol{\xi})$, bringing the whole IGA mesh from the parametric space $\hat{\Omega}$ into the physical space $\Omega$ and providing an explicit definition of the problem geometry (i.e., $G(\mathbf{x})$). Contrary to global nodes in FEA, the control points in IGA do not necessarily align with the mesh and may even be placed out of $\Omega$ [21, 25]. The restriction of $\mathbf{x}^g$ to each of the elements also enables the construction of the invertible *element geometric mapping* $\mathbf{x}^e$, which relies on the identification of the $n_b$ local basis functions and associated local control points defined over the element $e$ by means of a *connectivity array* as in FEA. Additionally, the connectivity array in IGA further accounts for the univariate basis functions that gave rise to the multivariate basis functions [21].

As the knot vectors are arbitrary, the elements in parametric space may have varying sizes. Thus, we further define a unique *quadrature space* $\tilde{\Omega}$, which is common to all elements, and an ancillary invertible element mapping $\phi^e(\tilde{\boldsymbol{\xi}})$ from $\tilde{\Omega}$ to each element in $\hat{\Omega}$. During the process of *formation* and *assembly*, we combine $\mathbf{x}^e$ and $\phi^e$ to perform the *pull-back* from $\Omega$ to $\tilde{\Omega}$ and calculate the integrals participating in matrices and vectors of the final system (see Eq. (16)). The rest of the assembly steps in IGA are essentially the same as in FEA (see [21]).

### 4.2.2 FEA and IGA for Image-Based Cancer Models

Let us now consider the reaction-diffusion cancer growth model in Eq. (4). Standard FEA and IGA approaches are *boundary-fitted*, i.e., the physical space $\Omega$ represents the patient's organ (see Fig. 4). Here, we will derive the weak form of Eq. (4) with the usual no-flux BC $\nabla N \cdot \mathbf{n} = 0$. We choose $\mathcal{U}, \mathcal{V} \subset \mathcal{H}^1$ and, using a Bubnov–Galerkin approach, we obtain $\mathcal{U} = \mathcal{V}$ [21, 43]. We define $\mathcal{U}^h \subset \mathcal{U}$ and $\mathcal{V}^h \subset \mathcal{V}$ and we choose a basis $\{B_A(\mathbf{x})\}_{A=1,\ldots,n_p}$ to span $\mathcal{U}^h$ and $\mathcal{V}^h$. Standard FEA and IGA approximate time-dependent functions $f(\mathbf{x}, t)$ as $f^h(\mathbf{x}, t) = \sum_{A=1}^{n_p} f_A(t) B_A(\mathbf{x})$. Then, the Galerkin discretization of the reaction-diffusion model's weak form is

$$\int_\Omega B_A \frac{\partial N^h}{\partial t} d\Omega + \int_\Omega \nabla B_A \cdot (D \nabla N^h) d\Omega - \int_\Omega B_A k^h N^h \left(1 - \frac{N^h}{\theta}\right) d\Omega = 0, \tag{17}$$

for all $A = 1, \ldots, n_p$ and where $k^h(\mathbf{x}) = \sum_{A=1}^{n_p} k_A B_A(\mathbf{x})$. To discretize and integrate in time, we propose the *generalized-$\alpha$ method* [17, 21, 48]. This approach relies on a partition of $[0, T]$ in subintervals $[t_n, t_{n+1}]$ with an arbitrary time step $\Delta t_n$, the discretization of $\partial N^h(\mathbf{x}, t)/\partial t = \sum_{A=1}^{n_p} \dot{N}_A(t) B_A(\mathbf{x})$, and the definition of vectors $\mathbf{N}^n = \mathbf{N}(t_n) = \{N_A(t_n)\}$ and $\dot{\mathbf{N}}^n = \dot{\mathbf{N}}(t_n) = \{\dot{N}_A(t_n)\}$. The generalized-$\alpha$ method is a predictor-multicorrector algorithm that provides $\mathbf{N}^{n+1}$ and $\dot{\mathbf{N}}^{n+1}$ given $\mathbf{N}^n$ and $\dot{\mathbf{N}}^n$. This method can be proven to be second-order accurate and $A$-stable by adequately choosing its defining parameters (see [20, 21]). The resulting system of nonlinear algebraic equations can be solved iteratively with Newton's method (see [20, 21]).

Finally, there are multiple approaches to generate *boundary-fitted* meshes in FEA and IGA [126]. *Unstructured meshes* are the usual strategy in FEA and consist of populating a geometric model of the patient's tumour-hosting organ generated from its segmentation with tetrahedral or hexahedral elements [1, 67, 115]. In IGA, a common approach is *parametric mapping*, whereby a known geometric model that is topologically equivalent to the organ's geometry is deformed to match the organ's segmentation [70, 71]. However, image-based cancer models may be more amenable to *immersed-boundary* methods [13, 24, 78, 85, 100]. These strategies rely on constructing the FEA/IGA mesh to align with the voxel grid and defining a function $g(\mathbf{x})$ accurately representing the organ's boundary. Hence, the function $g(\mathbf{x})$ enables the definition of the physical space $\Omega$ and a fictitious space over the rest of the mesh (see Fig. 4). The model PDEs are solved over the whole mesh but their parameters are weighted with $g(\mathbf{x})$, such that they take their usual value in the physical domain and a negligible value in the fictitious domain. Immersed-boundary methods may also rely on local refinement to improve the discretization of $g(\mathbf{x})$ [100].

## 5 Calibrating Image-Based Mathematical Oncology Models

Mathematical models of cancer define a *forward problem*, whose solution provides *state variables* (e.g., tumour cell density). In general, these models are parameterized by *unknown* biophysical parameters (and possibly initial conditions) that typically manifest substantial variability across subjects [68, 75, 99]. The estimation of these unknown variables (also called *inversion variables*) should be patient-specific and can be mathematically posed as an *inverse problem*, which aims at optimizing an objective function constrained by the model. Since image-based cancer models are usually represented by PDEs, the resulting inverse problem is formally a PDE-constrained optimization problem. In this section, we outline the general formulation of the inverse parameter estimation problem and discuss the standard methods to compute its solution, keeping in mind the ultimate goal of patient-specific tumour characterization and model prediction. Figure 5 illustrates the typical image-based inverse problem workflow in the context of brain tumours.

### 5.1 *Inverse Problems for Oncology Models*

Let $\mathcal{F}(\mathbf{p}, c) = 0$ denote a cancer model consisting of a PDE system, where $\mathbf{p}$ is a vector of unknown parameters and $c(\mathbf{x}, t)$ represents the state variables. The inverse problem seeks to estimate $\mathbf{p}$ such that the model state variables $c(\mathbf{x}, t)$ match given observational patient data (see Fig. 5). In oncology, the input data is usually a series of medical images (e.g., MRI, PET) at specific time instances $\{t_j\}_{j=1,\cdots,n_t}$ within a finite time horizon $(0, T]$. These define our observations $\hat{c}_j(\mathbf{x})$ of the state variables in the forward model $c(\mathbf{x}, t)$, e.g., tumour cell density (see Sect. 2.1). Then, the inverse problem can be mathematically formulated as

$$\min_{\mathbf{p}} \mathcal{J}(\mathbf{p}, c) = \frac{1}{2} \sum_{j=1}^{n_t} \int_{\Omega} \big(c(\mathbf{x}, t_j) - \hat{c}_j(\mathbf{x})\big)^2 d\mathbf{x} + \mathcal{R}(\mathbf{p}) \;\; \text{s.t.} \;\; \mathcal{F}(\mathbf{p}, c) = 0, \quad (18)$$

where $\Omega$ is the spatial domain representing the patient's tumour-harboring organ. The objective function $\mathcal{J}(\mathbf{p}, c)$ minimizes the mismatch between the predicted tumour and observed data at times $\{t_j\}_{j=1,\cdots,n_t}$ using an $L^2$ distance measure. Additionally, it balances this data fidelity/mismatch term with a regularization operator $\mathcal{R}(\mathbf{p})$.

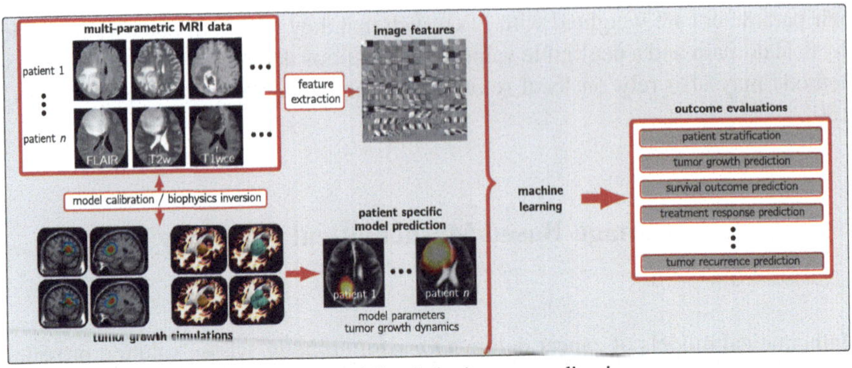


(a) Imaging workflow in brain tumor radiomics

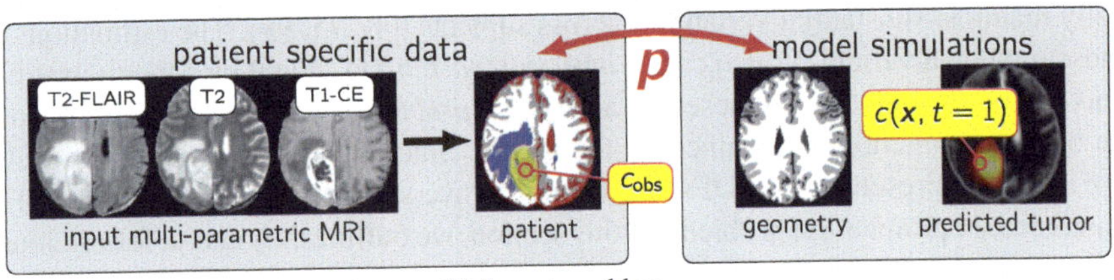


(b) Inverse problem

**Fig. 5** Workflow for an image-driven biophysical inverse problem. **a** The goal is to extract clinically relevant biomarkers from input multi-parametric MRI data that can predict and guide intervention so as to improve clinical outcome. Purely imaging-based approaches (feature extraction) can be integrated with biophysical priors (biophysical model inversion) to develop tools that can assist in treatment and prognosis. **b** The input data for patients with brain tumours (multi-parametric MRI scans) is at a single time point (treatment typically follows immediately after diagnosis) and is translated into model observables for tumour concentration $c_{\text{obs}}$. The inverse problem seeks to estimate unknown biophysical parameters $\mathbf{p}$ of a given mathematical model. The basic idea is to perform a number of model simulations with different parameters so that the model-predicted tumour matches the observed data. The panels in this figure are adapted from [73]

Designing algorithms for the efficient and effective numerical solution of such PDE-constrained optimization problems is a challenging task [2, 63]. While derivative-free optimization strategies are popular due to ease of implementation, they are typically associated with slow convergence and can become prohibitively expensive (especially if the dimensionality of $\mathbf{p}$ is large). Hence, optimization algorithms that utilize gradient information are preferable. In addition to improving convergence and computation time, these methods can reveal important characteristics of the objective function landscape, which can be exploited to design better algorithms and help us understand the sensitivities and ill-posedness inherent to PDE-constrained optimization problems. There exist multiple options for evaluating the gradient (and higher order derivatives) of the objective function, such as automatic differentiation, numerical approximation through finite differences, and adjoint-based methods. For oncology models, several groups have employed adjoints for inversion [19, 27, 29, 35, 57, 104]. Some efforts also employ Hessian information to accelerate convergence [29, 97]. Other strategies such as derivative-free optimization [15, 60, 76, 118] or finite difference approximations [39] have also been considered in literature, but are usually less effective than adjoint-based methods. For large-scale 3D inversion, parallel (distributed memory) algorithms have been considered in [29, 97, 104].

While the deterministic approaches just described are successful in estimating the optimal parameters to the minimization problem in Eq. (18), their utility could be limited due to uncertainties arising from modeling errors and noise in measurements/data. A probabilistic (Bayesian) formulation can mitigate this drawback by characterizing our confidence in the inversion variables **p** using probability density functions. This approach will be described in Sect. 6.2, as part of a comprehensive Bayesian framework for model selection. In the following section, we briefly describe adjoint-based inversion methods.

## 5.2 *Adjoint Methods for Inverse Problems*

The standard technique for solving the inverse problem posed in Eq. (18) is to introduce Lagrange multipliers $\lambda(\mathbf{x}, t)$, which are termed *adjoint variables* or simply *adjoints*, and construct the Lagrangian functional as

$$\mathcal{L}(c, \mathbf{p}, \lambda) = \mathcal{J}(\mathbf{p}, c) + \langle \mathcal{F}(\mathbf{p}, c), \lambda \rangle, \tag{19}$$

where $\langle . \rangle$ denotes an appropriate inner product (typically $L^2$) and $\mathcal{J}(\mathbf{p}, c)$ denotes the objective function defined in Eq. (18). By requiring stationarity with respect to the state, adjoint, and inversion variables, we arrive at the *first order optimality conditions* by taking the following variations:

$$\delta_\lambda \mathcal{L} = 0 \qquad \textit{(forward equations)}, \tag{20a}$$
$$\delta_c \mathcal{L} = 0 \qquad \textit{(adjoint equations)}, \tag{20b}$$
$$\delta_{\mathbf{p}} \mathcal{L} = 0 \qquad \textit{(inversion equations)}, \tag{20c}$$

where $\delta_z \mathcal{L}$ denotes the variation of $\mathcal{L}$ with respect to $z$. The forward equations are simply the tumour growth PDE model. The adjoint equations are linear PDEs in the adjoint variables backward in time. The inversion equations denote the PDE-constrained gradient of $\mathcal{J}(\mathbf{p}, c)$, which is set to zero at the local minimum.

For example, consider the reaction-diffusion cancer model in Eq. (4) with the usual no-flux BC $\nabla N \cdot \mathbf{n} = 0$ and two tumour cell density observations $\hat{N}_0(\mathbf{x})$ and $\hat{N}_1(\mathbf{x})$ derived from DW-MRI at $t = t_0 = 0$ and $t = t_1$ (see Sect. 2.1). We assume that $\hat{N}_0(\mathbf{x})$ are known initial conditions for Eq. (4), and use $\hat{N}_1(\mathbf{x})$ to calibrate $\mathbf{p} = \{D, k(\mathbf{x})\}$, where $D$ is a scalar constant and $k(\mathbf{x})$ is a spatial field, representing the tumour cell diffusion and proliferation rate, respectively (see Sect. 2.1). Then, we define the objective function $\mathcal{J}(\mathbf{p}, N)$ and the Lagrangian functional $\mathcal{L}(N, \mathbf{p}, \lambda)$ as

$$\mathcal{J}(\mathbf{p}, N) = \frac{1}{2} \int_{\Omega} \left(N(\mathbf{x}, t_1) - \hat{N}_1(\mathbf{x})\right)^2 d\mathbf{x} + \frac{a}{2} \left(D^2 + \int_{\Omega} k^2(\mathbf{x}) d\mathbf{x}\right), \tag{21}$$

$$\begin{aligned} \mathcal{L}(N, \mathbf{p}, \lambda) = \mathcal{J}(\mathbf{p}, N) + \int_0^{t_1} \int_{\Omega} \lambda(\mathbf{x}, t) \left( \frac{\partial N(\mathbf{x}, t)}{\partial t} - \nabla \cdot (D \nabla N(\mathbf{x}, t)) \right. \\ \left. - k(\mathbf{x}) N(\mathbf{x}, t) \left(1 - \frac{N(\mathbf{x}, t)}{\theta}\right)\right) d\mathbf{x} dt, \end{aligned} \tag{22}$$

where $a$ is a regularization parameter. Following Eq. (20b), the adjoint equation is

$$-\frac{\partial \lambda(\mathbf{x}, t)}{\partial t} = \nabla \cdot (D \nabla \lambda(\mathbf{x}, t)) + k(\mathbf{x}) \lambda(\mathbf{x}, t) \left(1 - \frac{2N(\mathbf{x}, t)}{\theta}\right), \tag{23}$$

which is a linear backward problem in time subject to the BC $\nabla \lambda \cdot \mathbf{n} = 0$ on $\partial \Omega$ and the terminal condition $\lambda(\mathbf{x}, t_1) = \hat{N}_1(\mathbf{x}) - N(\mathbf{x}, t_1)$. Following Eq. (20c), the inversion equations are given by

$$aD + \int_0^{t_1} \int_{\Omega} \nabla \lambda(\mathbf{x}, t) \cdot \nabla N(\mathbf{x}, t) d\mathbf{x} dt = 0, \tag{24}$$

$$ak(\mathbf{x}) - \int_0^{t_1} \lambda(\mathbf{x}, t) N(\mathbf{x}, t) \left(1 - \frac{N(\mathbf{x}, t)}{\theta}\right) dt = 0. \tag{25}$$

In general, Eq. (20) represents a large, non-linear, coupled system of PDEs, which can be significantly challenging to solve simultaneously. Instead, a standard approach involves a *reduced space* algorithm, which is an iterative strategy that reduces Eq. (20) to a system involving only the inversion variables. In each iteration, we use the current approximation to $\mathbf{p}$ to solve the state and adjoint equations to respectively get the current approximation to the state and adjoint variables. Then, we update our inversion variables using the inversion equations. This process is repeated until convergence, which is typically set by a user-defined threshold on the parameter update (i.e., the parameter gradient). In contrast to *full space* methods, reduced space methods present more tractable systems that can exploit existing PDE solvers for the state and adjoint equations and are better conditioned. We refer the reader to [2] for more details on adjoint methods in PDE-constrained optimization.

## 6 Model Selection and Identification of Relevant Parameters

Given the vast array of cancer growth models in the literature, it is not trivial to choose which is the best to represent the available data and to predict key quantities of interest (e.g., the tumour size, treatment efficacy, or percentage of necrosis) for a certain tumour type. In this section, we describe the Occam Plausibility Algorithm (OPAL), which has been proposed in [26] as an adaptive process for model selection and validation in the presence of uncertainties. The strategy relies on three key steps: sensitivity analysis to identify the relevant model parameters, model calibration of the relevant parameters, and calculation of model selection criteria. The OPAL can be referred to as *model agnostic* in that no single model is advocated; rather, the *best* model is selected based upon our model selection criteria. Details regarding each step of the OPAL are given in the next subsections.

## 6.1 Variance-Based Sensitivity Analysis

Sensitivity analysis quantifies how changes in parameter values affect the uncertainty in model output [95]. We can distinguish between *local* and *global* methods. Local methods compute the variation of the model output changing one parameter at a time (i.e., first-order effects) usually *via* derivation, but neglect the interactions between the parameters. In global methods, the contribution of each parameter along with its interactions with other parameters (i.e., higher-order effects) are taken into account, as all parameters are varied simultaneously over the entire parameter space. In this work, we present the variance-based global sensitivity analysis method, also known as the Sobol method [93, 102]. Details regarding local and other global methods can be found in [93].

Let $\mathbf{M}(\boldsymbol{\theta})$ be a model parameterized by $k$ parameters $\boldsymbol{\theta}$, which belong to a parameter space $\boldsymbol{\Theta} \subset \mathbb{R}^k$. The computational cost of the sensitivity analysis of model $\mathbf{M}(\boldsymbol{\theta})$ depends on the number of parameters $k$ and the sample size $N$, with the total number of model evaluations given by $N_T = N(k+1)$. There are several approaches to estimate the *total sensitivity index* for each parameter, which quantifies all effects of the parameter on the model output. Here, we present the strategy in [94], as it is known to demand a reduced sample size to converge.

First, we randomly generate two sampling matrices, $\mathbf{A}$ and $\mathbf{B}$, with size $N \times k$. Each row of these matrices represents a sampled value for the vector of parameter $\boldsymbol{\theta}$. Additionally, we create $k$ matrices $\mathbf{A}_{\mathbf{B}}^{(k)}$, where we copy the values from the matrix $\mathbf{A}$ and replace the values from column $k$ with the values from $\mathbf{B}$. For the case where $N = 1$, these matrices are given as: $\mathbf{A} = [\theta^a_{1,1} \;\; \theta^a_{1,2} \; \cdots \; \theta^a_{1,k}]$, $\mathbf{B} = [\theta^b_{1,1} \;\; \theta^b_{1,2} \; \cdots \; \theta^b_{1,k}]$, $\mathbf{A}_{\mathbf{B}}^{(1)} = [\theta^b_{1,1} \;\; \theta^a_{1,2} \; \cdots \; \theta^a_{1,k}]$, $\mathbf{A}_{\mathbf{B}}^{(2)} = [\theta^a_{1,1} \;\; \theta^b_{1,2} \; \cdots \; \theta^a_{1,k}]$, ..., and $\mathbf{A}_{\mathbf{B}}^{(k)} = [\theta^a_{1,1} \;\; \theta^a_{1,2} \; \cdots \; \theta^b_{1,k}]$.

Then, we run the forward model for each row in matrix $\mathbf{A}$ and all matrices $\mathbf{A}_{\mathbf{B}}^{(k)}$. The outputs of the model are stored in corresponding solution vectors for each matrix; i.e., $\mathbf{Y}_A$, $\mathbf{Y}^1_{AB}$, ..., $\mathbf{Y}^k_{AB}$. Finally, we compute the total sensitivity index $S_{T_i}$ for each parameter, which can be approximated [49, 93, 94] by

$$S_{T_i} = \frac{1}{2N\mathrm{Var}(\mathbf{Y}_A)} \sum_{j=1}^{N} \left( (\mathbf{Y}_A)_j - \left(\mathbf{Y}_{AB}^{(i)}\right)_j \right)^2, \tag{26}$$

where $\mathrm{Var}(\mathbf{Y}_A)$ is the variance of vector $\mathbf{Y}_A$. According to [93], a parameter $i$ can be considered non-influential if $S_{T_i} = 0$. In practice, we define a threshold $\epsilon$ and we identify a parameter $i$ as non-influential if $S_{T_i} < \epsilon$. The choice of $\epsilon$ is relative to other $S_{T_j}$ and problem-dependent. If $S_{T_i}$ is sufficiently small, then the parameter does not affect the quantities of interest, and the complexity of the model can be reduced by removing or fixing the parameter to any value within the uncertainty range [93].

## 6.2 Model Calibration

To characterize uncertainties in the observable data and the stochastic behavior of tumour growth, we follow the Bayesian statistical calibration procedure. This method captures these uncertainties by delivering a probabilistic distribution of the model parameters, instead of a single value for each of them [66, 67, 82, 83]. The basic ideas behind the Bayesian parameter estimation involve the following steps:

1. Select the observational data $\mathbf{D}$ to be used (e.g., baseline and follow-up MRI).
2. Establish the prior distribution of the model parameters $\pi_{prior}(\boldsymbol{\theta})$. In the cases where we do not have knowledge regarding the distribution of the parameters, and we can only estimate the range of these parameters, the usual approach is to assume a uniform prior distribution.
3. Construct the likelihood function, which, given the values assigned to the parameters $\boldsymbol{\theta}$, yields the probability of $\mathbf{D}$ being observed [83]. Assuming both the experimental error and the model inadequacy to be Gaussian, and the experimental data to be independent and identically distributed, the likelihood is given as

$$\pi_{like}(\mathbf{D}|\boldsymbol{\theta}) = \prod_{i=1}^{N_t} \frac{1}{\sqrt{2\pi\sigma^2}} \exp\left( -\frac{(D_i - Y_i(\boldsymbol{\theta}))^2}{2\sigma^2} \right), \tag{27}$$

   where $N_t$ is the number of data points, $\sigma$ is the standard deviation of the experimental error and model inadequacy, and $Y(\boldsymbol{\theta})$ is the model output.
4. Compute the posterior distribution of the parameters $\pi_{post}(\boldsymbol{\theta}|\mathbf{D})$ as

$$\pi_{post}(\boldsymbol{\theta}|\mathbf{D}) = \frac{\pi_{like}(\mathbf{D}|\boldsymbol{\theta})\pi_{prior}(\boldsymbol{\theta})}{\pi_{evid}(\mathbf{D})}, \tag{28}$$

   where $\pi_{evid}(\mathbf{D}) = \int_{\Theta} \left(\pi_{like}(\mathbf{D}|\boldsymbol{\theta})\pi_{prior}(\boldsymbol{\theta})\right) d\boldsymbol{\theta}$ is the model evidence. The resulting posterior distribution of the parameters allows the prediction of the quantities of interest taking into account the uncertainties in the parameters.

The posterior probability density function is, in general, non-Gaussian. Sampling schemes such as Markov Chain Monte Carlo (MCMC) methods can be used to evaluate posterior expectations [66, 67]. These stochastic methods can incur in a large computational cost, so efficient sampling strategies exploiting the problem structure are actively being investigated [23, 86, 87, 114].

## 6.3 Model Selection Criteria

Following the Bayesian framework used for calibration, we approach model selection by computing the *model plausibility* [10, 16, 54, 83, 87]. Given a set of $m$ models $\mathbf{M} = \{M_i(\boldsymbol{\theta}_i)\}_{i=1}^m$, the Bayes' rule in Eq. (28) can be rewritten assuming that probabilities are conditional on the model $M_i$ and the set $M$:

$$\pi_{post}(\boldsymbol{\theta}_i|\mathbf{D}, M_i, \mathbf{M}) = \frac{\pi_{like}(\mathbf{D}|\boldsymbol{\theta}_i, M_i, \mathbf{M})\pi_{prior}(\boldsymbol{\theta}_i|M_i, \mathbf{M})}{\pi_{evid}(\mathbf{D}|M_i, \mathbf{M})}. \tag{29}$$

The evidence of each model can be viewed as a likelihood for a discrete Bayesian calculation, yielding a new posterior called the model plausibility, $\pi_{plaus}$, given as

$$\pi_{plaus}(M_i|\mathbf{M}, \mathbf{D}) = \frac{\pi_{evid}(\mathbf{D}|, M_i, \mathbf{M})\pi_{prior}(M_i|\mathbf{M})}{\pi_{evid}(\mathbf{D}|\mathbf{M})}, \qquad 1 \le i \le m. \tag{30}$$

If we assume that all models are equally probable, $\pi_{prior}(M_i|\mathbf{M}) = \frac{1}{m}$. The sum of all model plausibilities is equal to one (i.e., $\sum_{i=1}^m \pi_{plaus}(M_i|\mathbf{M}, \mathbf{D}) = 1$). The model with highest plausibility is selected as the best model in $\mathbf{M}$ to capture the data.

Another popular method of model selection is the Akaike Information Criterion (AIC) [59]. In this method, the likelihood of the maximum likelihood estimator, $\hat{\boldsymbol{\theta}}$, is penalized according to the number of model parameters, i.e.,

$$AIC_i = -2\log \pi_{like}(\mathbf{D}|\hat{\boldsymbol{\theta}}) + 2k_i, \qquad 1 \le i \le m, \tag{31}$$

where $k_i$ the number of parameters in model $M_i$. In this case, the model with the lowest AIC is the best model in $\mathbf{M}$.

## 6.4 The Occam Plausibility Algorithm

OPAL has been proposed as a comprehensive framework that combines parameter sensitivity analysis, and model calibration, validation, selection, and prediction [26, 83]. Figure 6 shows a schematic representation of the main OPAL steps, which we outline in the following [26, 82]:

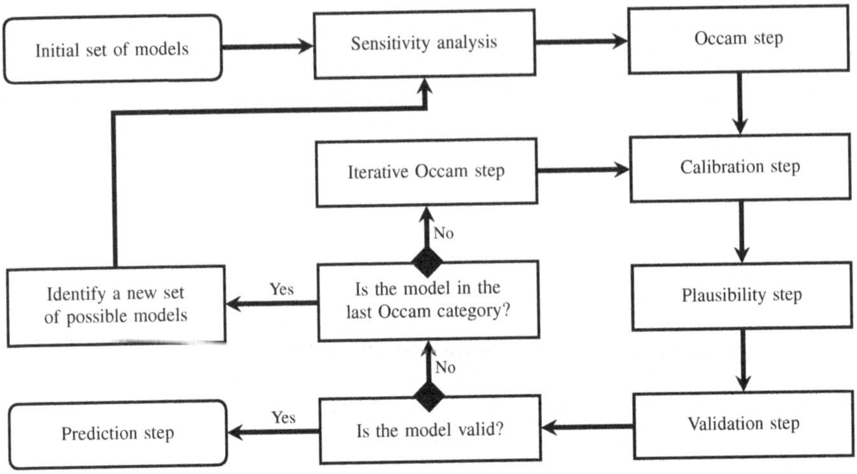


**Fig. 6** The Occam Plausibility Algorithm (OPAL). This framework brings together experimental data for model calibration and validation, sensitivity analysis, selection, and model prediction

1. Identify a set $\mathbf{M} = \{M_1(\boldsymbol{\theta}_1), M_2(\boldsymbol{\theta}_2), \ldots, M_m(\boldsymbol{\theta}_m)\}$ of parametric models, each with parameters $\boldsymbol{\theta}_i$ belonging to an appropriate parameter space $\boldsymbol{\Theta}_i$, $1 \leq i \leq m$.
2. Perform a sensitivity analysis to identify non-influential parameters. Based on this analysis, the values of these parameters are fixed to the mean value used in the sensitivity analysis. If there is a model in **M**, whose only difference to other model is given by the non-influential parameters, this model can be eliminated, yielding a reduced set $\bar{\mathbf{M}} = \{\bar{M}_1(\boldsymbol{\theta}_1), \bar{M}_2(\boldsymbol{\theta}_2), \ldots, \bar{M}_l(\boldsymbol{\theta}_l)\}$ of models, with $l \leq m$.
3. Divide the models in $\bar{\mathbf{M}}$ into "Occam Categories" according to their complexity (e.g., number of parameters). These categories are sorted in ascending order according to their complexity.
4. Calibrate the models in Category 1 using the calibration data $\mathbf{D}^c$.
5. Select the best model to represent the data in this category (e.g., the model with the highest plausibility).
6. Test the best model identified in the current category in a validation scenario, where the posterior from the calibration step is used as a prior and the distribution of the parameters are updated against the validation data $\mathbf{D}^v$. If the model is able to represent the data within a preset tolerance, the model is considered "valid". If not, we return to step 3 and move to the next Occam category. If we are not able to find a valid model, we need to return to step 1 and include new models.
7. After finding the "simplest" valid model, solve the forward model in the prediction scenario and compute the quantities of interest.

All of these steps are designed to consider uncertainties in the choice of model, the model parameters, the observational data, and the target quantities of interest. All uncertainties are generally characterized by probability densities.

# 7 Towards the Optimization of Personalized Treatment Plans

Several image-based mathematical models of cancer growth have shown promise in predicting treatment outcomes in a patient-specific manner, as discussed in Sects. 3.4 and 3.5. Those models could provide a means to determine optimal therapeutic regimens to treat a certain type of cancer in silico, which could then be investigated within a clinical trial in vivo. Hence, this computational approach seeks to help clinicians navigate the vast array of radiotherapy and drug combinations, dosing options, and treatment schedules and select optimal strategies, which are virtually impossible to assess in clinical trials. Ultimately, cancer models could also serve as a *digital twin* for the patient's tumour, thereby enabling the pathological assessment, monitoring, and design of optimal therapeutic regimens for the individual patient in silico. In this section, we discuss the use of image-based predictive tumour growth models accounting for the therapeutic regimen and associated tumour response for the discovery of optimal therapeutic regimens and the design of patient-specific optimal treatment strategies.

## *7.1 Potential to Select Treatment Plans for Individual Patients*

Selecting a treatment regimen for a patient is a complex process. Oncologists use decision tree algorithms to select therapeutics for each patient considering, for example, tumour grade and cell markers [12]. However, the determination of the optimal dosing regimens for these therapies is vastly underinvestigated. This limitation follows from the impossibility to test all the potential dosing strategies within a clinical trial. Additionally, regimens may be altered by the treating oncologist due to considerations like side effects and quality of life for the patient, where doses may be skipped, dosages decreased, and/or supportive medications prescribed. However, these changes are made only with a limited knowledge of their effects on the treatment outcome for any given patient.

As patients present with varying physiologies and sensitivities, the one-size-fits all approach is clearly not optimal for all patients. Mathematical models of tumour growth and treatment response can help us predict therapeutic efficacy accounting for each patient's specific tumour dynamics and, potentially, select the best therapeutic regimen for each individual patient. For instance, Jarrett et al. [52] used the model in Eq. (9) to investigate alternative dosing regimens of cytotoxic therapies for breast cancer in silico. First, the model was parameterized patient-wise using MRI data collected prior to the start of therapy and after one drug cycle as indicated in Sect. 3.5. The resulting personalized model was then simulated to the time of completion of the prescribed therapeutic regimen, and the simulated tumour growth was compared to the actual tumour response measured by MRI for each patient. The model predictions were found to be highly correlated with actual tumour response ($N = 13$, CCC $> 0.90$, $p < 0.01$ for total cellularity, total volume, and longest axis), so the model was considered valid to reproduce the effects of chemotherapy in breast cancer. Their validated model was then used to explore alternative therapeutic regimens, which were defined patient-wise by fixing the total dose prescribed in standard regimen and varying frequency and dosage. The authors indicated that an additional 0–46% reduction (median=17%) in total cellularity may have been achievable across the patient cohort ($N = 13$) compared to the standard chemotherapeutic regimens that the patients were prescribed. The dosing regimens that the model predicted to reduce/control each tumour were also found to significantly outperform standard regimens for tumour control ($p < 0.001$), thereby supporting the claim that standard regimens may not be the most effective for every patient.

## 7.2 *Optimal Control Theory for Personalized Treatment Planning*

Consider a dynamical system involving a set of variables $u(t)$ and controls $z(t)$, which are functions describing external forces that can alter the system dynamics. *Optimal control theory* (OCT) was developed to determine the solution of the system that achieves a particular outcome by adequately adjusting the controls. The mathematical formulation of the optimal control problem consists of minimizing or maximizing an objective functional $J(u, z)$. Thus, given a particular dynamical system over a certain time interval $[0, T]$, applying OCT largely consists of determining the objective functional, problem-specific constraints, and a method for solving the OCT problem. The general form of the objective functional for OCT is

$$J(u, z) = \Phi[u(t_j), z(t_j), t_j] + \int_{t_0}^{t_f} L(u(t), z(t), t)dt, \tag{32}$$

where $\Phi[u(t_j), z(t_j), t_j]$ includes target values of the variables and the controls at specific times $\{t_k\}_{k=1,\dots,n_t}$, while $L(u(t), z(t), t)$ accounts for the target dynamics of the variables and the controls over $[0, T]$. $\Phi[u(t_j), z(t_j), t_j]$ has several names in OCT literature, including *endpoint cost* in minimization problems and *terminal payoff* in maximization problems. The formulation of the objective function can also be divided into three canonical types: *endpoint control*, which only includes $\Phi[u(t_j), z(t_j), t_j]$; *bang-bang control*, which only features a linear $L(u(t), z(t), t)$; and *continuous control*, which only has a quadratic $L(u(t), z(t), t)$. The objective function can also be constructed by combining these canonical types. The term *bang-bang* refers to the usual dynamics of optimal control $z(t)$ for this type of functional, which switches between the maximum admissible value and $z(t) = 0$ (i.e., no effect). Additionally, the quadratic term in the continuous control is not usually motivated by problem-dependent phenomena, but it ensures that the optimization problem is *convex*. Hence, the optimal control problem has key mathematical features, including the existence of a global minimum.

In the context of cancer, we can apply OCT to obtain optimal treatment strategies by using a mathematical model to simulate tumour growth and therapeutic response as a dynamical system, setting the treatment as a control, and selecting clinically-relevant treatment outcomes in the objective functional. Here we will briefly discuss the formulation of the OCT problem using a simplification of the PDE model of breast cancer chemotherapy in Eq. (9), whereby we directly model the total number of tumour cells $n(t)$ using the ordinary differential equation (ODE) [56, 124]:

$$\frac{dn(t)}{dt} = kn(t)\left(1 - \frac{n(t)}{\theta}\right) - \alpha z(t)n(t) \tag{33}$$

where $k$ describes global tumour cell proliferation, $\theta$ is the tissue carrying capacity, $\alpha$ models chemotherapy efficacy, and $z(t)$ represents the dynamics of the concentration of drug(s) in the plasma, which is used to derive the initial concentration of drug in the tissue $C_{drug}(\mathbf{x}, t)$ in Eq. (9) [50]. We chose this ODE model for the sake of simplicity and because ODE models have been commonly used in the cancer OCT literature [5, 51, 62, 96, 107]. However, the following ideas could also be applied to the PDE model in Eq. (9) by defining $n(t) = \int_\Omega N(\mathbf{x}, t)d\mathbf{x}$ (see [3]).

A primary goal of OCT problems for cancer treatment is to minimize the tumour burden only at the completion of the therapy, which can be formulated by the endpoint control functional

$$J_1(n) = n(t_f). \tag{34}$$

This optimal control problem requires additional constraints for the therapeutic regimen $z(t)$. For example, these can limit the maximum dose by setting $z(t) < z_{max}$ and/or the total maximum dose by imposing $\int_0^T z(t)dt < z_{tot}$, which is termed an *isoperimetric constraint* (see Fig. 7). The limits $z_{max}$ and $z_{tot}$ are drug-dependent and may be patient-specific (e.g., quality of life, comorbidities). However, the optimization of $J_1(n)$ is insensitive to the dynamics of the drug concentration; i.e., it does not formally adapt the drug regimen to the tumour burden. Another limitation of endpoint control formulations is that they do not consider the potential growth of the tumour during therapy, which may be relevant to select actionable therapeutic regimens depending on the type of cancer.

To address this limitation, we can extend the objective functional in Eq. (34) with a bang-bang term accounting for $z(t)$:

$$J_2(u, z) = w_1 n(t_f) + w_2 \int_0^T z(t)dt. \tag{35}$$

where $w_1$ and $w_2$ are problem specific weights that can be included in the objective function to give greater or lesser importance to the different terms in the objective functional during optimization. For example, if a particular treatment has significant adverse side effects, $w_2 > w_1$ may be enforced to focus the problem on minimizing the total drug dose. Similarly, an objective function that includes the continuous control of the therapy can be written as

$$J_3(u, z) = w_1 n(t_f) + \frac{w_2}{2} \int_0^T (z(t))^2 dt. \tag{36}$$

Furthermore, the objective function can also account for the tumour growth over the entire treatment period:

$$J_4(u, z) = \frac{1}{2} \int_0^T w_1 (n(t))^2 + w_2 (z(t))^2 dt. \tag{37}$$

The main difference between the optimal treatment solution found for objective function $J_2$ versus $J_3$ and $J_4$ is that $J_2$ assumes that a constant dose is given over a certain interval, while $J_3$ and $J_4$ assume that the drug concentration $z(t)$ can change over time. These differences are shown in Fig. 7. In principle, an on-off (bang-bang) control would be more clinically relevant than the continuous control, as patients are not usually treated over time in a continuous manner. However, this type of control would result in a sudden drop in $z(t)$ after the conclusion of each drug cycle (see Fig. 7), whereas in reality $z(t)$ decays with an approximately exponential trend. With the introduction of take-home infusion pumps for chemotherapy [125], evaluating continuous control may be a more plausible avenue of investigation. However, the optimal $z(t)$ obtained under continuous control may also be unachievable with the current drugs for cancer treatment (e.g., due to incompatible pharmacokinetics). Ultimately, the optimal drug concentrations $z(t)$ obtained with these functionals can render valuable information to guide the design of clinically-feasible optimal therapeutic strategies. Another future goal could even be the synthesis of new drugs or the adaptation of current drug compounds to match target dynamics emanating from the combination of OCT, cancer modeling, and pharmacodynamics [44, 72, 101].

Finally, optimal control problems using $J_2$, $J_3$, and $J_4$ may include clinically-relevant constraints to further focus the solution, including the limitations to $z(t)$ discussed for $J_1$. For instance, we can limit the tumour burden at any time by imposing $n(t) < n_r$, where $n_r$ is an arbitrary threshold. Additional constraints can also limit the frequency $f$ of doses, for example, by setting $f \leq T/n_d$, where $n_d$ is the maximum number of doses per treatment period. Larger systems of equations that may account for the healthy and immune cell populations may also require additional constraints and/or incorporate these other variables into the optimal control problem itself. Beyond biological concerns are the logistical, monetary, and psychological costs that may also be considered for an optimal regimen. The reader is referred to [5, 51, 62, 96, 107] for further details on applying OCT to cancer models.

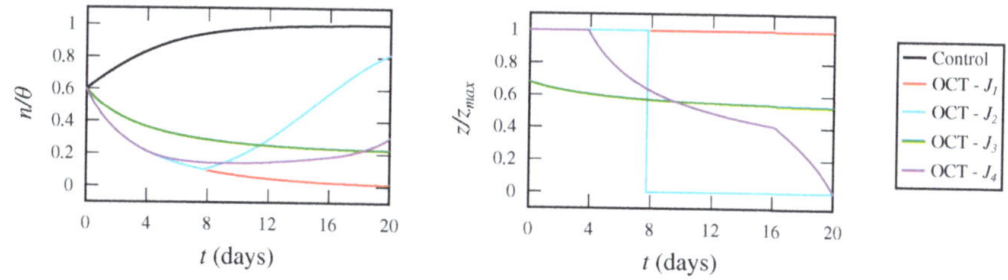


(a) Optimal control without isoperimetric constraint

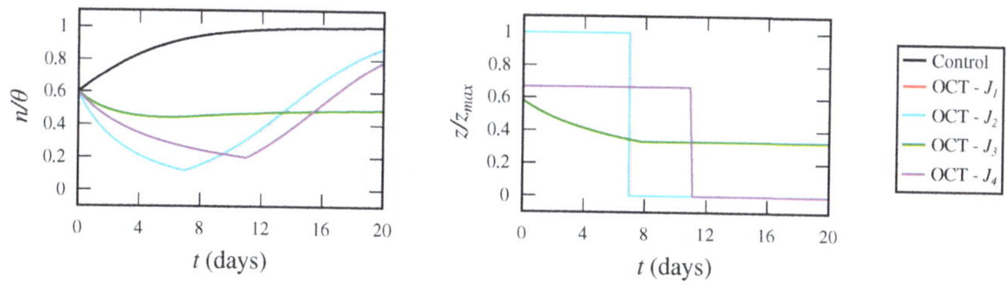


(b) Optimal control with isoperimetric constraint

**Fig. 7** Optimal control problem solutions for the ODE model of cancer chemotherapy in Eq. (33). **a** Without an isoperimetric constraint, the optimal control problem using $J_1$ yields $z(t) = z_{max}$, which would probably induce acute toxicity in patients. The optimal $z(t)$ for the bang-bang control using $J_2$ produces the characteristic stepwise solution, while the optimal $z(t)$ obtained with the continuous controls in $J_3$ and $J_4$ show a continuous decay. The drug regimens calculated with $J_1$, $J_3$ and $J_4$ successfully control tumour growth, while for $J_2$ we observe regrowth once $z(t)$ drops to zero. **b** If we impose the isoperimetric constraint, the optimal $z(t)$ for $J_1$ and $J_2$ coincide and produce a similar regimen to the bang-bang solution without isoperimetric constraint. The optimal $z(t)$ obtained with the continuous controls in $J_3$ and $J_4$ now take lower concentration values, which translate into a limited tumour control. Balancing the maximum drug delivered during treatment, the maximum reduction in tumour burden, and the relative weights in the formulations of $J_2$, $J_3$, and $J_4$ is a challenge in OCT applications in cancer, which may render very different results

## 8 Barriers to Success

One of the principal issues preventing the successful translation of image-based computational modeling technologies to routine clinical practice is access to proper data. Calibration and validation of image-based cancer models require individual quantitative imaging data at multiple time points during the course of surveillance or treatment. The acquisition of frequent and rich imaging datasets might be feasible in preclinical settings, thereby enabling the realization of controlled studies to thoroughly assess model validity. However, such controlled studies are extremely challenging in a clinical scenario for two key reasons. First, the timely and effective treatment of the patient and their quality of life are the utmost priorities. Thus, a controlled validation study would require the acquisition of minimal datasets to ensure a reliable calibration of the model, while ensuring that data collection is not a burden for patients and that those in the model-informed arm will be closely monitored in case immediate action is warranted during the course of the controlled study. Second, quantitative imaging data are not routinely collected in the current clinical protocols. Standard-of-care imaging primarily focuses on delineating tumour boundaries for staging and planning interventions (e.g., biopsy, surgery, and radiotherapy). However, anatomical imaging fundamentally limits modeling approaches as it does not quantitatively characterize the unique heterogeneous nature of each patient's tumour. In addition, standard-of-care images are often acquired only before and following the completion of therapy. This is partly due to financial constraints, hospital workflow, scanner availability, and patient burden. Thus, carefully designed studies are needed to determine a clinically-feasible strategy to collect sufficient quantitative imaging data enabling an accurate parameterization of predictive models of cancer. For instance, abbreviated imaging protocols may fit quantitative, research-focused scans into routine clinical visits alleviating the need for separate research scans. Additionally, many quantitative imaging data types can now be acquired in the community setting (i.e., away from major hospitals or oncology centers), which may be more convenient for patients [103].

From a modeling viewpoint, a central challenge is balancing model complexity, a rational use of computational resources, data requirements to ensure an accurate calibration, and predictive accuracy. The dynamics of cancer growth and treatment response is extremely complex, involving a multitude of biophysical processes interacting at various spatial and temporal scales [74]. Mathematical models of cancer are built upon a series of relevant biophysical phenomena, whose selection, formulation, and calibration ultimately determines the predictive power of the model [90, 122, 123]. The accurate modeling of some of these phenomena (e.g., angiogenesis [113]) would require complex equations at multiple scales, a large number of parameters requiring extensive spatiotemporal data, and advanced numerical methods that may incur in a large computational cost. However, such models would be incompatible with the constraints on quantitative imaging data availability and patient care during the controlled validation studies discussed above.

Thus, clinically-oriented image-based models usually require conservative modeling assumptions, which enables the description of cancer growth and treatment response at organ scale using simple modeling paradigms and involving a minimal set of parameters whose calibration is feasible with existing quantitative imaging techniques [122]. However, these model assumptions may incur substantial errors that ultimately limit the predictive power of the model. Moreover, some biophysical model fields (e.g., tumour cell density) are not directly observed. Instead, the observed imaging data must be preprocessed to create a proxy to the biophysical observables. Such *pseudo-correspondences* introduce additional uncertainties and can significantly affect the reconstruction results [29, 32, 61, 108]. Likewise, the phenomenological nature of macroscopic cancer models introduces parameters without direct biological counterparts (e.g., tumour cell diffusivity, drug efficacy), which further complicates model assessment and validation.

Nevertheless, the future development of image-based cancer models requires the initial validation of simpler models because this enables the identification of key improvements in model formulation to refine predictive accuracy. In addition, the success of simpler models also provides justification to collect further data to calibrate more complex biophysical mechanisms in subsequent model extensions. For example, this has been the process behind the extension of the logistic model in Eq. (3) to the reaction-diffusion model in Eq. (4), then to a mechanically-coupled model in Sect. 3.2, and later to a vasculature-coupled or drug-informed model in Sects. 3.3 and 3.5, respectively.

Biophysical inversion is a promising strategy to calibrate predictive models of cancer, but also presents several challenges. Complex, typically nonlinear and time-dependent, PDE-based models often result in ill-conditioned and non-convex optimization problems, which require sophisticated numerical algorithms to stabilize the inversion, such as multiresolution continuation, parameter continuation, and regularization schemes [29, 98, 99, 104]. Data scarcity can exacerbate the ill-posedness. Mitigating this issue can entail imposing additional modeling assumptions and regularization strategies [104]. The noise arising from various sources in imaging data also complicates model inversion. Further modeling priors and structure-exploiting algorithms can help mitigate some of these issues. Other mathematical considerations such as the choice of mismatch function (e.g., $L^2$ loss, cross-entropy loss) and regularization models can further complicate the inversion. Inversion methods may also require several forward model evaluations, so specialized solvers are needed to prevent prohibitive computational costs.

OPAL is an attractive methodology to decide potential model extensions by comprehensively assessing the improvement in model predictions against the increase in model complexity. In particular, OPAL can guide the modeler to select the *best* valid model representing a certain experimental or clinical setting while accounting for uncertainties in both data and model parameters. However, OPAL may also face certain challenges. For example, computationally expensive models might require a more efficient approach for sensitivity analysis than the Sobol method, such as the elementary effects method or metamodels [93]. Another major difficulty in model selection is that every model in the initial set of models might be invalid (i.e., they do not satisfy the validation criteria), which would require to extend such set with further models.

Finally, application of OCT to mathematical models of tumour growth and treatment response is a promising strategy for the optimization of therapeutic regimens in silico [52]. However, the reliability and plausibility of solutions generated by OCT methods depend on several factors including the validity of the model, the accurate definition of the objective functional, the uncertainty in the parameters and data, the application of clinically-relevant constraints, and the accuracy in solving the OCT problem itself. Additionally, implementation of OCT approaches within the clinical trial system is even more complicated than model validation. Beyond the challenges on data availability, model assumptions, and biophysical inversion described above, a controlled clinical study to validate OCT-generated therapies would involve the test of novel computationally-derived regimen protocols in patients. This requires a close monitoring of treatment response, toxicity, and patient well-being, which can be extremely complex to balance and maintain during such clinical study. Thus, robust preclinical evidence showing the advantages of OCT for the design of therapeutic regimens is required before advancing to clinical scenarios.

## 9 Conclusion

Integrating quantitative data obtained from biomedical imaging with mechanism-based mathematical modeling represents a significant departure from current paradigms in cancer biology and oncology. More specifically, this approach is fundamentally different from the current trend in modeling which emphasizes applying the methods of artificial intelligence to extremely large data sets. However, statistical inference—though enormously powerful—relies on properties of large populations that can frequently obscure important characteristics (or conditions) that are specific to individual patients and may drive the development of their disease or their response to therapy. The high-consequence decisions present in clinical oncology simply must be based on more than data analytics. These decisions must incorporate biophysical processes within a rigorous, mathematical framework that can be calibrated with patient-specific data to make patient-specific predictions. The transformation from population-based care to patient-based care is inevitable, and the intimate integration of quantitative imaging, mechanism-based mathematical modeling, and efficient computational methods enabling precise in silico tumour forecasts is a very promising avenue to achieve this important goal.

**Acknowledgements** GL was partially supported by a Peter O'Donnell Jr. Postdoctoral Fellowship from the Oden Institute for Computational Engineering and Sciences at The University of Texas at Austin and acknowledges funding from the European Union's Horizon 2020 research and innovation programme under the Marie Skłodowska-Curie grant agreement No. 838786. EABFL is supported by the Oncological Data and Computational Sciences collaboration between the Oden Institute for Computational Engineering and Sciences, The University of Texas MD Anderson Cancer Center, and Texas Advanced Computing Center. TEY is a CPRIT Scholar in Cancer Research and acknowledges funding from CPRIT RR160005, NCI U24CA226110, NCI R01CA235800, and NCI U01CA253540.

## References

1. Agosti, A., Giverso, C., Faggiano, E., Stamm, A., Ciarletta, P.: A personalized mathematical tool for neuro-oncology: a clinical case study. Int. J. Non Linear Mech. **107**, 170–181 (2018)
2. Akcelik, V., Biros, G., Ghattas, O., Hill, J., Keyes, D., van Bloemen, Wanders B.: Parallel algorithms for PDE constrained optimization (Chap. 16). In: Parallel Processing for Scientific Computing, vol. 20, pp. 291–322. SIAM, Philadelphia, Pennsylvania, US (2006)
3. Almeida, L., Bagnerini, P., Fabrini, G., Hughes, B.D., Lorenzi, T.: Evolution of cancer cell populations under cytotoxic therapy and treatment optimisation: insight from a phenotype-structured model. ESAIM:M2AN **53**(4), 1157–1190 (2019)

4. Anderson, A.W., Xie, J., Pizzonia, J., Bronen, R.A., Spencer, D.D., Gore, J.C.: Effects of cell volume fraction changes on apparent diffusion in human cells. Magn. Reson. Imaging **18**(6), 689–695 (2000)
5. Aniţa, S., Arnăutu, V., Capasso, V.: An Introduction to Optimal Control Problems in Life Sciences and Economics. From Mathematical Models to Numerical Simulation with MATLAB®. Birkhauser (2011)
6. Atuegwu, N.C., Arlinghaus, L.R., Li, X., Welch, E.B., Chakravarthy, A.B., Gore, J.C., Yankeelov, T.E.: Integration of diffusion-weighted MRI data and a simple mathematical model to predict breast tumor cellularity during neoadjuvant chemotherapy. Magn. Reson. Med. **66**(6), 1689–1696 (2011)
7. Baldock, A., Rockne, R., Boone, A., Neal, M., Bridge, C., Guyman, L., Mrugala, M., Rockhill, J., Swanson, K.R., Trister, A.D., Hawkins-Daarud, A., Corwin, D.M.: From patient-specific mathematical Neuro-Oncology to Precision Medicine. Front. Oncol. **3**, 62 (2013)
8. Barnes, S.L., Sorace, A.G., Loveless, M.E., Whisenant, J.G., Yankeelov, T.E.: Correlation of tumor characteristics derived from DCE-MRI and DW-MRI with histology in murine models of breast cancer. NMR Biomed. **28**(10), 1345–1356 (2015)
9. Baumann, M., Krause, M., Overgaard, J., Debus, J., Bentzen, S.M., Daartz, J., Richter, C., Zips, D., Bortfeld, T.: Radiation oncology in the era of precision medicine. Nat. Rev. Cancer **16**(4), 234 (2016)
10. Beck, J.L.: Bayesian system identification based on probability logic. Struct. Control. Health Monit. **17**(7), 825–847 (2010)
11. Benzekry, S., Lamont, C., Beheshti, A., Tracz, A., Ebos, J.M.L., Hlatky, L., Hahnfeldt, P.: Classical mathematical models for description and prediction of experimental tumor growth. PLoS Comput. Biol. **10**(8), e1003,800 (2014)
12. Bevers, T.B., Anderson, B.O., Bonaccio, E., Buys, S., Daly, M.B., Dempsey, P.J., Farrar, W.B., Fleming, I., Garber, J.E., Harris, R.E., Heerdt, A.S., Helvie, M., Huff, J.G., Khakpour, N., Khan, S.A., Krontiras, H., Lyman, G., Rafferty, E., Shaw, S., Smith, M.L., Tsangaris, T.N., Williams, C., Yankeelov, T., Network, N.C.C.: NCCN clinical practice guidelines in oncology: breast cancer screening and diagnosis. J. Natl. Compr. Cancer Netw. **7**(10), 1060–1096 (2009)
13. Burman, E., Claus, S., Hansbo, P., Larson, M.G., Massing, A.: CutFEM: discretizing geometry and partial differential equations. Int. J. Numer. Methods Eng. **104**(7), 472–501 (2015)
14. Castell, F., Cook, G.J.R.: Quantitative techniques in 18FDG PET scanning in oncology. Br. J. Cancer **98**(10), 1597–601 (2008)
15. Chen, X., Summers, R.M., Yoa, J.: Kidney tumor growth prediction by coupling reaction-diffusion and biomechanical model. IEEE Trans. Biomed. Eng. **60**(1), 169–173 (2012)
16. Chow, G.C.: A comparison of the information and posterior probability criteria for model selection. J. Econom. **16**(1), 21–33 (1981)
17. Chung, J., Hulbert, G.: A time integration algorithm for structural dynamics with improved numerical dissipation: the generalized-$\alpha$ method. J. Appl. Mech. **60**(2), 371–375 (1993)
18. Clatz, O., Sermesant, M., Bondiau, P.Y., Delingette, H., Warfield, S.K., Malandain, G., Ayache, N.: Realistic simulation of the 3-D growth of brain tumors in MR images coupling diffusion with biomechanical deformation. IEEE Trans. Med. Imaging **24**(10), 1334–1346 (2005)
19. Colin, T., Iollo, A., Lagaert, J.B., Saut, O.: An inverse problem for the recovery of the vascularization of a tumor. J. Inverse Ill Posed Probl. **22**(6), 759–786 (2014)
20. Colli, P., Gomez, H., Lorenzo, G., Marinoschi, G., Reali, A., Rocca, E.: Mathematical analysis and simulation study of a phase-field model of prostate cancer growth with chemotherapy and antiangiogenic therapy effects. Math. Models Methods Appl. Sci. **30**(07), 1253–1295 (2020)
21. Cottrell, J.A., Hughes, T.J.R., Bazilevs, Y.: Isogeometric Analysis: Toward Integration of CAD and FEA. Wiley, New York (2009)
22. Douglas, B.G., Fowler, J.F.: The effect of multiple small doses of X rays on skin reactions in the mouse and a basic interpretation. Radiat. Res. **66**(2), 401–426 (1976)
23. Drzisga, D., Gmeiner, B., Rüde, U., Scheichl, R., Wohlmuth, B.: Scheduling massively parallel multigrid for multilevel Monte Carlo methods. SIAM J. Sci. Comput. **39**(5), S873–S897 (2017)

24. Düster, A., Parvizian, J., Yang, Z., Rank, E.: The finite cell method for three-dimensional problems of solid mechanics. Comput. Methods Appl. Mech. Eng. **197**(45–48), 3768–3782 (2008)
25. Farin, G.: NURBS Curves and Surfaces: From Projective Geometry to Practical Use, 2nd edn. A. K. Peters Ltd. (1999)
26. Farrell, K., Oden, J.T., Faghihi, D.: A Bayesian framework for adaptive selection, calibration, and validation of coarse-grained models of atomistic systems. J. Comput. Phys. **295**, 189–208 (2015)
27. Feng, X., Hormuth, D.A., Yankeelov, T.E.: An adjoint-based method for a linear mechanically-coupled tumor model: application to estimate the spatial variation of murine glioma growth based on diffusion weighted magnetic resonance imaging. Comput. Mech. **63**(2), 159–180 (2019)
28. Gambhir, S.S.: Molecular imaging of cancer with positron emission tomography. Nat. Rev. Cancer **2**(9), 683–93 (2002)
29. Gholami, A., Mang, A., Biros, G.: An inverse problem formulation for parameter estimation of a reaction-diffusion model of low grade gliomas. J. Math. Biol. **72**(1), 409–433 (2016)
30. Gillies, R.J., Schornack, P.A., Secomb, T.W., Raghunand, N.: Causes and effects of heterogeneous perfusion in tumors. Neoplasia **1**(3), 197–207 (1999)
31. Gomez, H., van der Zee, K.G.: Computational phase-field modeling. In: Stein, E., de Borst, R., Hughes, T.J.R. (eds.) Encyclopedia of Computational Mechanics, 2nd edn., pp. 1–35 (2017)
32. Harpold, H.L.P., Alvord, E.C., Swanson, K.R.: The evolution of mathematical modeling of glioma proliferation and invasion. J. Neuropathol. Exp. Neurol. **66**(1), 1–9 (2007)
33. Hathout, L., Patel, V., Wen, P.: A 3-dimensional DTI MRI-based model of GBM growth and response to radiation therapy. Int. J. Oncol. **49**(3), 1081–1087 (2016)
34. Helmlinger, G., Netti, P.A., Lichtenbeld, H.C., Melder, R.J., Jain, R.K.: Solid stress inhibits the growth of multicellular tumor spheroids. Nat. Biotech. **15**(8), 778–783 (1997)
35. Hogea, C., Davatzikos, C., Biros, G.: An image-driven parameter estimation problem for a reaction-diffusion glioma growth model with mass effects. J. Math. Biol. **56**(6), 793–825 (2008)
36. Hormuth, D.A., Weis, J.A., Barnes, S., Miga, M.I., Quaranta, V., Yankeelov, T.E.: Biophysical modeling of in vivo glioma response after whole-brain radiation therapy in a murine model of brain cancer. Int. J. Radiat. Oncol. Biol. Phys. **100**(5), 1270–1279 (2018)
37. Hormuth, D.A., Jarrett, A.M., Feng, X., Yankeelov, T.E.: Calibrating a predictive model of tumor growth and angiogenesis with quantitative MRI. Ann. Biomed. Eng. **47**(7), 1539–1551 (2019)
38. Hormuth, D.A., Jarrett, A.M., Yankeelov, T.E.: Forecasting tumor and vasculature response dynamics to radiation therapy via image based mathematical modeling. Radiat. Oncol. **15**(1), 4 (2020)
39. Hormuth, D.A., II., Weis, J.A., Barnes, S.L., Miga, M.I., Rericha, E.C., Quaranta, V., Yankeelov, T.E.: Predicting in vivo glioma growth with the reaction diffusion equation constrained by quantitative magnetic resonance imaging data. Phys. Biol. **12**(4), 046,006 (2015)
40. Hormuth, D.A., II., Weis, J.A., Barnes, S.L., Miga, M.I., Rericha, E.C., Quaranta, V., Yankeelov, T.E.: A mechanically coupled reaction-diffusion model that incorporates intra-tumoural heterogeneity to predict in vivo glioma growth. J. R. Soc. Interface **14**(128), 20161,010 (2017)
41. Hormuth, D.A., II., Jarrett, A.M., Lima, E.A.B.F., McKenna, M.T., Fuentes, D.T., Yankeelov, T.E.: Mechanism-based modeling of tumor growth and treatment response constrained by multiparametric imaging data. JCO Clin. Cancer Inform. **3**, 1–10 (2019)
42. Hormuth, D.A., II., Sorace, A.G., Virostko, J., Abramson, R.G., Bhujwalla, Z.M., Enriquez-Navas, P., Gillies, R., Hazle, J.D., Mason, R.P., Quarles, C.C., Weis, J.A., Whisenant, J.G., Xu, J., Yankeelov, T.E.: Translating preclinical MRI methods to clinical oncology. J. Magn. Reson. Imaging **50**(5), 1377–1392 (2019)
43. Hughes, T.J.R.: The Finite Element Method: Linear Static and Dynamic Finite Element Analysis. Dover Publications (2000)

44. Iyengar, R., Zhao, S., Chung, S.W., Mager, D.E., Gallo, J.M.: Merging systems biology with pharmacodynamics. Sci. Transl. Med. **4**(126), 126ps7 (2012)
45. Jain, R.K., Safabakhsh, N., Sckell, A., Chen, Y., Jiang, P., Benjamin, L., Yuan, F., Keshet, E.: Endothelial cell death, angiogenesis, and microvascular function after castration in an androgen-dependent tumor: Role of vascular endothelial growth factor. Proc. Natl. Acad. Sci. USA **95**(18), 10,820-10,825 (1998)
46. Jain, R.K., di Tomaso, E., Duda, D.G., Loeffler, J.S.: Sorensen AG, Batchelor TT: Angiogenesis in brain tumours. Nat. Rev. Neurosci. **8**(8), 610–622 (2007)
47. Jain, R.K., Martin, J.D., Stylianopoulos, T.: The role of mechanical forces in tumor growth and therapy. Annu. Rev. Biomed. Eng. **16**, 321–346 (2014)
48. Jansen, K.E., Whiting, C.H., Hulbert, G.M.: A generalized-$\alpha$ method for integrating the filtered Navier–Stokes equations with a stabilized finite element method. Comput. Methods Appl. Mech. Eng. **190**(3–4), 305–319 (2000)
49. Jansen, M.J.: Analysis of variance designs for model output. Comput. Phys. Commun. **117**(1–2), 35–43 (1999)
50. Jarrett, A.M., Hormuth, D.A., Barnes, S.L., Feng, X., Huang, W., Yankeelov, T.E.: Incorporating drug delivery into an imaging-driven, mechanics-coupled reaction diffusion model for predicting the response of breast cancer to neoadjuvant chemotherapy: theory and preliminary clinical results. Phys. Med. Biol. **63**(10), 105,015 (2018)
51. Jarrett, A.M., Faghihi, D., Hormuth, D.A., II., Lima, E.A.B.F., Virostko, J., Biros, G., Patt, D., Yankeelov, T.E.: Optimal control theory for personalized therapeutic regimens in oncology: background, history, challenges, and opportunities. J. Clin. Med. **9**(5), 1314 (2020)
52. Jarrett, A.M., Hormuth, D.A., Wu, C., Kazerouni, A.S., Ekrut, D.A., Virostko, J., Sorace, A.G., DiCarlo, J.C., Kowalski, J., Patt, D., Goodgame, B., Avery, S., Yankeelov, T.E.: Evaluating patient-specific neoadjuvant regimens for breast cancer via a mathematical model constrained by quantitative magnetic resonance imaging data. Neoplasia **22**(12), 820–830 (2020)
53. Jbabdi, S., Mandonnet, E., Duffau, H., Capelle, L., Swanson, K.R., Pélégrini-Issac, M., Guillevin, R., Benali, H.: Simulation of anisotropic growth of low-grade gliomas using diffusion tensor imaging. Magn. Reson. Med. **54**(3), 616–624 (2005)
54. Jeffreys, H.: The Theory of Probability, 3rd edn. Oxford University Press, Oxford (1998)
55. Jiang, X., Li, H., Xie, J., McKinley, E.T., Zhao, P., Gore, J.C., Xu, J.: In vivo imaging of cancer cell size and cellularity using temporal diffusion spectroscopy. Magn. Reson. Med. **78**(1), 156–164 (2017)
56. Johnson, K.E., Howard, G.R., Morgan, D., Brenner, E., Gardner, A.L., Durrett, R.E., Mo, W., Al'Khafaji, A.M., Sontag, E., Jarrett, A.M., Yankeelov, T.E., Brock, A.: Integrating multimodal data sets into a mathematical framework to describe and predict therapeutic resistance in cancer (2020)
57. Knopoff, D.A., Fernández, D.R., Torres, G.A., Turner, C.V.: Adjoint method for a tumor growth PDE-constrained optimization problem. Comput. Math. Appl. **66**(6), 1104–1119 (2013)
58. Koh, D.M., Collins, D.J.: Diffusion-weighted MRI in the body: applications and challenges in oncology. Am. J. Roentgenol. **188**(6), 1622–1635 (2007)
59. Konishi, S., Kitagawa, G.: Information Criteria and Statistical Modeling, 1st edn. Springer, New York (2008)
60. Konukoglu, E., Clatz, O., Menze, B.H., Stieltjes, B., Weber, M.A., Mandonnet, E., Delingette, H., Ayache, N.: Image guided personalization of reaction-diffusion type tumor growth models using modified anisotropic Eikonal equations. IEEE Trans. Med. Imaging **29**(1), 77–95 (2010)
61. Lê, M., Delingette, H., Kalpathy-Cramer, J., Gerstner, E.R., Batchelor, T., Unkelbach, J., Ayache, N.: Personalized radiotherapy planning based on a computational tumor growth model. IEEE Trans. Med. Imaging **36**(3), 815–825 (2017)
62. Lenhart, S., Workman, J.T.: Optimal Control Applied to Biological Models. Chapman and Hall/CRC (2007)
63. Leugering, G., Benner, P., Engell, S., Griewank, A., Harbrecht, H., Hinze, M., Rannacher, R., Ulbrich, S. (eds.): Trends in PDE Constrained Optimization. Springer (2014)

64. LeVeque, R.J.: Finite Difference Methods for Ordinary and Partial Differential Equations: Steady-State and Time-Dependent Problems. SIAM (2007)
65. Li, X., Welch, E.B., Arlinghaus, L.R., Chakravarthy, A.B., Xu, L., Farley, J., Loveless, M.E., Ia, M., Kelley, M.C., Meszoely, I.M., Means-Powell, J.A., Abramson, V.G., Grau, A.M., Gore, J.C., Yankeelov, T.E.: A novel AIF tracking method and comparison of DCE-MRI parameters using individual and population-based AIFs in human breast cancer. Phys. Med. Biol. **56**(17), 5753–5769 (2011)
66. Lima, E.A.B.F., Oden, J.T., Hormuth, D.A., Yankeelov, T.E., Almeida, R.C.: Selection, calibration, and validation of models of tumor growth. Math. Models Methods Appl. Sci. **26**(12), 2341–2368 (2016)
67. Lima, E.A.B.F., Oden, J.T., Wohlmuth, B., Shahmoradi, A., Hormuth, D.A., II., Yankeelov, T.E., Scarabosio, L., Horger, T.: Selection and validation of predictive models of radiation effects on tumor growth based on noninvasive imaging data. Comput. Methods Appl. Mech. Eng. **327**, 277–305 (2017)
68. Lipkova, J., Angelikopoulos, P., Wu, S., Alberts, E., Wiestler, B., Diehl, C., Preibisch, C., Pya, T., Comps, S., Hadjidoukas, P., Leemput, K.V., Koumoutsakos, P., Lowengrub, J., Menze, B.: Personalized radiotherapy planning for glioma using multimodal Bayesian model calibration. IEEE Trans. Med. Imaging **38**(8), 1875–1884 (2019)
69. Liu, Y., Sadowski, S.M., Weisbrod, A.B., Kebebew, E., Summers, R.M., Yao, J.: Patient specific tumor growth prediction using multimodal images. Med. Image Anal. **18**(3), 555–566 (2014)
70. Lorenzo, G., Scott, M.A., Tew, K., Hughes, T.J.R., Zhang, Y.J., Liu, L., Vilanova, G., Gomez, H.: Tissue-scale, personalized modeling and simulation of prostate cancer growth. Proc. Natl. Acad. Sci. U.S.A. **113**(48), E7663–E7671 (2016)
71. Lorenzo, G., Hughes, T.J.R., Dominguez-Frojan, P., Reali, A., Gomez, H.: Computer simulations suggest that prostate enlargement due to benign prostatic hyperplasia mechanically impedes prostate cancer growth. Proc. Natl. Acad. Sci. U.S.A. **116**(4), 1152–1161 (2019)
72. Lüpfert, C., Reichel, A.: Development and application of physiologically based pharmacokinetic-modeling tools to support drug discovery. Chem. Biodivers. **2**(11), 1462–1486 (2005)
73. Mang, A., Bakas, S., Subramanian, S., Davatzikos, C., Biros, G.: Integrated biophysical modeling and image analysis: application to neuro-oncology (2020). arXiv:2002.09628
74. Marusyk, A., Polyak, K.: Tumor heterogeneity: causes and consequences. Biochim. Biophys. Acta Rev. Cancer **1805**(1), 105–117 (2010)
75. Massey, S.C., White, H., Whitmire, P., Doyle, T., Johnston, S.K., Singleton, K.W., Jackson, P.R., Hawkins-Daarud, A., Bendok, B.R., Porter, A.B., Vora, S., Sarkaria, J.N., Hu, L.S., Mrugala, M.M., Swanson, K.R.: Image-based metric of invasiveness predicts response to adjuvant temozolomide for primary glioblastoma. PLoS ONE **15**(3), 1–18 (2020)
76. Mi, H., Petitjean, C., Dubray, B., Vera, P., Ruan, S.: Prediction of lung tumor evolution during radiotherapy in individual patients with PET. IEEE Trans. Med. Imaging **33**(4), 995–1003 (2014)
77. Miller, K.D., Siegel, R.L., Lin, C.C., Mariotto, A.B., Kramer, J.L., Rowland, J.H., Stein, K.D., Alteri, R., Jemal, A.: A Cancer treatment and survivorship statistics. CA Cancer J. Clin. **66**(4), 271–289 (2016)
78. Mittal, R., Iaccarino, G.: Immersed boundary methods. Annu. Rev. Fluid Mech. **37**(1), 239–261 (2005)
79. Mortimer, J.E., Bading, J.R., Park, J.M., Frankel, P.H., Carroll, M.I., Tran, T.T., Poku, E.K., Rockne, R.C., Raubitschek, A.A., Shively, J.E., Colcher, D.M.: Tumor uptake of (64)Cu-DOTA-Trastuzumab in patients with metastatic breast cancer. J. Nucl. Med. **59**(1), 38–43 (2018)
80. Nagelkerke, A., Bussink, J., Rowan, A.E., Span, P.N.: The mechanical microenvironment in cancer: how physics affects tumours. Semin. Cancer Biol. **35**, 62–70 (2015)
81. Neal, M.L., Trister, A.D., Cloke, T., Sodt, R., Ahn, S., Baldock, A.L., Bridge, C.A., Lai, A., Cloughesy, T.F., Mrugala, M.M., Rockhill, J.K., Rockne, R.C., Swanson, K.R.: Discriminating

survival outcomes in patients with glioblastoma using a simulation-based, patient-specific response metric. PLoS ONE **8**(1), e51,951 (2013)
82. Oden, J.T., Lima, E.A.B.F., Almeida, R.C., Feng, Y., Rylander, M.N., Fuentes, D., Faghihi, D., Rahman, M.M., DeWitt, M., Gadde, M., Zhou, J.C.: Toward predictive multiscale modeling of vascular tumor growth. Arch. Comput. Methods Eng. **23**(4), 735–779 (2016)
83. Oden, J.T., Babuška, I., Faghihi, D.: Predictive computational science: computer predictions in the presence of uncertainty. In: Stein, E., de Borst, R., Hughes, T.J.R. (eds.) Encyclopedia of Computational Mechanics, pp. 1–26. Wiley (2018)
84. Padhani, A.R., Liu, G., Mu-Koh, D., Chenevert, T.L., Thoeny, H.C., Takahara, T., Dzik-Jurasz, A., Ross, B.D., Van Cauteren, M., Collins, D., Hammoud, D.A., Rustin, G.J.S., Taouli, B., Choyke, P.L.: Diffusion-weighted magnetic resonance imaging as a cancer biomarker: consensus and recommendations. Neoplasia **11**(2), 102–125 (2009)
85. Parvizian, J., Düster, A., Rank, E.: Finite cell method: h- and p-extension for embedded domain problems in solid mechanics. Comput. Mech. **41**(1), 121–133 (2007)
86. Petra, N., Martin, J., Stadler, G., Ghattas, O.: A computational framework for infinite-dimensional Bayesian inverse problems, Part II: stochastic Newton MCMC with application to ice sheet flow inverse problems. SIAM J. Sci. Comput. **36**(4), A1525–A1555 (2014)
87. Prudencio, E., Cheung, S.H.: Parallel adaptive multilevel sampling algorithms for the Bayesian analysis of mathematical models. Int. J. Uncertain. Quantif. **2**(3), 215–237 (2012)
88. Rajendran, J.G., Krohn, K.A.: F-18 fluoromisonidazole for imaging tumor hypoxia: imaging the microenvironment for personalized cancer therapy. Semin. Nucl. Med. **45**(2), 151–162 (2015)
89. Rockne, R.C., Trister, A.D., Jacobs, J., Hawkins-Daarud, A.J., Neal, M.L., Hendrickson, K., Mrugala, M.M., Rockhill, J.K., Kinahan, P., Krohn, K.A., Swanson, K.R.: A patient-specific computational model of hypoxia-modulated radiation resistance in glioblastoma using (18)F-FMISO-PET. J. R. Soc. Interface **12**(103), 20141,174 (2015)
90. Rockne, R.C., Hawkins-Daarud, A., Swanson, K.R., Sluka, J.P., Glazier, J.A., Macklin, P., Hormuth, D.A., Jarrett, A.M., Lima, E.A.B.F., Tinsley Oden, J., Biros, G., Yankeelov, T.E., Curtius, K., Al Bakir, I., Wodarz, D., Komarova, N., Aparicio, L., Bordyuh, M., Rabadan, R., Finley, S.D., Enderling, H., Caudell, J., Moros, E.G., Anderson, A.R.A., Gatenby, R.A., Kaznatcheev, A., Jeavons, P., Krishnan, N., Pelesko, J., Wadhwa, R.R., Yoon, N., Nichol, D., Marusyk, A., Hinczewski, M., Scott, J.G.: The 2019 mathematical oncology roadmap. Phys. Biol. **16**(4), 41,005 (2019)
91. Roque, T., Risser, L., Kersemans, V., Smart, S., Allen, D., Kinchesh, P., Gilchrist, S., Gomes, A.L., Schnabel, J.A., Chappell, M.A.: A DCE-MRI driven 3-D reaction-diffusion model of solid tumour growth. IEEE Trans. Med. Imaging **37**(3), 724–732 (2018)
92. Ruuth, S.J.: Implicit-explicit methods for reaction-diffusion problems in pattern formation. J. Math. Biol. **34**(2), 148–176 (1995)
93. Saltelli, A., Ratto, M., Andres, T., Campolongo, F., Cariboni, J., Gatelli, D., Saisana, M., Tarantola, S.: Global Sensitivity Analysis: The Primer. Wiley, New York (2008)
94. Saltelli, A., Annoni, P., Azzini, I., Campolongo, F., Ratto, M., Tarantola, S.: Variance based sensitivity analysis of model output. Design and estimator for the total sensitivity index. Comput. Phys. Commun. **181**(2), 259–270 (2010)
95. Saltelli, A., Aleksankina, K., Becker, W., Fennell, P., Ferretti, F., Holst, N., Li, S., Wu, Q.: Why so many published sensitivity analyses are false: a systematic review of sensitivity analysis practices. Environ. Model. Softw. **114**, 29–39 (2019)
96. Schättler, H., Ledzewicz, U.: Optimal Control for Mathematical Models of Cancer Therapies. Springer, Berlin (2015)
97. Scheufele, K., Mang, A., Gholami, A., Davatzikos, C., Biros, G., Mehl, M.: Coupling brain-tumor biophysical models and diffeomorphic image registration. Comput. Methods Appl. Mech. Eng. **347**, 533–567 (2019)
98. Scheufele, K., Subramanian, S., Mang, A., Biros, G., Mehl, M.: Image-driven biophysical tumor growth model calibration (2019). arXiv:1907.07774

99. Scheufele, K., Subramanian, S., Biros, G.: Fully automatic calibration of tumor-growth models using a single mpMRI scan. IEEE Trans. Med. Imaging (2020). https://doi.org/10.1109/TMI.2020.3024264
100. Schillinger, D., Ruess, M.: The finite cell method: a review in the context of higher-order structural analysis of CAD and image-based geometric models. Arch. Comput. Methods Eng. **22**(3), 391–455 (2015)
101. Shi, J., Kantoff, P.W., Wooster, R., Farokhzad, O.C.: Cancer nanomedicine: progress, challenges and opportunities. Nat. Rev. Cancer **17**(1), 20–37 (2017)
102. Sobol, I.M.: Global sensitivity indices for nonlinear mathematical models and their Monte Carlo estimates. Math. Comput. Simul. **55**(1–3), 271–280 (2001)
103. Sorace, A.G., Wu, C., Barnes, S.L., Jarrett, A.M., Avery, S., Patt, D., Goodgame, B., Luci, J.J., Kang, H., Abramson, R.G., Yankeelov, T.E., Virostko, J.: Repeatability, reproducibility, and accuracy of quantitative MRI of the breast in the community radiology setting. J. Magn. Reson. Imaging **48**(3), 695–707 (2018)
104. Subramanian, S., Scheufele, K., Mehl, M., Biros, G.: Where did the tumor start? An inverse solver with sparse localization for tumor growth models. Inverse Prob. **36**(4), 045,006 (2020)
105. Sundgren, P.C., Dong, Q., Gómez-Hassan, D., Mukherji, S.K., Maly, P., Welsh, R.: Diffusion tensor imaging of the brain: review of clinical applications. Neuroradiology **46**(5), 339–350 (2004)
106. Swan, A., Hillen, T., Bowman, J.C., Murtha, A.D.: A patient-specific anisotropic diffusion model for brain tumour spread. Bull. Math. Biol. **80**(5), 1259–1291 (2018)
107. Swan, G.W.: Role of optimal control theory in cancer chemotherapy. Math. Biosci. **101**(2), 237–284 (1990)
108. Swanson, K.R., Rostomily, R.C., Alvord, E.C.: A mathematical modelling tool for predicting survival of individual patients following resection of glioblastoma: a proof of principle. Br. J. Cancer **98**(1), 113–119 (2008)
109. Swanson, K.R., Rockne, R.C., Claridge, J., Ma, C., Alvord, E.C., Anderson, A.R.A.: Quantifying the role of angiogenesis in malignant progression of gliomas: in silico modeling integrates imaging and histology. Cancer Res. **71**(24), 7366–7375 (2011)
110. Swierniak, A., Kimmel, M., Smieja, J.: Mathematical modeling as a tool for planning anticancer therapy. Eur. J. Pharmacol. **625**(1), 108–121 (2009)
111. Therasse, P., Arbuck, S.G., Eisenhauer, E.A., Wanders, J., Kaplan, R.S., Rubinstein, L., Verweij, J., Van Glabbeke, M., van Oosterom, A.T., Christian, M.C., Gwyther, S.G.: New guidelines to evaluate the response to treatment in solid tumors. J. Natl. Cancer Inst. **92**(3), 205–216 (2000)
112. Vaupel, P., Mayer, A.: Hypoxia in cancer: significance and impact on clinical outcome. Cancer Metastasis Rev. **26**(2), 225–239 (2007)
113. Vilanova, G., Colominas, I., Gomez, H.: Computational modeling of tumor-induced angiogenesis. Arch. Comput. Methods Eng. **24**(4), 1071–1102 (2017)
114. Wang, K., Bui-Thanh, T., Ghattas, O.: A randomized maximum a posteriori method for posterior sampling of high dimensional nonlinear Bayesian inverse problems. SIAM J. Sci. Comput. **40**(1), A142–A171 (2018)
115. Weis, J.A., Miga, M.I., Arlinghaus, L.R., Li, X., Chakravarthy, A.B., Abramson, V., Farley, J., Yankeelov, T.E.: A mechanically coupled reaction-diffusion model for predicting the response of breast tumors to neoadjuvant chemotherapy. Phys. Med. Biol. **58**(17), 5851–5866 (2013)
116. Whisenant, J.G., Ayers, G.D., Loveless, M.E., Barnes, S.L., Colvin, D.C., Yankeelov, T.E.: Assessing reproducibility of diffusion-weighted magnetic resonance imaging studies in a murine model of HER2+ breast cancer. Magn. Reson. Imaging **32**(3), 245–249 (2014)
117. Withers, H.R.: Biologic basis for altered fractionation schemes. Cancer **55**(9 Suppl), 2086–2095 (1985)
118. Wong, K.C.L., Summers, R.M., Kebebew, E., Yoa, J.: Pancreatic tumor growth prediction with elastic-growth decomposition, image-derived motion, and FDM-FEM coupling. IEEE Trans. Med. Imaging **36**(1), 111–123 (2017)

119. Woolf, D.K., Beresford, M., Li, S.P., Dowsett, M., Sanghera, B., Wong, W.L., Sonoda, L., Detre, S., Amin, V., Ah-See, M.L., Miles, D., Makris, A.: Evaluation of FLT-PET-CT as an imaging biomarker of proliferation in primary breast cancer. Br. J. Cancer **110**(12), 2847–2854 (2014)
120. Xu, J., Vilanova, G., Gomez, H.: Phase-field model of vascular tumor growth: three-dimensional geometry of the vascular network and integration with imaging data. Comput. Methods Appl. Mech. Eng. **359**(112), 648 (2020)
121. Yankeelov, T.E., Gore, J.C.: Dynamic contrast enhanced magnetic resonance imaging in oncology: theory, data acquisition, analysis, and examples. Curr. Med. Imaging Rev. **3**(2), 91–107 (2009)
122. Yankeelov, T.E., Atuegwu, N., Hormuth, D.A., Weis, J.A., Barnes, S.L., Miga, M.I., Rericha, E.C., Quaranta, V.: Clinically relevant modeling of tumor growth and treatment response. Sci. Transl. Med. **5**(187), 187ps9 (2013)
123. Yankeelov, T.E., Quaranta, V., Evans, K.J., Rericha, E.C.: Toward a science of tumor forecasting for clinical oncology. Cancer Res. **75**(6), 918–923 (2015)
124. Yin, A., Moes, D.J.A.R., van Hasselt, J.G.C., Swen, J.J., Guchelaar, H.J.: A review of mathematical models for tumor dynamics and treatment resistance evolution of solid tumors. CPT Pharmacometrics Syst. Pharmacol. **8**(10), 720–737 (2019)
125. Zahnd, D., Aebi, S., Rusterholz, S., Fey, M.F., Borner, M.M.: A randomized crossover trial assessing patient preference for two different types of portable infusion-pump devices. Ann. Oncol. **10**(6), 727–729 (1999)
126. Zhang, Y.J.: Geometric Modeling and Mesh Generation from Scanned Images. Mathematical and Computational Imaging Sciences Series, CRC Press, Chapman & Hall/CRC (2018)

# The Effect of Over-Feeding in a Computational Model of Tumour Growth

**Pan Pantziarka, Lina Ghibelli, and Albrecht Reichle**

## 1 Introduction

This work explores the effects of excess nutrient in an agent-based lattice model of tumour growth. The Non-physiological Evolutionary Algorithm for Tumour Growth (NEATG) model is a platform that has previously been used to explore aspects of tumour regrowth following cytotoxic treatment and the impact of tissue-cell communicative breakdown (anakoinosis) [1]. The model has been shown to recapitulate real-world tumour growth dynamics and display emergent behaviours in line with in vitro tumour systems. This work is motivated by a chance observation during regression testing of the model after refactoring of the code. During this series of tests a number of model runs had shown excessive run-times due to unexpectedly high rates of tumour growth. On investigation it was ascertained that a single model parameter had been mistyped and that the nutrient supply had been set at a supra-physiological level. This study expands on this chance finding to explore the impact of over-feeding on tumour growth dynamics and its relation to response to cytotoxic treatment.

The NEATG platform is coded in the Java programming language and is designed to explore the effects of different treatment scenarios on tumour growth. It consists of an agent-based model of cells, representing both cancer and non-cancer populations,

P. Pantziarka (✉)
The George Pantziarka TP53 Trust, London KT1 2JP, UK
e-mail: anticancer.org.uk@gmail.com

Anticancer Fund, Strombeek-Bever, 1853 Brussels, Belgium

L. Ghibelli
Dipartimento di Biologia, Universita' di Roma Tor Vergata, Roma, Italy

A. Reichle
Department of Hematology and Oncology, University Hospital Regensburg, Regensburg, Germany

 
I. Balaz and A. Adamatzky (eds.), *Cancer, Complexity, Computation*, Emergence, Complexity and Computation 46,
https://doi.org/10.1007/978-3-031-04379-6_4

on a rectangular lattice. Each grid element of the lattice can contain a configurable number of cells in competition for nutrient and growth factors. Cells contain genes which encode for fitness, as well as phenotypic features which are also taken into account for fitness. The model is evolutionary in that genetic changes can occur at cell division and cells can also migrate from one element of the lattice to another when over-crowding leads to competition. In addition to the core model, NEATG also contains a reporting engine so that detailed information can be recorded during run-time. More details of the model and its behaviour are reported in [2].

The theory of anakoinosis views tumours as complex adaptive systems with aberrant signalling/communication between tumour cell populations and the local tissue environment [3, 4]. As a treatment paradigm is posits that communicative reprogramming is an essential ingredient for successful tumour control—and has been tested in multiple clinical trials across a range of cancer types. This hypothesis was incorporated into an enhanced version of the NEATG model which we called NEATG_A, and which was reported in a prior study [1]. In this version of the model communication between cells and the tissue was modelled as a simple handshake protocol between cells and the local lattice grid element. Simply put, communicative dysfunction created a permissive local environment which more easily enabled tumour growth. Anakoinosis treatment reprogrammed this handshaking algorithm and reduced the permissive environment, correcting aberrant cell-tissue communication. Results from the model showed that a combination of cytotoxic treatment and anakoinosis treatment was more successful than either single treatment in slowing the accelerated tumour regrowth following initial treatment—a finding in line with results from clinical trials of anakoinosis treatment.

As detailed in previous studies, at each time step cells are able to consume nutrient and gene factors and are able to change state based on their internal state and the state of the grid element in which they are located. Details of this process are outlined in the two previous studies already mentioned. The model in this study is the NEATG_A model with a number of minor enhancements in reporting and CPU performance. The initial impetus for this study arose in the context of the regression testing of changes in the platform to ensure that code refactoring had not changed the underlying behaviour of the model. During this series of tests one of the input parameters to a treatment scenario script file was mistyped and the nutrient supply was multiplied by 10, resulting in an excessive run-time. This was caused by a very high rate of cancer cell growth. In previous studies the impact of nutrient supply had focused on the effects of restricting nutrient supply. It had been shown that restricting nutrient supply caused a reduction in normal cell counts compared to the number under homeostatic conditions (i.e. with adequate nutrient supply). It had been noted that a slightly increased nutrient supply did not have significant effects on cell counts, but no investigation had been performed to assess the impact of very high excess nutrient supply. This work, therefore, explores the effects of very high nutrient supply on tumour growth dynamics, on the interaction of over-feeding with cytotoxic treatments and the relationship with anakoinosis.

## 2 Methods

Cells in the NEATG family of models consume nutrient that is supplied to each element of the lattice grid. Each grid element can contain a variable number of cells, some of which are Malignant cells and some are Normal (non-transformed) cells. Each cell has a nutrient store which can conserve nutrient that is not consumed and the amount of nutrient consumed per time period is controlled by the nutrient rate for that cell. The nutrient rate is a mutable characteristic of a cell, and therefore may increase or decrease in daughter cancer cells during cell division. When the nutrient store is depleted, or a cell's lifetime (number of clock ticks) has counted down to zero, then a cell will undergo a change of state, as shown in the lifecycle in Fig. 1.

Each grid element is seeded with an initial population of Normal cells, under homeostatic conditions and with no Malignant cells present this population remains stable as cells multiply, complete their lifetimes and undergo apoptosis. However, Malignant cells can disrupt this and multiply at a greater rate therefore increasing the number of cells such that cell competition takes place when the population exceeds the maximum carrying capacity. Nutrient is represented as anr integer value and is divided between all cells in a grid element according to the following distribution function:

$$N_i = S\frac{T_i}{\sum_{p=1}^{P} T_p}$$

where $N_i$ is the nutrient supplied to the $i$th cell, $T$ is the target value (nutrient demand) for a cell and $S$ the nutrient supplied to the grid element, and $P$ the number of cells in the grid element.

Cell lifetime is the number of time periods or clock-ticks that define the duration of a cell's existence. In this work all cells were defined to have a lifetime of 100 clock-ticks, although as a mutable property of a cell this could evolve over time in Malignant cells. Typical model runs were for 1500 or more clock-ticks/iterations. Other key cell parameters include the number of genes and individual gene parameters as described previously. In this work the same 3-gene structure and settings were used as in previous studies as the behaviours associated with these parameters are now well-characterised.

The NEATG_A model implemented a simple handshaking protocol to represent cell-tissue communication, in this case communication between a cell and the grid element in which it is located. This communication consists of a cell reporting a signature which is compared to a signature contained at the grid element to check for differences—defined as the Levenshtein distance between signatures. The signatures are derived from a combination of genomic and phenotypic data such that two cells with the same genomes and phenotypes have the same signature, for example all Normal cells, but any changes, for example a gene mutation, will cause a difference between the mutated and non-mutated version of the cell. The Levenshtein distance is a numeric measurement of the degree of change [5]. The handshake protocol

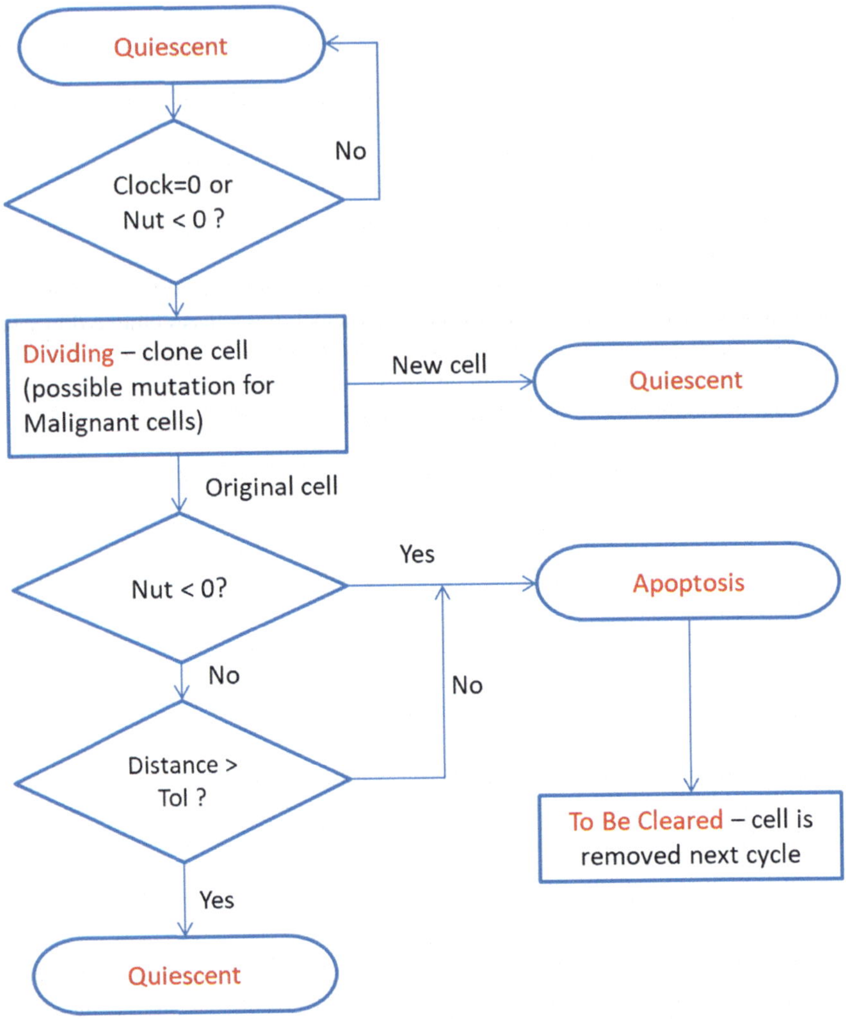


**Fig. 1** Cell fate program—note the same cycle is used for both Malignant and Normal cells. At cell division Malignant cells may probabilistically undergo a mutational event. The Tol figure corresponds to the degree of permissiveness of the host environment

implements a 'go-no-go' decision point when cell division takes place such that if a the distance between the cell signature and the grid level signature exceeds a given value, termed the *Tolerance*, then cell division is inhibited. Systems in which the Tolerance value is greater are therefore more tolerant of genomic change than systems in which the Tolerance value is lower and genomic change is constrained. The theory of anakoinosis posits that cancer exists where there is communicative dysfunction, and that correcting or reprogramming this aberrant cell-tissue signalling is essential for the successful treatment of cancer.

**Table 1** Key NEATG_A parameters used in this series of experiments

| Parameter | Effect | Value |
|---|---|---|
| Maximum cell count | The maximum carrying capacity of a grid element—when the number of cells reaches this value cell competition takes place | 10 |
| Nutrient rate | The amount of nutrient consumed by a cell per clock tick | 1 |
| Nutrient supply | Amount of nutrient supplied to a grid element—shared between cells using the distribution function previously described | 10 |
| Nutrient target | Nutrient demand for a cell | 10 |
| Invasion rate | The probability of a cell migrating to a neighbouring grid element during cell division | 0.10 |
| Mutation rate | Probability that a cell division will lead to a mutation event | 0.05 |
| Tolerance | The maximum allowed Levenshtein distance between tissue and cell signature (i.e. a measure of the permissiveness of the system) | 9 |
| Width/Length | The number of grid elements (width x length) | 50, 50 |

Note that these values of nutrient target, nutrient supply and the maximum cell count ensure that in homeostatic conditions all cells receive sufficient nutrient to meet their metabolic demands (i.e. nutrient supply = nutrient rate). A key objective of this work is to investigate tumour growth in situations where the nutrient supply far exceeds the nutrient rate

Our previous work has established a set of key parameters that produce homeostasis in the absence of Malignant cells (e.g. cell metabolism, cell turnover etc.) and tumour growth when seeded with one or more Malignant cells (e.g. growth and spread of Malignant cells, mutational events leading to the emergence of clonal subpopulations etc.). Key parameters are shown in Table 1.

In this study we perform a series of experiments to assess the impact of excessive nutrient supply on cancer growth and how this interacts with cytotoxic treatment and anakoinosis treatment. All experiments were carried out on a Windows 10 system and the 64-bit OpenJDK Java run-time environment (version 11.0.6). Statistical analysis was performed using R (release 4.0.3) and RStudio Desktop (version 1.4.1103). The Akaike Information Criterion (AIC) was used to select whether regression models should include interaction terms [6].

# 3 Results

## 3.1 *Malignant Cell Growth*

To assess the influence of both nutrient availability and Tolerance, and their interaction, on tumour growth a series of model runs were performed varying each of the factors. The nutrient supply was varied from 10 to 100 in increments of 10, and for each value of nutrient the Tolerance was varied from 0 to 9 in increments of 1. Each combination of nutrient supply and Tolerance ($10 \times 10$) was run 10 times, with a

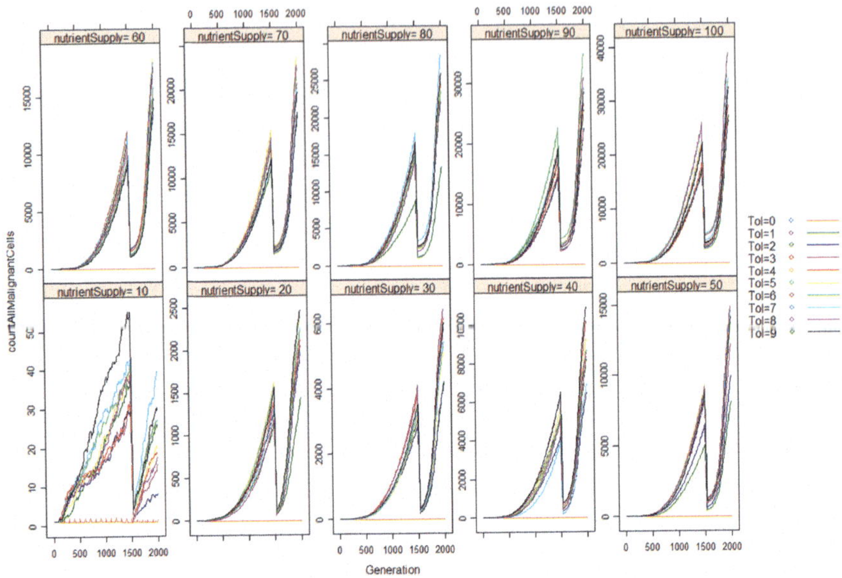

**Fig. 2** Average Malignant cell growth for 2000 clock-ticks for different Tolerance and nutrient supply settings—with cytotoxic treatment

duration of 2000 clock-ticks per run. A cytotoxic treatment scenario was initiated at clock-tick = 1500 which persisted for 25 clock-ticks. Each run was terminated at clock-tick = 2000. The plot of the average Malignant cell count growth for different Tolerance values and nutrient supply values is shown in Fig. 2. Note the y-axis varies per chart.

The associations between nutrient supply and Tolerance are shown at clock-tick = 1499 (immediately before cytotoxic treatment is applied) in the left panel of Fig. 3. Greater values of Tolerance and higher amounts of nutrient lead to sharply increased Malignant cell counts.

The count of Malignant cells alone does not tell us of the degree of spread of tumour within the grid. If we define a tumour grid element as one that includes one or more Malignant cells we can also look at how Tolerance value and nutrient supply are associated with the spread of tumour cells throughout the grid (in this case a 50 × 50 grid). This is shown in the right panel of Fig. 3.

Multiple regression analysis was used to test if the Tolerance, nutrient supply and their interaction significantly predicted the count of Malignant cells at time points clock-tick = 1000 and clock-tick = 1499. At clock-tick 1000 the regression showed a highly significant relationship that explained 53% of the variance (Adjusted R-squared = 0.53, F(3, 997) = 375.6, P < 2.2e-16). At clock-tick = 1499 the regression showed a higher explanatory value with an adjusted R-squared value of 67% (Adjusted R-squared = 0.67, F(3, 997) = 685.9, P < 2.2e-16).

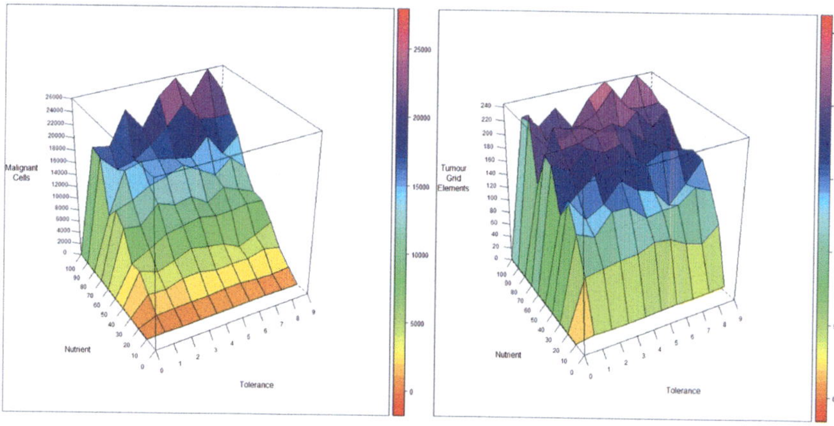


**Fig. 3** Left panel: Relationship between malignant cell count and nutrient supply and Tolerance value. Right panel: Association of number of tumour grid elements with Tolerance value and nutrient supply

In addition to the increase in Malignant cell numbers and invasiveness across grid elements we would also expect to see evolutionary change associated with increasing Tolerance and nutrient supply. The left panel of Fig. 4 shows the average number of distinct genotypes, including extinct genotypes, which have existed by clock-tick = 1499 for different values of Tolerance and nutrient Supply. Both greater Tolerance and nutrient supply values lead to a larger exploration of the genotype space than more restrictive Tolerance environments or with more constrained nutrient supply. A multiple regression model indicated a highly statistically significant relationship (Adjusted R-squared = 0.65, F(3,997) = 627.9, P < 2.2e-16).

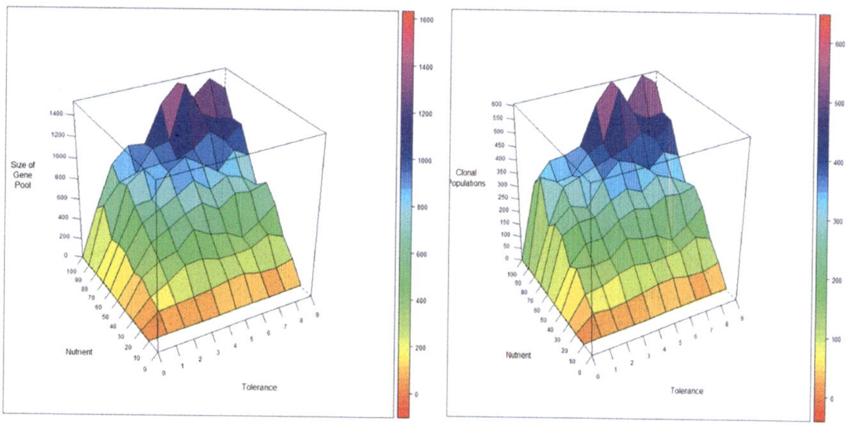


**Fig. 4** Left panel: Number of distinct genotypes at clock-tick = 1499. Right panel: Number of clonal subpopulations at click-tick = 1499

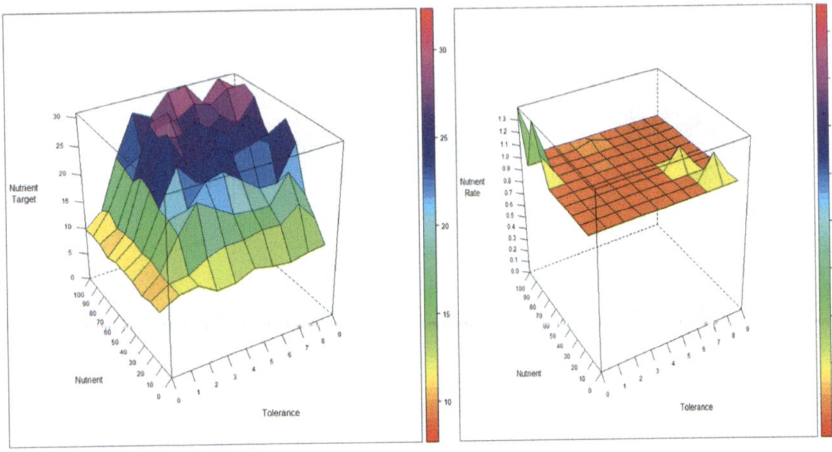


**Fig. 5** Left panel: Changes in nutrient target (i.e. nutrient demand) and changes in nutrient supply and Tolerance. Right panel: Change in nutrient rate as a function of nutrient supply and Tolerance

Intra-tumour heterogeneity is another, and arguably more clinically relevant, measure of evolutionary change within a tumour system. This is shown in the right panel of Fig. 4. Regression analysis of the number of clonal subpopulations as a function of Tolerance and nutrient Supply, and their interaction, shows that the model is able to predict 70% of the variance (Adjusted R-squared = 0.70, $F(3,997) = 763.2$, $P < 2.2e\text{-}16$).

Given that nutrient demand, in the form of the variable nutrient Target, is a mutable factor in Malignant cells we can also investigate the relationship between this value and the increasing Tolerance and nutrient supply values. Figure 5 shows how the nutrient target value rises from a baseline of 10 in response to increases both in nutrient supply and in Tolerance. In other words Malignant cells become greedier in response to increasing nutrient supply and increasing Tolerance.

The nutrient rate parameter is also mutable and is roughly analogous to the metabolic rate of a cell as it encodes how much nutrient a cell consumes in each clock-tick. As shown in the right panel of Fig. 5 the average nutrient rate of Malignant cells shows little change in response to increases in nutrient supply or Tolerance. Greedier cells are therefore not consuming additional nutrient but are using it to boost fitness and therefore their chance to proliferate in competition with other cells.

A series of runs were also performed in which the nutrient supply was reduced to below the baseline homeostatic value of 10. In the majority of runs there was little or no Malignant cell growth and a high degree of cell atrophy as non-Malignant cells died off due to lack of nutrient (data not shown).

## 4 Cytotoxic Treatment

The NEATG systems feature a number of different treatment scenarios which can be used to model the effects of treatment on malignant cell numbers, the degree of invasion, genetic heterogeneity etc. The most basic treatment option is loosely modelled on maximum tolerated dose chemotherapy. In this cytotoxic scenario apoptosis is induced in Malignant cells which are close to cell division. There is a degree of collateral damage in that some non-Malignant cells which can also be affected—although Malignant cells are more susceptible.

The kill rate we have defined as the percentage difference between the number of Malignant cells before the commencement of cytotoxic treatment, at clock-tick = 1499, and the first clock-tick after the cessation of treatment (i.e. at clock-tick = 1526). A kill rate of 100% means all Malignant cells have undergone apoptosis and there are no remaining Malignant cells, and a negative kill rate means that there has actually been an increase in Malignant cell numbers despite the treatment.

Figure 6 shows how the kill rate varies as a function of both the Tolerance and the nutrient supply. It is clear that at zero Tolerance the treatment has little effect but beyond that the treatment is effective in reducing Malignant cell numbers by over 80%.

Another view of the kill rate is shown in the left panel of Fig. 7. The coloured surface shows the average number of Malignant cells remaining after treatment has ended. Superimposed is the surface plot showing the number of Malignant cells at clock-tick = 1499, immediately prior to treatment commencing.

A multiple regression model showed a high correlation between Tolerance, nutrient supply and their interaction and the number of Malignant cells killed by the treatment. The adjusted R-squared = 0.70, F(3,997) = 787.1, P < 2.2e-16). However,

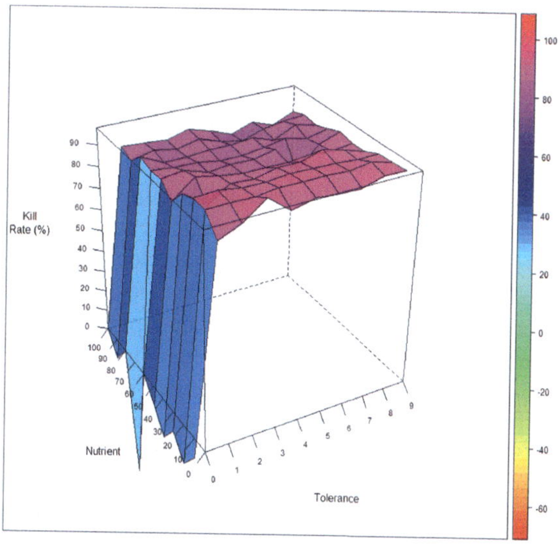


**Fig. 6** Kill rate, (percentage of Malignant cells killed by cytotoxic treatment), as a function of Tolerance and nutrient supply

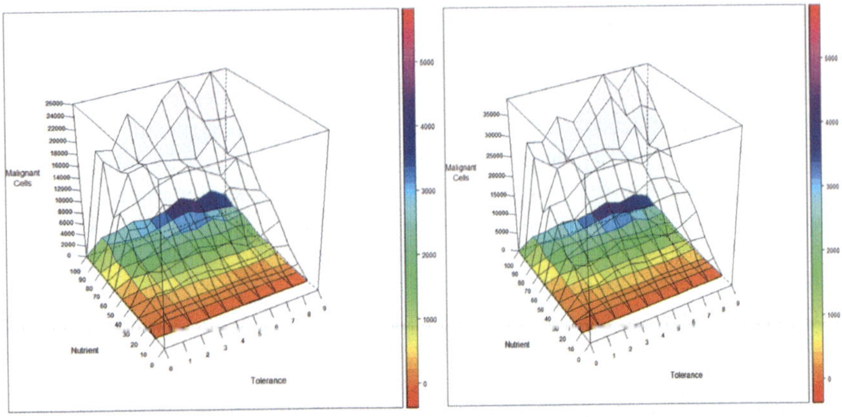


**Fig. 7** Left panel: Number of Malignant cells after cytotoxic treatment (in colour), superimposed is the Malignant cell count prior to treatment (at clock-tick = 1499). Right panel: Malignant cell growth after end of cytotoxic treatment. In colour—number of Malignant cells at treatment end (clock-tick = 1526), superimposed is the Malignant cell count at clock-tick = 2000

when looking at the kill rate, AIC model selection showed that a regression model that did not include the interaction term was a better fit for the data. The adjusted R-squared = 0.20, F(3,997) = 127.7, P < 2.2e-16).

Tumour repopulation occurs when the number of Malignant cells begins to increase again after the cessation of treatment. Of course, an ideal treatment would remove all Malignant cells, but in practice tumour repopulation is a common occurrence in the real world. We define the repopulation rate as the percentage increase in Malignant cells between the end of treatment, at clock-tick = 1526, and an arbitrary time point post-treatment. In these experiments model runs were for 2000 clock-ticks, therefore the repopulation rate is defined as the percentage increase in Malignant cells between clock-tick = 1526 and clock-tick = 2000.

Tumour repopulation is clearly shown in the right panel of Fig. 7. Notable is the very high increase in post-treatment regrowth with high values of Tolerance and nutrient supply. In terms of the cell growth post-treatment, a linear regression model assessing the influence of the Tolerance value, nutrient Supply and their interaction showed a high level of correlation. The adjusted R-squared = 0.74, F(3,996) = 942.4, P < 2.2e-16. The repopulation rate is defined as the rate of increase of Malignant cells between the end of treatment at clock-tick = 1526 and our final time point at clock-tick = 2000. Using AIC the model selected did not include a term for the interaction of Tolerance and nutrient supply and the model, although statistically significant, had low explanatory value. The adjusted R-squared = 0.03, F(3,997) = 17.3, P = 4.2e-08. It is also notable that in this model, in contrast to all others, the coefficient for nutrient supply was not statistically significant.

A regression model of the final count of Malignant cells, from clock-tick = 0 to clock-tick = 2000, so passing through the treatment period, shows a very strong relationship between the overall growth of Malignant and Tolerance, nutrient supply and their interaction. Adjusted R-squared = 0.72, F(3,996) = 849.7, P < 2.2e-16.

## 5 Cytotoxic Treatment + Nutrient Reduction (NR)

In this set of experiments the treatment regimen consists of cytotoxic treatment in combination with a normalisation of nutrient supply to correct the overfeeding. The aim of this experiment is to reduce the rate of tumour regrowth following cytotoxic treatment. As before the cytotoxic treatment is initiated at clock-tick = 1500 and ends at clock-tick = 1525. The nutrient normalisation reverts the nutrient supply to 10, which is the optimum, and commences at clock-tick = 1500 and returns to the pre-treatment level at clock-tick = 1600. The nutrient restriction treatment is for a longer period than the cytotoxic treatment in that informal testing had shown that very short-term nutrient treatments had marginal effects compared to the more drastic step of directly inducing cell death via cytotoxic treatment. The growth curves are shown in Fig. 8.

In comparison to the treatment in Fig. 2 we can see that the nutrient restriction delays the start of the accelerated regrowth that occurs at the end of the cytotoxic treatment for the duration of the nutrient restriction. This delay leads to a statistically significant reduced malignant cell growth at the final time-point at clock-tick = 2000 (comparison of the means of the Malignant cell count in the two treatments, across all values of Tolerance and nutrient supply, shows a lower mean in the cytotoxic + nutrient restriction treatment, P = 3.564e-05). However, it could be argued that there is a longer duration between the end of the cytotoxic treatment and clock-tick = 2000. If we therefore compare Malignant cell counts for a duration of 250 clock-ticks after

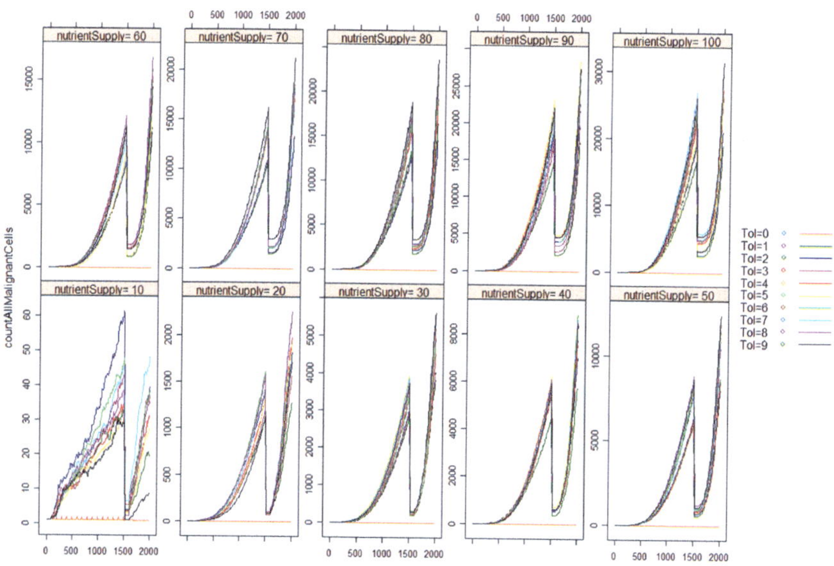


**Fig. 8** Average Malignant cell growth for 2000 clock-ticks for different Tolerance and nutrient supply settings—with cytotoxic treatment + nutrient restriction

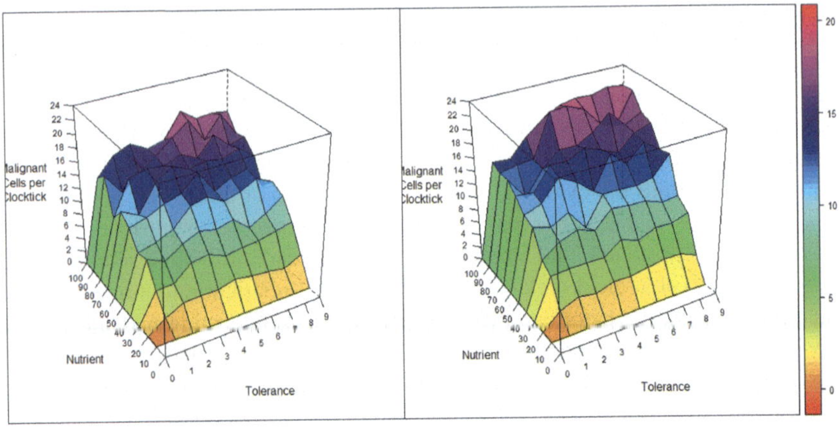


**Fig. 9** Gradient of Malignant cell repopulation for the first 250 clock-ticks after the end of treatment. Left panel, cytotoxic treatment, right panel cytotoxic + nutrition restriction

the cessation of treatment, then there is in fact a higher Malignant mean cell count in the cytotoxic + nutrient restriction group (4029 versus to 3572 for the cytotoxic treatment, P = 0.0026). The gradient of the Malignant cell repopulation for the first 250 clock-ticks after the end of treatment is shown, for both treatments, in Fig. 9. The addition of nutrient restriction delays the beginning of population regrowth but when nutrient levels return to pre-treatment levels the regrowth is more aggressive.

## 6 Cytotoxic Treatment + Tolerance Normalisation (TN)

In this treatment arm the Tolerance level is normalised during the treatment period, commencing at the same time as the cytotoxic treatment (clock-tick = 1500), and continuing for 100 clock-ticks whereas the cytotoxic treatment completes at clock-tick = 1525. The over-feeding is not altered in this treatment arm. The growth curves are shown in Fig. 10. Direct comparison of the final Malignant cell counts at clock-tick = 2000 between this treatment and cytotoxic + nutrient reduction shows a statistically significant difference, with this treatment arm having much lower final cell counts (9342 versus 11,427, P = 4.477e-07). However, comparison of Figs. 8 and 10 shows a very similar story in terms of the shapes of the curve—steep decline when cytotoxic therapy begins followed by a plateau during the combination treatment and then an accelerated repopulation.

Figure 11 shows the comparison of the rate of tumour cell repopulation between cytotoxic + nutrient reduction and cytotoxic + Tolerance normalisation treatments. The slower rate of accelerated repopulation in the right panel is striking.

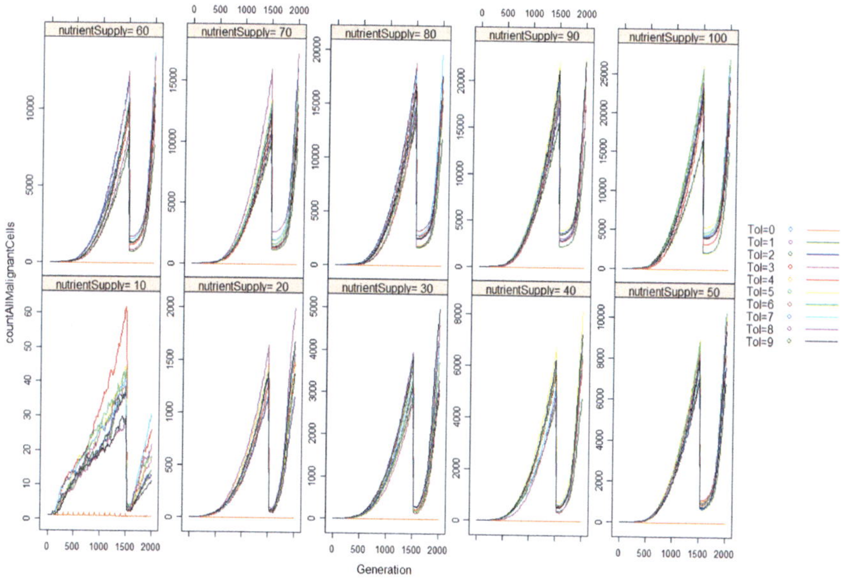


**Fig. 10** Average Malignant cell growth for 2000 clock-ticks for different Tolerance settings—with cytotoxic treatment and Tolerance normalisation

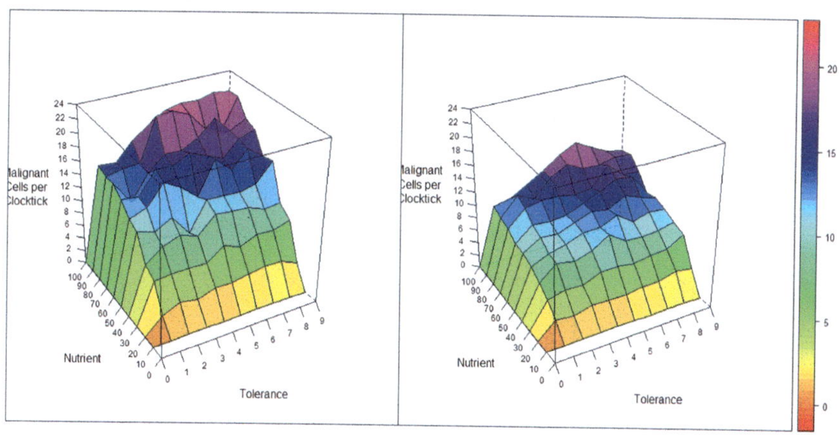


**Fig. 11** Gradient of Malignant cell repopulation for the first 250 clock-ticks after the end of treatment. Left panel, cytotoxic treatment + nutrient reduction, right panel cytotoxic + Tolerance normalisation

# 7 Cytotoxic Treatment + Nutrient Reduction + Tolerance Normalisation (NR-TN)

This treatment schedule combines nutrient restriction and Tolerance normalisation commencing at clock-tick = 1500 and runs concurrently with the nutrient restriction and reverts, along with the nutrient supply, to the pre-treatment level at clock-tick = 1600. The growth curves for different values of nutrient supply and Tolerance are shown in Fig. 12. In comparison to Figs. 2 and 8, Fig. 12 shows a reduction in final Malignant cell count and an extended period before regrowth accelerates. The difference in mean cell counts at clock tick = 2000 between this treatment and the NR treatment is significant, (9153 vs 11,427, P = 2.356e-08), although the difference between this and TN, although lower, does not reach significance (9153 vs 9342, P = 0.602).

Comparison of the repopulation rates between the NR and TN treatments for 250 clock-ticks post-treatment, from clock-tick = 1600 to clock-tick = 1850, is shown in Fig. 13. Clearly the addition of Tolerance normalisation strongly inhibits, but does not completely abolish, tumour regrowth in this model. The improvement in response is clear both in absolute Malignant cell counts and in the comparison of the trajectory of regrowth.

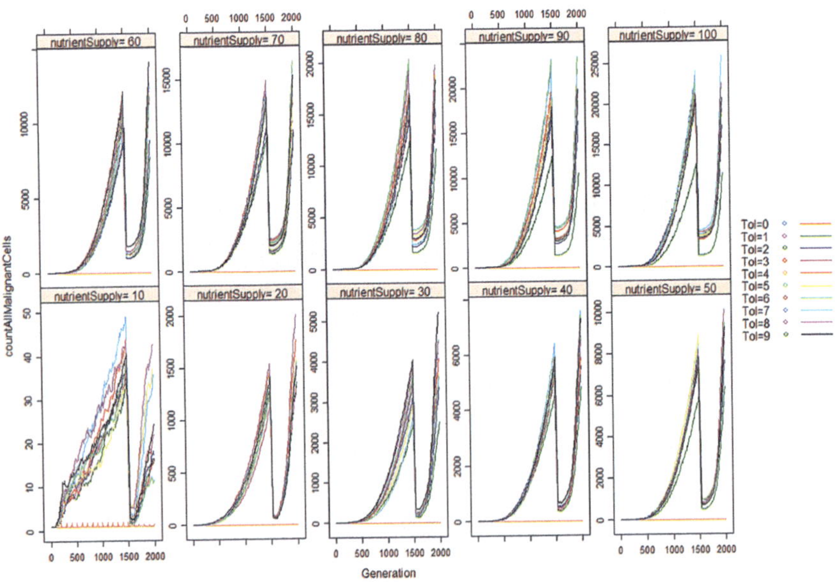


**Fig. 12** Average Malignant cell growth for 2000 clock-ticks for different Tolerance and nutrient supply settings—with cytotoxic treatment, nutrient restriction and Tolerance normalisation

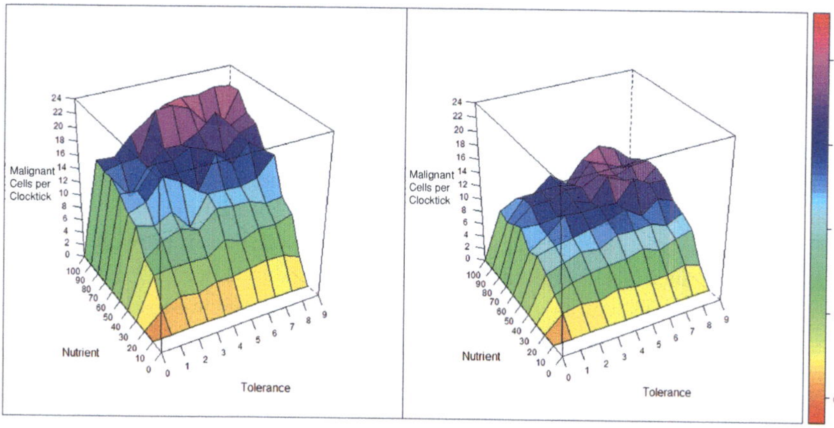


**Fig. 13** Gradient of Malignant cell repopulation for the first 250 clock-ticks after the end of treatment. Left panel, cytotoxic treatment + nutrient restriction, right panel cytotoxic + nutrition restriction + Tolerance normalisation

## 8 Discussion

In previous works we have shown that the NEATG models display clinically relevant emergent behaviours related to cancer growth dynamics and responses to treatment. Previous analyses have also shown that results are in line with in vitro laboratory data from a panel of cancer cell lines [1]. Furthermore, NEATG_A was the first computational model of anakoinosis, a new treatment paradigm that features communicative reprogramming of aberrant cell-tissue communication.

In this new work we have explored the role of over-feeding, both in terms of cancer cell growth and in terms of response to treatments. The question of nutrition and cancer is one that is enormously complex but also of great interest to clinicians and patients alike. While the image of cancer cells as addicted to glucose and therefore vulnerable to metabolic change is a starting point for many ‘anticancer diets’, the reality is that cancer cells display high levels of metabolic plasticity and respond to nutrient challenge in ways that are still being elucidated.

In our results the clear impact of excess nutrient supply on the growth and spread of cancer cells is shown in Figs. 2 and 3. The increased growth associated with higher nutrient supply is tempered but not abolished by lower values of Tolerance. The combination of high Tolerance and high nutrient supply leads to the highest levels of tumour growth. Notably, a multiple regression model predicts Malignant cell numbers at clock-tick = 1499, immediately prior to treatment, with an adjusted R-squared of 0.67. It is also notable that the model includes a term for the interaction of nutrient supply and Tolerance. This is as we would predict, a higher Tolerance value, representing greater levels of cell-tissue communicative dysfunction, means that there is more scope for cancer cells to experience genetic change in response to cell competition. The finding that greater Tolerance and nutrient supply is also associated with higher rates of intra-tumour heterogeneity, as shown in Fig. 4, is in line with this. One aspect of this genetic change is in the increased nutrient demand that is displayed, although there is no concomitant increase in metabolic rate as represented by the nutrient rate, as shown in Fig. 5.

These results show that in response to higher Tolerance levels and increased nutrient supply, Malignant cell numbers increase, cell competition grows and drives genetic change and increases the Malignant cell avidity for nutrient.

The cytotoxic treatment strategy, which we have modelled in previous studies, leads to rapid reductions in Malignant cell numbers, as shown in Fig. 2. However, in line with our previous results, the initial cytotoxic insult does not lead to complete clearance of all Malignant cells and there is a very fast cancer cell repopulation. Figure 6 shows that there is little variation in the tumour kill rate related to either Tolerance or nutrient supply values. However, there is an effect on the degree and speed of regrowth, as shown in Fig. 7. A linear regression model of Tolerance and nutrient supply as predictors of Malignant cell numbers had a high R-squared = 0.74, and again included an interaction term.

Given the impact of high nutrient supply on both pre- and post-treatment cancer cell growth, we next investigated the impact of restricting the nutrient supply as part of a treatment strategy. In this scenario the nutrient supply was restricted to the optimal value of 10 for a period of 100 clock-ticks, running from the initiation of cytotoxic treatment. At treatment end the nutrient supply reverts to the pre-treatment level. The growth curves are shown in Fig. 8. Here again we see a well-defined collapse of Malignant cell numbers, a period of little growth and then a very aggressive regrowth. A comparison of the slope of the regrowth between cytotoxic and cytotoxic + nutrient restriction treatments, in Fig. 9, shows in fact that the regrowth is faster for the second strategy. In other words restricting the nutrient supply, keeping the Tolerance value unchanged, has no effect on outcomes. Indeed, in comparing the Malignant cell counts at a common time point, (250 clock-ticks after the end of treatment), there is a statistically significant higher mean value in the cytotoxic + nutrient restriction datasets.

In contrast, the combination of cytotoxic treatment and Tolerance normalisation, with nutrient supply unchanged during the treatment, does show both a lower count of Malignant cells at the end of the experiment, as shown in Fig. 10, and a reduced rate of regrowth in Fig. 11. The mean Malignant cell counts also show a statistically significant reduction. Tolerance normalisation, the very essence of the anakoinosis treatment paradigm, can be effective even when the over-feeding continues.

In the final experiment the treatment strategy combines cytotoxic treatment, nutrient restriction and Tolerance normalisation. Here, in addition to restricting the nutrient supply as before, the Tolerance value was normalised to Tol = 0 for the duration of the treatment and then reverted to the pre-treatment level at treatment end. Although the growth curves, Fig. 12, are similar to the previous curves we should note the lower Malignant cell counts (y-axes). Comparison of the slopes of the regrowth curves, shown in Fig. 13, indicates a clear difference and confirms the slowing of the rate of tumour regrowth when adding control of Tolerance to the treatment mix. While it clearly does not abrogate the regrowth, it does slow it down in comparison to cytotoxic treatment and cytotoxic treatment + nutrient restriction. However, the improvement between this treatment option and cytotoxic + Tolerance normalisation is marginal. It is of course possible that over a longer period the NR + TN treatment strategy would show more significant improvement over TN alone.

This is a non-physiological model and does not directly address biological mechanisms. A major limitation is that the over-feeding of cells serves only to increase the relative fitness of those cells. While these cells display increased avidity for nutrient, which we may say is analogous to glucose or glutamine addiction in cancer cells, there are no other biological correlates being modelled. Neither does the model incorporate the complex metabolic plasticity that has been observed in tumours, with multiple metabolic compartments and the complex shuttling of nutrients and by-products between cancer and stromal cell populations [7, 8]. Another limitation is that the model does not fully explore the interactions between increased nutrient supply and cell-tissue communicative dysfunction. While there is an interaction term in many of the linear regressions used in the analysis, the model itself is not structured to explicitly explore the relationship of these two factors. Intuitively we would expect that over-feeding has effects both at both cell and tissue levels, for example over-feeding may increase oxidative stress leading to chronic inflammation and a concomitant negative effect on tissue homeostasis [9].

However, within these limits the model does display behaviour that is relevant to cancer. It suggests that an excess of nutrients is a driver of tumour growth, particularly when there is a high degree of Tolerance. Furthermore, it suggests that accelerated tumour regrowth—the phoenix rising phenomenon—is also more pronounced with higher nutrient availability. This is part of the rationale for many so-called anticancer diets. It is an idea that has much intuitive appeal in terms of 'starving cancer cells of the nutrients they need to survive'—however, this model suggests that the effects will be limited without addressing the issue of cell-tissue communicative dysfunction. Biologically there is much evidence that fast growing cancers display high levels of metabolic plasticity and can use alternative substrates to satisfy the metabolic needs associated with increased proliferation and the low availability of glucose. Our results suggest that nutrient restriction alone may have little impact on cancer growth or regrowth following cytotoxic insult.

These results have implications for those looking at dietary interventions as a way of controlling tumour growth. It is therefore important that our results are validated in vitro—for example by assessing the impact of overfeeding on cancer cell growth dynamics and response to chemotherapeutic agents. The use of anakoinosis treatment to reprogram cell-tissue communication networks should also be explored in the context of research on anticancer diets and nutrient control in cancer. It should also be noted that pioglitazone, a key therapeutic drug used in anakoinosis treatment, is used primarily to treat type II diabetes [10]. It is possible that the use of nutrient restriction in combination with anakoinosis treatment may emerge as a new strategy to explore clinically.

## References

1. Pantziarka, P., Ghibelli, L., Reichle, A.: A computational model of tumor growth and anakoinosis. Front. Pharmacol. **10**(March), 287 (2019)
2. Pantziarka, P.: Emergent properties of a computational model of tumour growth. PeerJ **4**, e2176 (2016)
3. Heudobler, D., Rechenmacher, M., Lüke, F., Vogelhuber, M., et al.: Clinical efficacy of a novel therapeutic principle anakoinosis. Front. Pharmacol. **9**(November), 1357 (2018)
4. Hart, C., Vogelhuber, M., Wolff, D., Klobuch, S., et al.: Anakoinosis: communicative reprogramming of tumor systems - for rescuing from chemorefractory neoplasia. Cancer Microenviron.: Off. J. Int. Cancer Microenviron. Soc. **8**(2), 75–92 (2015)
5. Levenshtein, V.I.: Binary codes capable of correcting deletions, insertions, and reversals. In: Soviet Physics Doklady, pp. 707–710 (1966)
6. Sakamoto, Y., Ishiguro, M., Kitagawa, G.: Akaike Information Criterion Statistics, vol. 81(10.5555), pp. 26853. D. Reidel, Dordrecht, The Netherlands (1986)
7. Salem, A.F., Whitaker-Menezes, D., Lin, Z., Martinez-Outschoorn, U.E., et al.: Two-compartment tumor metabolism: autophagy in the tumor microenvironment and oxidative mitochondrial metabolism (OXPHOS) in cancer cells. Cell cycle (Georgetown, Tex.) **11**(13), 2545–56.
8. Martinez-Outschoorn, U.E., Balliet, R.M., Rivadeneira, D.B., Chiavarina, B., et al.: Oxidative stress in cancer associated fibroblasts drives tumor-stroma co-evolution: A new paradigm for understanding tumor metabolism, the field effect and genomic instability in cancer cells. Cell Cycle (Georgetown, Tex.) **9**(16), 3256–3276 (2010)
9. Rezende, L.P., Galheigo, M.R.U., Landim, B.C., Cruz, A.R., et al.: Effect of glucose and palmitate environment on proliferation and migration of PC3-prostate cancer cells. Cell Biol. Int. **43**(4), 373–383 (2019)
10. Filipova, E., Uzunova, K., Kalinov, K., Vekov, T.: Effects of pioglitazone therapy on blood parameters, weight and BMI: a meta-analysis. Diabetol. Metab. Syndr. **9**, 90 (2017)

# Modeling Tumour Growth with a Modulated Game of Life Cellular Automaton Under Global Coupling

**Vladimir García-Morales, José A. Manzanares, and Javier Cervera**

**Abstract** We propose a coupled map lattice that simulates the Gompertz tumour growth model used in cancer diagnosis and is able to reproduce the long-time parameter correlations that have been found in previous studies. The coupled map lattice is a modulated Game of Life cellular automaton model that includes only two free parameters. Parameter $\gamma$ governs the strength of a global coupling of the cells due to the confinement pressure. Parameter $\kappa$ governs the strength of the intercellular coupling which modulates the local dynamic rules of Conway's Game of Life.

## 1 Introduction

Tumours are self-organizing, complex systems, that must be studied and treated as such on a micro- and macroscopic level [1]. Macroscopically, a widely-used phenomenological model of tumour growth is Gompertz model, which describes the evolution of the tumour volume $V$ as a function of time $t$ as [2–7]

$$V(t) = V_0 \exp\left(\frac{A}{B}(1 - \exp(-Bt))\right) \tag{1}$$

where $V_0$ is the volume at time $t = 0$, and $A$ and $B$ are parameters. At times $t \ll 1/B$, Eq. (1) reduces to $V = V_0 \exp(At)$ (exponential growth). At the later stages of tumour growth, $t \gg 1/B$, the volume attains a maximum value $V_{\max} := V_0 \exp(A/B)$. A remarkable empirical correlation found experimentally by Norton et al. [5] is that parameters $A$ and $B$ seem to be correlated as

$$\exp A = a \ln B + b \tag{2}$$

V. García-Morales (✉) · J. A. Manzanares · J. Cervera
Departament de Física de la Terra i Termodinàmica, Universitat de València, 46100 Burjassot, Spain
e-mail: vladimir.garcia@uv.es

 
I. Balaz and A. Adamatzky (eds.), *Cancer, Complexity, Computation*,
Emergence, Complexity and Computation 46,
https://doi.org/10.1007/978-3-031-04379-6_5

where $a$ and $b$ are constants that depend on the type of tumour. This correlation has the following important implication in cancer diagnosis: By determining the coefficient $A$ from early stages of tumour development (in which the growth is exponential), parameter $B$ dictating the characteristic time $\tau_c = 1/B$ that marks the transition between early and late stages of tumour development can be predicted from Eq. (2) as $B = \exp[(-b + \exp A)/a]$. Consequently, the maximum tumour volume $V_{\max}$ to be attained at later stages of growth can be predicted, as well as the whole function $V(t)$ in Eq. (1).

The Gompertz equation is useful for describing tumour growth, but it does not reflect any microscopic mechanisms [8]. This fact has motivated some researchers to devise cellular automata models to explain Gompertzian tumour growth as an emergent behavior. In a pioneering work, Qi et al. [8] considered a probabilistic cellular automaton with four different types of cells over a von Neumann neighborhood dependent on six parameters (three of which could be fixed by experiments, the remaining being free). One of the free parameters $d_c$ was meant to account for the effect of the confinement mechanical pressure: a tumour can only expand if its inner pressure is larger than the external pressure, a fact that dynamically constrains the maximum tumour size. However, $d_c$ was related to the distance of the cells to the origin, placed at the center of the lattice, therefore treating the center of the lattice in a different way to the rest of the lattice (the translational symmetry characteristic of any cellular automaton being broken). If we consider that a tumour can equally originate at any other region of the lattice, such symmetry breaking introduces, in our opinion, an undesirable feature since it not only limits the number of physically meaningful initial conditions (requiring cancerous cells to be necessarily at the center of the lattice for a tumour to grow): it also introduces a preferred reference frame that is unjustified. This feature is also present in the work by Kansal et al. [9] who presented a probabilistic 3D cellular automaton model depending on four free parameters on a Voronoi lattice, which is more dense at the center where the tumour was initially placed. The above mentioned works, [8, 9], discussed two different kinds of tumours actually found in biological systems. Qi et al. considered the most common type in which the tumour can be active not only at its surface but also in its interior, e.g., the mouse carcinoma KHT and the spontaneous carcinoma C3H in mice [8]. Kansal et al. considered activity of the tumour only at its surface, the interior of the tumour being composed of necrotic, inactive cells, as it is found in the glioblastoma multiforme in the brain [9]. Both models satisfactorily account for Gompertzian growth as described by Eq. (1). However, the empirical correlation shown in Eq. (2), which seems to be apparent in realistic tumour growth, was not addressed in Refs. [8, 9].

In this work, we present a 2D coupled-map lattice model based on cellular automata dynamics [10, 11] under global coupling. Our model has only two free parameters, predicts Gompertzian tumour growth given by Eq. (1) as well as the correlation between the parameters, Eq. (2), found in Ref. [5]. The local dynamical rules of the model are similar to those of Conway's Game of Life [12–16] but are modulated by a coupling parameter $\kappa$. The other free parameter $\gamma$ accounts for the global confinement pressure and is called the strength of the global coupling.

## 2 Model

We represent a cell in the multicellular ensemble as a site in a 2D square lattice with $\Omega = n^2$ sites, where $n$ is the number of sites on a side of the lattice. The dimensionless, continuous variable $u_t^{i,j}$ represents the dynamical state of the cell at site $(i, j)$ at time $t$; hereinafter we refer to the state of the cell as the state of the site. We say that the site $(i, j)$ is normal when $u_t^{i,j} \leq 0.15$ and that it is cancerous when $u_t^{i,j} > 0.15$. A lattice where most sites are found in state 0 is said to be normal, i.e. composed of normal non-cancerous cells. The total number of abnormal cells in the lattice is counted as

$$N_{ab} = \sum_{i=1}^{n} \sum_{j=1}^{n} H(u_t^{i,j} - 0.15) \tag{3}$$

where $H(x)$ is the Heaviside step function ($H(x) = 1$ if $x > 0$ and $H(x) = 0$ otherwise). The tumour volume is $V = N_{ab} v_0$, where $v_0$ is the "volume" of a single cell in the lattice.

The biological signals that couple individual cells to their local multicellular environment may contribute to normalization and are modeled using a continuous parameter $\kappa$ that accounts for a weak coupling between cells. This parameter quantifies the weakening of the local dynamical rules due to the limited coupling of each individual cell to its local neighborhood. For different values of $\kappa$, different kinds of tumour are found, being able to reproduce tumours that resemble those in [8] and others that resemble those in [9], the type of tumours depending on the values of the parameter $\kappa$. In general, low values of $\kappa$ tend to enforce the local rules over the ensemble while high values of $\kappa$ are associated with limited intercellular communication [10]. Importantly, our model does not break any translation invariance. A tumour can start anywhere, not necessarily the center. Although the model is simple and deterministic, its output is unpredictable because it incorporates Conway's Game of Life as the cellular automaton governing the local dynamics in the limit $\kappa \to 0$. The deterministic character of the model, and the smoothing and loosening of the local dynamical rules, as the parameter $\kappa$ is varied, are helpful to understand its intricate dynamics with the tools of bifurcation analysis and the qualitative theory of smooth nonlinear dynamical systems.

Parameter $\gamma$ accounts for a global coupling. The local dynamics is coupled to the global average of the dynamical state of the cells in the lattice

$$u_t = \left\langle u_t^{i,j} \right\rangle := \frac{1}{\Omega} \sum_{i=1}^{n} \sum_{j=1}^{n} u_t^{i,j}. \tag{4}$$

We note that a tumour would grow without being stabilized if $\gamma = 0$. A non-vanishing global coupling is necessary for tumour volume stabilization. The way the latter enters in our model is analogous to the one found in models of spatiotemporal pattern formation in chemical systems [17]: the pressure has an stabilizing effect of the homogeneous mode and creates a global coupling that is proportional to the lattice average of a local order parameter. This stabilizing effect is responsible, in our scenario, for the inhibition of the tumour growth.

## 2.1 Lattice and States

We consider square (Moore) neighborhoods of $3 \times 3$ sites. Using periodic boundary conditions in the lattice, the neighborhood of the site $(i, j)$ is formed by the sites $(i+k, j+m)$ where $k$ and $m$ can take the values $-1$, $0$ and $1$. If a cancerous cell is in a stable state and is homogeneously surrounded in its neighborhood by stable cancerous cells with approximately the same value of their dynamical state, we say that the cell is necrotic.

The evolution of the state of site $(i, j)$ at discrete time steps is given by

$$u_{t+1}^{i,j} = \mathcal{B}_\kappa\left(3 - s_t^{i,j}, \frac{1}{2}\right) + u_t^{i,j}\mathcal{B}_\kappa\left(4 - s_t^{i,j}, \frac{1}{2}\right) - \gamma u_t \tag{5}$$

where $\kappa$ is the coupling parameter, $\gamma$ is the global coupling strength parameter, and

$$s_t^{i,j} := \sum_{k,m=-1}^{1} u_t^{i+k,j+m} \tag{6}$$

is the neighborhood sum. The $\mathcal{B}_\kappa$ function [11] of real variables $x$ and $y$ is

$$\mathcal{B}_\kappa(x, y) := \frac{1}{2}\left[\tanh\left(\frac{x+y}{\kappa}\right) - \tanh\left(\frac{x-y}{\kappa}\right)\right] \tag{7}$$

For all finite values of $x$ and $y$, the $\mathcal{B}_\kappa$ function satisfies the limits [11]:

$$\lim_{\kappa\to\infty} \mathcal{B}_\kappa(x, y) = 0 \qquad \lim_{\kappa\to\infty} \frac{\mathcal{B}_\kappa(x, y)}{\mathcal{B}_\kappa(0, y)} = 1 \tag{8}$$

$$\lim_{\kappa\to 0} \mathcal{B}_\kappa(x, y) = \mathcal{B}(x, y) = \frac{1}{2}\left(\frac{x+y}{|x+y|} - \frac{x-y}{|x-y|}\right) = \begin{cases} \operatorname{sgn} y & \text{if } |x| < |y| \\ \frac{\operatorname{sgn} y}{2} & \text{if } |x| = |y| \\ 0 & \text{if } |x| > |y| \end{cases} \tag{9}$$

where we have introduced the function $\mathcal{B}(x, y)$, which allows a universal map for cellular automata to be formulated [18].

In the limit $\kappa \to 0$ and in the absence of global coupling $\gamma = 0$, Eq. (5) becomes

$$u_{t+1}^{i,j} = \mathcal{B}\left(3 - s_t^{i,j}, \frac{1}{2}\right) + u_t^{i,j}\mathcal{B}\left(4 - s_t^{i,j}, \frac{1}{2}\right) = \begin{cases} 1 & \text{if } s_t^{i,j} = 3 \\ u_t^{i,j} & \text{if } s_t^{i,j} = 4 \\ 0 & \text{otherwise} \end{cases} \tag{10}$$

## 2.2 Relation to Conway's Game of Life

For Boolean initial conditions and $\kappa \to 0$ this model is similar to Conway's Game of Life [10, 18]:

1. *Any site in state* 1 *with fewer than two nearest neighbors in state* 1 *takes state* 0 *at the next time step.* The rule establishes the normalizing effect of the local neighborhood when normal cells predominate.
2. *Any site in state* 1 *with two or three nearest neighbors in state* 1 *remains in state* 1 *at the next time step.* The rule assumes that the normalization effect of the local neighborhood is lost when sufficient abnormal cells are present.
3. *Any site in state* 0 *with three nearest neighbors in state* 1 *changes to state* 1 *at the next time step.* The rule considers the promotion from a normal to an abnormal state.
4. *Any site in state* 1 *with more than three nearest neighbors in state* 1 *changes to state* 0 *at the next time step.* The rule establishes a limit to abnormal cell expansion, e.g. because of finite available resources, representing a change from positive to negative cooperativity.

Note that a predominantly normal neighborhood may constitute a normalizing microenvironment for a cell because of the abnormal cell underpopulation (rule 1). On the contrary, a significantly abnormal neighborhood may impair the normalization effect and promote the abnormal state (rules 2 and 3). In the case of abnormal cell overcrowding, however, limited proliferation could arise because of the competition for finite resources (rule 4) [10].

When $\kappa \neq 0$, Eq. (5) produces a more 'fuzzy' dynamics, and $u_t^{i,j}$ is a continuous variable. Further, we have:

5. *The coupling between sites due to the local rules 1 to 4 is modulated by the parameter* $\kappa \geq 0$. This parameter incorporates the collective influence of biological phenomena such as the stochastic intercellular diffusion of signaling molecules, the intrinsically probabilistic gene expression, and the individual cell heterogeneity. These noisy phenomena should weaken rules 1 to 4, which hold exactly only in the limit $\kappa \to 0$.
6. *A global coupling whose strength is given by a parameter* $\gamma \in [0, 1]$ *competes against the local inhomogeneities tending to stabilize a global homogeneous mode of normal cells.* The term $-\gamma u_t$ in Eq. (5) does not contribute significantly to the local dynamics if most cells in the lattice are normal, i.e., if $u_t \approx 0$. If many cells are abnormal, the term $-\gamma u_t < 0$ has a normalizing effect on an abnormal cell at position $(i, j)$. The increased impact of the global coupling as the number of abnormal cells grow eventually stops the tumour growth. This global coupling represents the effect of the mechanical pressure.

When $\gamma > 0$ is increased, as shown in the next section, the homogeneous state consisting of normal cells tends to be more stable and the abnormal cells tend to be confined on a smaller region, where they tend to cluster and coalesce.

## 2.3 Mean-Field Approximation

Insight for the globally coupled case $\gamma \neq 0$ can be obtained using the analytical mean-field approximation in the absence of global coupling ($\gamma = 0$)

$$u_{t+1} = \mathcal{B}_\kappa\left(3 - 9u_t, \frac{1}{2}\right) + u_t\mathcal{B}_\kappa\left(4 - 9u_t, \frac{1}{2}\right) \tag{11}$$

where it is considered that all cells have approximately the same average value $u_t$, Eq. (4). The mean-field approximation provides a good description of the dynamics when $\kappa$ is sufficiently large and is away of the cellular automaton limit. Let

$$f(u_t; \kappa) := \mathcal{B}_\kappa\left(3 - 9u_t, \frac{1}{2}\right) + u_t\mathcal{B}_\kappa\left(4 - 9u_t, \frac{1}{2}\right). \tag{12}$$

Then, a homogeneous fixed point $u^*$ of Eq. (11) satisfies $u_{t+1} = u_t = u^*$, that is,

$$u^* = f(u^*; \kappa). \tag{13}$$

There is a homogeneous fixed point $u^*_{\text{norm}} \approx 0$ for every value of $\kappa$. In the cases $\kappa \to 0$ and $\kappa \to \infty$, we have, exactly, $u^*_{\text{norm}} = 0$. For finite non-vanishing $\kappa$ values, the numerical solution of Eq. (13) shows that, in every case $u^*_{\text{norm}} < 0.15$. We call this the normalized homogeneous state, since all cells are normal, non-cancerous ones, at this fixed point.

Let $u_t = u^*_{\text{norm}} + \xi_t$, where $\xi_t$ is a small perturbation around the fixed point. The evolution of the small perturbation is governed by the map

$$\xi_{t+1} = \left.\frac{\mathrm{d}f(x; \kappa)}{\mathrm{d}x}\right|_{x=u^*_{\text{norm}}} \xi_t \tag{14}$$

which is obtained by expanding Eq. (11) around $u^*_{\text{norm}}$ in powers of $\xi$ and retaining only the first term. Clearly, the perturbation from the normalized homogeneous state can only grow if

$$\left.\frac{\mathrm{d}f(x; \kappa)}{\mathrm{d}x}\right|_{x=u^*_{\text{norm}}} > 1. \tag{15}$$

The mean-field approximation with global coupling is

$$u_{t+1} = f(u_t; \kappa) - \gamma u_t, \tag{16}$$

an equation that has fixed points $u^*_\gamma$ given by

$$u^*_\gamma = \frac{f(u^*_\gamma; \kappa)}{1 + \gamma}. \tag{17}$$

As before, there is a normalized homogeneous fixed point $u^*_{\gamma,\text{norm}}$ and linearization of Eq. (16) around it yields, for a small perturbation,

$$\xi_{t+1} = \left[ \left.\frac{\mathrm{d}f(x;\kappa)}{\mathrm{d}x}\right|_{x=u^*_{\gamma,\text{norm}}} - \gamma \right] \xi_t. \tag{18}$$

A perturbation can only grow if

$$\left.\frac{\mathrm{d}f(x;\kappa)}{\mathrm{d}x}\right|_{x=u^*_{\gamma,\text{norm}}} > 1 + \gamma \tag{19}$$

where we see the stabilizing effect of the non-vanishing global coupling respect to the $\gamma = 0$ case: The larger the value of $\gamma$ the less likely condition (19) is satisfied because $\mathrm{d}f(x;\kappa)/\mathrm{d}x|_{x=u^*_{\gamma,\text{norm}}} \approx \mathrm{d}f(x;\kappa)/\mathrm{d}x|_{x=u^*_{\text{norm}}}$.

### 2.4 *Initial Conditions for Simulations*

The lattice dynamics is studied using numerical simulations over a lattice with $159 \times 159 = 25281$ cells. We shall consider an initial condition consisting of a random distribution of 0 and 1 values in a $25 \times 25$ central square region ($N_{\text{ab},0} = 123$ abnormal cells), the rest of the lattice being at state 0.

## 3 Results and Discussion

### 3.1 *Spatiotemporal Dynamics in the Absence of Global Coupling ($\gamma = 0$)*

To get insight in the full model with $\gamma \neq 0$, we first study the model in the absence of global coupling. For $\gamma = 0$ the model reduces to *Model II* discussed in our recent work [10].

Figure 1 shows the snapshots obtained for an inhomogeneous region occupying initially a central cluster. If $\kappa \gtrsim 1.9$ a homogeneous normal state is obtained at long times. As $\kappa$ is decreased, the central inhomogeneity can grow unboundedly. Domain formation and oscillations are observed within the growing inhomogeneity. The number of abnormal cells $N_{ab}$ grows until the whole lattice is occupied by abnormal cells.

The dynamics of the model can be analyzed further using the mean field approximation in the absence of global coupling, Eq. (11). We describe next the bifurcation diagram of Eq. (11) as $\kappa$ is decreased from $\kappa \geq 2$ to 0:

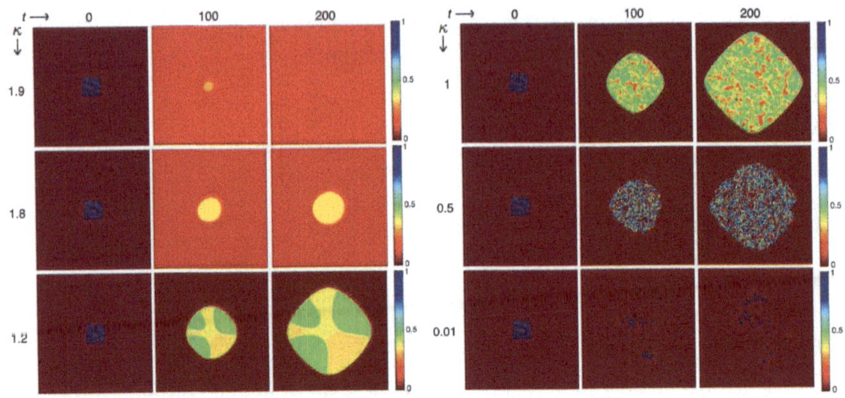


**Fig. 1** Spatiotemporal evolution of the cell states $u_t^{i,j}$ obtained by iterating Eq. (5) in a multicellular ensemble of $159 \times 159 = 25281$ cells for six different $\kappa$ values. The initial ($t = 0$) state is the same for all cases and consists of a random distribution of 0 and 1 values in a central square region of the lattice of $25 \times 25$ size, the rest of the lattice being at state 0 [10]

- A bifurcation is encountered at $\kappa \approx 1.95$, which is close to the value $\kappa \approx 1.9$ found in the numerical simulations of the exact dynamics, Eq. (5). The system abruptly splits into two branches leading to the bistable regime A (Fig. 2A). The two states correspond to a homogeneous abnormal (upper branch) and normal (lower branch) states. Remarkably, the system would tend to a normal homogeneous state when $\kappa \to 0$ only if the lower branch in Fig. 2A were followed, i.e., if the initial conditions were always constituted by a majority of normal cells, with a tiny and dilute proportion of abnormal cells.
- A bifurcation of the upper branch, representing the abnormal homogeneous state, is found at $\kappa \approx 1.35$ leading to two branches of abnormal states that perform period-2 oscillations. Further period doubling bifurcations between abnormal states are then observed at $\kappa \approx 1.2$ leading through a period-doubling cascade into chaos which is most prominent at $\kappa = 1$ (regime B in Fig. 2A). To substantiate this observation, we calculated the Lyapunov exponent $\lambda$ showing that it is positive exactly in this regime [10] for initial conditions corresponding to an abnormal homogeneous state.
- In regimes C and D of Fig. 2A, the mean-field approximation, Eq. (11), fails because it can no longer be assumed that all neighborhoods are well described by an average cell value. Equation (5) needs to be considered in these regimes. Noise is high in regime C (see Fig. 1 for $\kappa = 0.5$) and more degrees of freedom are involved here in the spatiotemporal dynamics invalidating the mean field approximation.

The bifurcation diagram (Fig. 2) explains the pattern formation observed in Fig. 1 for $1 \leq \kappa \leq 1.9$: the upper branch with bifurcations corresponds to the abnormal region and the lower branch to the normal homogeneous region. The bifurcation

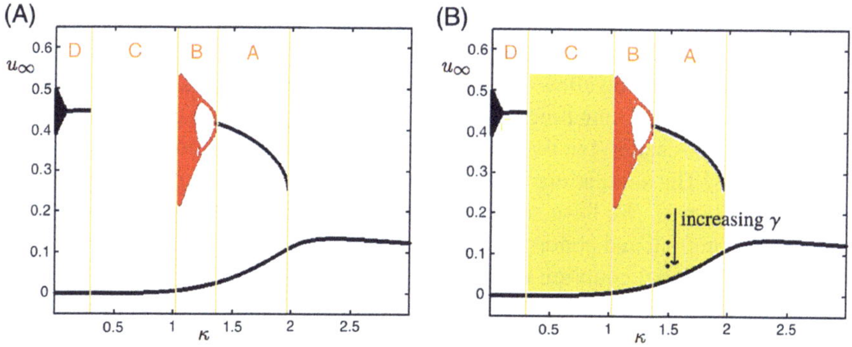


**Fig. 2** Bifurcation diagram calculated from the asymptotic behavior of Eq. (11) in the absence **A** and in presence **B** of global coupling. The black curves correspond to stationary states $u_\infty$ obtained at large times. The red points, following the branch of abnormal cell states, indicate the period doubling bifurcation cascades into chaos. The yellow region in **B** correspond to states that can be stabilized by means of a nonzero global coupling. These states (points marked) do not represent homogeneous states but the coexistence of normal and abnormal phases so that the global average value is $u_\infty$. This phase coexistence corresponds to tumours of a fixed volume in the long time limit for initial conditions in which the abnormal cells are surrounded by a large homogeneous region of normal cells. Increasing $\gamma$ leads to a larger fraction of normal cells in the coexisting phases

diagram also clarifies why oscillations occur only in the region of abnormal cells. Furthermore, as we see in the next section the bifurcation diagram allows to understand the spatiotemporal dynamics of the multicellular ensemble in the presence of a global coupling.

## 3.2 *Spatiotemporal Dynamics in the Presence of Global Coupling ($\gamma > 0$)*

As mentioned above, if there is initially a bounded region of abnormal cells in the lattice, the number of abnormal cells $N_{ab}$ grows indefinitely until the whole lattice is filled with abnormal cells.

As we have explained in Sect. 2.2 the global coupling ($0 < \gamma \leq 1$) tends to stabilize a homogeneous state of normal cells. We now consider the introduction of a global coupling to the multicellular ensemble, aided by the insights provided by the bifurcation diagram in Fig. 2A. We note that the local spatiotemporal dynamics found in absence of global coupling is not significantly altered by the global coupling if: (1) the majority of the cells is in the normal state so that $u_t < 0.15$; (2) $\gamma$ is small. The condition (1) is easily met if we consider initial conditions in which a tiny cluster of cells in the abnormal state is initially present. For the initial conditions employed in the simulations, we have that, at time $t = 0$, $\gamma u_0 = 123\gamma/25281 = 0.0049\gamma \leq 0.0049$, which makes the contribution of the global coupling to the spatiotemporal dynamics of the ensemble governed by Eq. (5) initially negligible.

Regions containing abnormal cells can coexist in a stable manner with regions containing only normal cells if a global coupling is added. When a global coupling is introduced, points in the yellow-shaded region of the bifurcation diagram in Fig. 2B can be stabilized in the long time limit. Points within the shaded region do not represent homogeneous states, but two-phase coexistence between regions of normal and abnormal cells. The simulations in Fig. 3 for increasing $\gamma$ lie in regime A in Fig. 2B (four points shown). We have seen that in regime A there are two different stable branches, the normal and abnormal branches. Depending on the initial condition, in the absence of global coupling the trajectory is attracted to any of these branches. If we introduce a nonzero global coupling, both branches can coexist, and there are homogeneous regions in the lattice that can remain in the normal state and other connected regions in the abnormal state. If we start from a small cluster of abnormal cells surrounded by a homogeneous region of normal cells, we expect at initial times a behavior similar to the one found in Fig. 1 for $\kappa = 1.8$ and $\gamma = 0$. However, the growth of abnormal cells will stop if $\gamma \neq 0$: the contribution to the local dynamics due to the global coupling $-\gamma u_t$ grows and, for a sufficiently large contribution, normal cells at the boundary of the tumour are stabilized, precluding the growth of the latter.

In Fig. 3A the evolution of the number of abnormal cells $N_{\text{ab}}(t)$ is shown for $\kappa = 1.5$ and the values of $\gamma$ indicated on the curves. In Fig. 3B we show the results obtained by Gompertz's law

$$N_{\text{ab}}(t) = N_{\text{ab},0} \exp\left(\frac{A}{B}(1 - \exp(-Bt))\right) \tag{20}$$

for $N_{\text{ab},0} = 123$ and $A = 0.0211$, $B = 0.0047$ (curve a), $A = 0.0174$, $B = 0.0047$ (curve b), $A = 0.0140$, $B = 0.0044$ (curve c) and $A = 0.0096$, $B = 0.0035$ (curve d). Excellent agreement is obtained in all instances of the spatiotemporal evolution of the model and Gompertz's law.

The snapshots of the stable, stationary tumours obtained at time $t = 2000$ (Fig. 3C) show that the homogeneous region of normal cells grows and the region of cancerous cells shrinks when the global coupling strength increases. These tumours are like those described in Ref. [9]: the cells within the tumour are necrotic and only the cells at the perimeter of the tumour remain active. Regardless of the initial condition on a bounded region in the lattice, the tumour grows to a spherical shape. Even when the lattice is square and the underlying local dynamics is inherited from Conway's Game of Life, the smoothing parameter $\kappa$ is responsible for a spatiotemporal evolution that is characteristic of a partial differential equation rather than that of a cellular automaton. As a consequence, isotropy is achieved.

The tumours reported in Ref. [8] behaved differently. Initially, the tumour quickly grows to a size that remains approximately constant while its perimeter fluctuates. The interior remains active for any long time even when the average number of abnormal cells saturates to a constant value, and those abnormal cells remain in a connected region of the same total area (disregarding small fluctuations). From the bifurcation diagram, Fig. 2A, we observed that, in regime C, in absence of global coupling noise

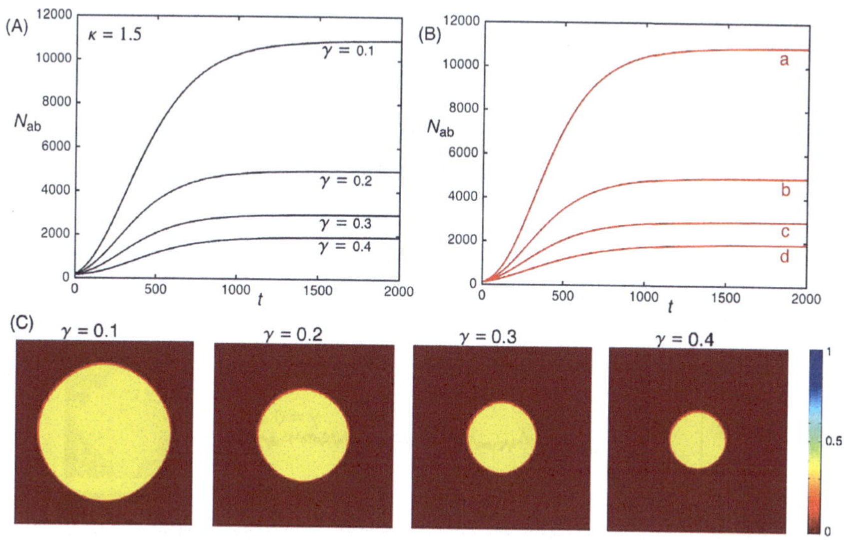


**Fig. 3** Evolution of the number of abnormal cells $N_{ab}(t)$ on a lattice of $\Omega = 25281$ cells. **A** Results obtained from the model, Eq. (5), for $\kappa = 1.5$ and the values of $\gamma$ indicated below the curves, by using Eq. (3) to count the abnormal cells. **B** Results obtained by the Gompertz law, Eq. (20) for $N_{ab,0} = 123$ and $A = 0.0211$, $B = 0.0047$ (curve a), $A = 0.0174$, $B = 0.0047$ (curve b), $A = 0.0140$, $B = 0.0044$ (curve c) and $A = 0.0096$, $B = 0.0035$ (curve d). **C** Snapshots at time $t = 2000$ of the stable, stationary tumours. This tumour behavior is similar to that reported in Ref. [9].

is high in the branch of abnormal cells and the mean field approximation breaks down: more degrees of freedom are needed to describe the branch of abnormal cells that becomes essentially inhomogeneous. In Fig. 1 it is observed that for $0.3 \leq \kappa \leq 1$ a cluster of active abnormal cells grows from an initial condition where only a few cells are abnormal. The cluster of active abnormal cells against the homogeneous region of normal cells grows until the whole lattice is filled in absence of global coupling. For $\gamma > 0$ it is now possible to stabilize a significant part of the normal region, leading to saturation of the tumour growth. Therefore, tumours like those reported in Ref. [8] can be predicted by our model for $\kappa$ parameter values in the range $0.3 \leq \kappa \leq 1$ and $\gamma > 0$. In Fig. 4 we show that $\kappa = 0.9$ and $\gamma = 0.9$ describe a tumour growth behavior similar to that of Ref. [8].

In Fig. 5, the evolution of the number of abnormal cells $N_{ab}(t)$ is shown for $\kappa = 0.9$ and the values of $\gamma$ indicated on the curves. The black curves are the results obtained from the model, Eq. (5). The red curves represent Gompertz's law, Eq. (20) for $N_{ab,0} = 123$ and $A = 0.0458$, $B = 0.0105$ (curve a), $A = 0.0481$, $B = 0.0120$ (curve b) and $A = 0.0524$, $B = 0.0150$ (curve c). Excellent agreement is obtained in both instances of the averaged spatiotemporal evolution of the model and Gompertz's law. The snapshots of the stable, stationary tumours obtained at time $t = 2000$ (Fig. 5B) show that the homogeneous region of normal cells grows and the region of

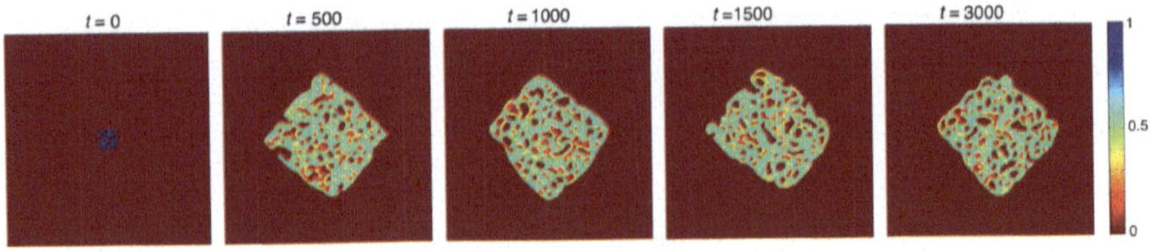


**Fig. 4** Snapshots of the spatiotemporal evolution of the cell states obtained for $\kappa = 0.9$ and $\gamma = 0.9$ at the times indicated on the panels. The initial state is also shown. This tumour behavior is similar to that reported in Ref. [8]

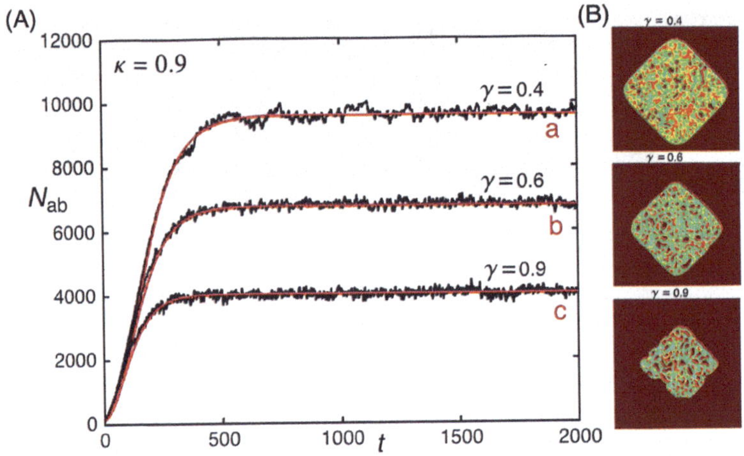


**Fig. 5** Evolution of the number of abnormal cells $N_{ab}(t)$ on a lattice of $\Omega = 25281$ cells. **A** Results obtained from the model, Eq. (5), for $\kappa = 0.9$ and the values of $\gamma$ indicated on the curves (black curves). The red curves are the representations of Gompertz's law, Eq. (20), for $N_{ab,0} = 123$ and $A = 0.0458$, $B = 0.0105$ (curve a), $A = 0.0481$, $B = 0.0120$ (curve b) and $A = 0.0524$, $B = 0.0150$ (curve c). **B** Snapshots of the tumours at time $t = 2000$. Although their interior is not stationary but dynamical, the area occupied by the tumour is constant and remains connected

cancerous cells shrinks when the global coupling strength increases. These tumours are like those described in Ref. [8]: not only the cells in the perimeter of the tumour are active, but also those in the interior. The tumour grows up to a maximal size and remains bounded, although the fluctuations of both the interior of the tumour and its perimeter persist. This is the case, regardless of the initial condition on a bounded region in the lattice.

For $\kappa = 0.9$ and increasing global coupling, $\gamma = 0.4$, 0.5, 0.6, 0.7, 0.8 and 0.9, the parameters $A$ and $B$ calculated by fitting Gompertzian curves, Eq. (20) are obtained. A linear correlation with coefficient of determination $R^2 = 0.995$ is obtained between $\exp A$ and $\ln B$, such that $\exp A \approx 0.020 \ln B + 1.138$. We thus observe that the parameters calculated for the Gompertzian curves that fit the results from our model satisfy the linear correlation reported by Norton et al. [5] for experimental biological instances of tumour growth (Fig. 6).

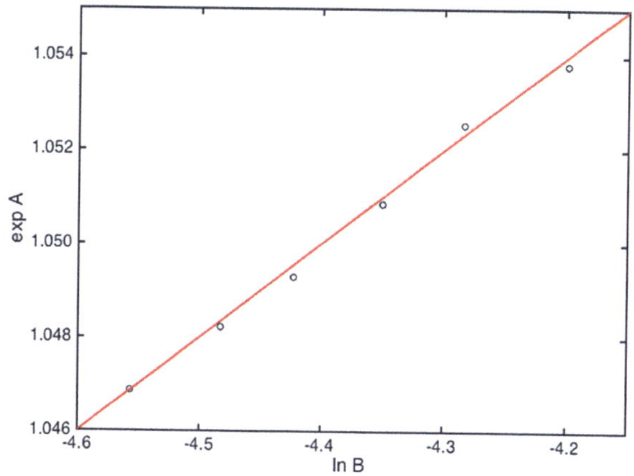


**Fig. 6** The parameters $A$ and $B$ of the Gompertzian curves fitting the spatiotemporal evolution of the tumour dynamics predicted by the model, Eqs. (3) and (5) for $\kappa = 0.9$ and $\gamma = 0.4$, 0.5, 0.6, 0.7, 0.8 and 0.9 show an excellent agreement with the empirical relation found in Ref. [5], Eq. (2). From these simulation results, the parameter values in Eq. (2) are determined as $a = 0.020 \pm 0.002$ and $b = 1.138 \pm 0.008$

Parameters $a$ and $b$ in Eq. (2) depend on the type of cancer under consideration. In our model, these parameters exclusively depend on $\kappa$, a parameter which, as we have seen, yields different types of tumours depending on the regime found in the bifurcation diagram in Fig. 2.

Values of $\kappa$ in the interval $1.35 \leq \kappa \leq 1.95$ (regime A in Fig. 2), yields tumours composed of necrotic, inactive cancerous cells like those found in Ref. [9]. Values of $\kappa$ in the range $1 \leq \kappa \leq 1.35$ (regime B in Fig. 2) yield tumours bounded by a circular shape as those in the interval $1.35 \leq \kappa \leq 1.95$ but their interior is now active, presenting coherent periodic oscillations in different domains of the tumour ($1.18 \leq \kappa \leq 1.35$) or chaotic, aperiodic behavior ($1 \leq \kappa \leq 1.15$). Values of $\kappa$ in the interval $0.3 \leq \kappa \leq 1$ (regime C in Fig. 2) yields tumours composed of active cancerous cells like those found in Ref. [8]. The difference between these tumours and those found in Regime B is that the perimeter of the tumour is not a circumference but a complex, dynamically-changing, shape. Finally, values of $\kappa$ in the range $0 \leq \kappa \leq 0.3$ (regime D in Fig. 2) correspond to the cellular automaton limit of the map and do not yield connected tumours.

## 4 Conclusions

A coupled map lattice that simulates Gompertz tumour growth as an emergent behavior has been presented. The model, Eq. (5), depends only on two free parameters, $\kappa$ and $\gamma$, that control the strengths of the intercellular and global couplings, respectively. The model displays the correlations between parameters in the Gompertz tumour growth model [5], an important feature used in cancer diagnosis to predict long-time behavior from early stages of tumour development. Previous cellular automata models [8, 9, 19] have not addressed these correlations and depend on a larger number of parameters.

As the intercellular coupling parameter $\kappa$ is varied, different types of tumours are modelled. On one hand, for $\kappa$ large, the model describes tumours that are only active at the surface, the interior of the tumour being mostly composed of necrotic, inactive cells [9]. These tumours, with homogeneous interior, are well described by the mean-field approximation of the model, Eq. (11). As $\kappa$ is lowered from $\kappa = 1.35$, the degree of inner activity of the tumour can be switched from coherent and periodic to aperiodic and chaotic. As $\kappa$ is further lowered below $\kappa = 1$, more degrees of freedom become active, the mean-field approximation, Eq. (11), breaks down, and the model reproduces tumours that are active and noisy not only at its surface but also in its interior [8]. On the contrary, previous models accounted only for one type of tumour.

Growth inhibition and tumour volume saturation is described by a global coupling that can represent the contribution of the mechanical pressure, limited amount of nutrients, chemical inhibitors, etc. Importantly, the tumour growth is not here artificially constrained as in previous models in which the dynamics of the tumour was made explicitly dependent on the radius or the volume of the tumour in an ad hoc manner. Geometric features of the tumour do not enter in the model as explicit variables but arise as an emergent phenomenon as a consequence of the spatiotemporal dynamics. This feature is important because, in contrast with previous approaches, the tumour does not need to start 'at the center of the lattice' but can start from any other place. The model is translationally invariant: the local dynamics is the same in every site of the lattice.

Compared to previous cellular automata models applied to tumour growth that are all probabilistic, our model is deterministic and it is given by a simple map, Eq. (5), that can be iterated on a lattice. However, in spite of being deterministic, our coupled map lattice incorporates Conway's Game of Life as a cellular automaton limit ($\kappa \to 0$) and, therefore, yields complex unpredictable behavior for arbitrary initial conditions. Quite remarkably, for $\kappa > 1$ the model yields isotropic growth in spite of taking place on a square lattice.

Gompertzian growth behavior has been found for all different kinds of tumour, independently of the value of the parameter $\kappa$. Thus, our model suggests that Gompertzian growth has a universal character, as pointed out by previous macroscopic approaches [5, 7].

## References

1. Kraus, M., Wolf, B.: Emergence of self-organization in tumor cells: relevance for diagnosis and therapy. Tumor Biol. **14**, 338–353 (1993). https://doi.org/10.1159/000217849

2. Brunton, G.F., Wheldon, T.E.: Prediction of the complete growth pattern of human multiple myeloma from restricted initial measurements. Cell Tissue Kinet. **10**, 591–594 (1977). https://doi.org/10.1111/j.1365-2184.1977.tb00316.x
3. Steel, G.G.: Growth Kinetics of Tumours. Clarendon Press, Oxford (1977)
4. Klein, C.A.: Parallel progression of primary tumours and metastases. Nat. Rev. Cancer **9**, 302–312 (2009). https://doi.org/10.1038/nrc2627
5. Norton, L., Simon, R., Brereton, H.D., Bogden, A.E.: Predicting the course of Gompertzian growth. Nature **264**, 542–545 (1976). https://doi.org/10.1038/264542a0
6. Norton, L.: A Gompertzian model of human breast cancer growth. Cancer Res. **48**, 7067–7071 (1988)
7. Benzekry, S., Lamont, C., Beheshti, A., Tracz, A., Ebos, J.M.L., et al.: Classical mathematical models for description and prediction of experimental tumor growth. PLoS Comput. Biol. **10**(8), e1003800 (2014). https://doi.org/10.1371/journal.pcbi.1003800
8. Qi, A.-S., Zheng, X., Du, C.-Y., An, B.-S.: A cellular automaton model of cancerous growth. J. Theor. Biol. **161**, 1–12 (1993). https://doi.org/10.1006/jtbi.1993.1035
9. Kansal, A.R., Torquato, S., Harsh, G.R., IV., Chiocca, E.A., Deisboeck, T.S.: Simulated brain tumor growth dynamics using a three-dimensional cellular automaton. J. Theor. Biol. **203**, 367–382 (2000). https://doi.org/10.1006/jtbi.2000.2000
10. García-Morales, V., Manzanares, J.A., Mafé, S.: Weakly coupled map lattice models for multicellular patterning and collective normalization of abnormal single-cell states. Phys. Rev. E **95**, 042324 (2017). https://doi.org/10.1103/physreve.95.042324
11. Garcia-Morales, V.: From deterministic cellular automata to coupled map lattices. J. Phys. A.: Math. Theor. **49**, 295101 (2016). https://doi.org/10.1088/1751-8113/49/29/295101
12. Berlekamp, E.R., Conway, J.H., Guy, R.K.: Wining Ways for Your Mathematical Plays, vol. 2. Academic Press, New York (1982)
13. Adamatzky, A. (ed.): Game of Life Cellular Automata. Springer, New York (2010)
14. Adachi, S., Peper, F., Lee, J.: The Game of Life at finite temperature. Phys. D **198**, 182 (2004). https://doi.org/10.1016/j.physd.2004.04.010
15. Chua, L.O., Roska, T., Venetianer, P.L.: The CNN is universal as the Turing machine. IEEE Trans. Circuits Syst. I **40**, 289 (1993). https://doi.org/10.1109/81.224308
16. Pazienza, G.E., Gomez-Ramirez, E., Vilasi's-Cardona, X.: Polynomial cellular neural networks for implementing the Game of Life. In: Marques de Sá, J., Alexandre, L.A., Duch, W., Mandic, D.P. (eds.) Proceedings of ICANN 2007. LNCS, vol. 4668, pp. 914–923. Springer, Berlin (2007). https://doi.org/10.1007/978-3-540-74690-4-9
17. Mertens, F., Imbihl, R., Mikhailov, A.: Breakdown of global coupling in oscillatory chemical reactions. J. Chem. Phys. **99**, 8668 (1993). https://doi.org/10.1063/1.465590
18. García-Morales, V.: Universal map for cellular automata. Phys. Lett. A **376**, 2645 (2012). https://doi.org/10.1016/j.physleta.2012.07.021
19. Deutsch, A., Dormann, S.: Cellular Automaton Modeling of Biological Pattern Formation: Characterization, Applications and Analysis. Birkhäuser, Boston (2005)

# Pinning Control to Regulate Cellular Response in Cancer for the p53-Mdm2 Genetic Regulatory Network

**Oscar J. Suarez, Carlos J. Vega, Edgar N. Sanchez, Guanrong Chen, Ana E. González-Santiago, Otoniel Rodríguez-Jorge, Alma Y. Alanis, and Esteban A. Hernandez-Vargas**

**Abstract** This chapter presents the pinning control strategy applied to the p53-Mdm2 regulatory network to achieve a target cellular response in cancer cells. The network, regulated by p300 and HDAC1, is modeled by a system of 14 ordinary differential equations. Based on this model, an oncogenic condition of Mdm2 over-expression is derived and the functional suppression of p53 is simulated as one possible carcinogenic behavior. Then, control action is introduced to induce an increased expression of p53 levels and downregulation of Mdm2 and p53 activation using only

This work is supported by CONACYT, Mexico, Project 257200.

O. J. Suarez
Mechatronics Engineering Department, Universidad de Pamplona, Pamplona, Colombia

C. J. Vega
School of Management, Universidad del Rosario, Bogota, Colombia

E. N. Sanchez
Electrical Engineering Department, Centro de Investigación y de Estudios Avanzados del Instituto Politécnico Nacional, Guadalajara, Mexico
e-mail: edgar.sanchez@cinvestav.mx

G. Chen (✉)
Electrical Engineering Department, City University of Hong Kong, Hong Kong, China
e-mail: eegchen@cityu.edu.hk

A. E. González-Santiago
Biomedical Sciences Department, Centro de Investigación Multidisciplinario en Salud, Universidad de Guadalajara, Tonalá, Mexico

O. Rodríguez-Jorge
Escuela de Estudios Superiores de Axochiapan, Universidad Autónoma del Estado de Morelos, Axochiapan, Mexico

A. Y. Alanis
Computer Sciences Department, Universidad de Guadalajara, Guadalajara, Mexico

E. A. Hernandez-Vargas
Frankfurt Institute for Advanced Studies, Frankfurt, Germany



I. Balaz and A. Adamatzky (eds.), *Cancer, Complexity, Computation*, Emergence, Complexity and Computation 46,
https://doi.org/10.1007/978-3-031-04379-6_6

three controlled states (pinned nodes) of the network for the first case and two states for the second case, respectively. The analytic results are simulated, which illustrate the feasibility of the proposed control scheme.

**Keywords** Cancer · Complex network · Pinning control · p53 · Mdm2 · p300 · HDAC1

## 1 Introduction

Complex networks theory has been intensively investigated in interdisciplinary areas such as systems biology, social science, engineering technology, among others [5, 7, 45, 57]. Complex networks theory is used as a quantitative method, mathematically grounded in graph theory, to visualize and analyze complex systems. A complex system is a set of dynamical agents interacting in a common environment, such that system attributes that cannot be observed at the individual agent level.

Genetic regulatory networks can be modeled and analyzed from data produced by experimental techniques and computational methods [66]. Nodes and edges form these networks. Nodes usually represent genes, both coding (mRNA) and non-coding (miRNAs, lncRNAs, etc.), or regulatory elements like enhancers, proteins, molecules, etc. Edges represent various regulatory connections (activation, repression, modulation, etc.), and they can be weighted to describe the strengths of the regulatory relationship (establishing regulatory properties) [14, 47].

The p53-Mdm2 (Mouse double minute 2 homolog) genetic regulatory network is one of the most explored biological networks, which provides adequate information for determining cell behavior in response to damage or stress. $TP53$ (p53) is an essential regulatory gene that allows the cell to execute tasks such as cell cycle arrest, DNA repair induction, senescence, or cell death [8, 61]. p53 is referred to as "the guardian of the genome" due to its responses to different levels of stress, maintaining genome stability by preventing mutation, and determining cell fate decisions, and is one of the potential targets of antitumour therapy [28, 42, 64]. When cells present irreparable defects, survival signals are inhibited, and the effector proteins of cell fate promote cell death mechanisms [16]. If the mechanisms of response to DNA damage are intact, the damaged cell will repair its DNA and avoid carcinogenesis. On the other hand, cell death control can avoid passing damaged DNA to the following cell generation [29]. Numerous antitumour strategies have been proposed in recent years, including promoting and stimulating the antitumour activities of p53 and its regulatory network, thereby inducing cell death [54]. Mdm2 has been identified as a p53 interacting protein, which binds to p53 to induce protein-mediated degradation by proteasome, leading to low p53 protein stability and resulting in decreased p53 levels. In consequence, p53 transcriptional activity is prevented [22, 28, 42]. In this sense, the disruption of the protein-protein interaction between p53 and its negative regulator Mdm2 comprises one of the most studied cancer therapy studies.

In this chapter, a deterministic model published in [2], describing the p53-Mdm2 network regulated by p300 (acetyltransferase) and HDAC1 (deacetylase) under ionizing gamma-radiation, is adopted. Apart from a case of large p53 inactivation caused by Mdm2 overexpression, which represses *TP53* transactivation activity and mediates p53 degradation as in [24, 44], a simple and effective control technique named pinning control [12] is applied, which uses a small number of controllers to produce global effects for network modulation. The task of the proposed scheme corresponds to the carcinogenesis case characterized by *MDM2* overexpression [46].

This chapter contributes to the ongoing research, where control methodologies are applied to complex networks to regulate the system dynamics, targeting the controllability of biological networks. The novelty lies in developing an effective controller, which guarantees modulation in the p53-Mdm2 network regulated by p300 and HDAC1, using a discontinuous feedback control law combined with pinning control approach.

## 2 Methods

### 2.1 The p53-Mdm2 Genetic Regulatory Network

The p53-Mdm2 network illustrated in Fig. 1 coordinates a series of tumour suppression activities in response to several stress-associated signals, resulting mainly in the transcription of numerous genes involved in cell differentiation, cell cycle arrest/senescence, DNA repair, and cell death (apoptosis). Under some circumstances such as genotoxic damage, p53-mediated transcriptional regulation is activated to promote behaviors like apoptosis response through Bax, Fas, PARP1, PAI, among others. However, p53 also promotes the transcription of genes that can inhibit apoptosis, an opportunity to repair DNA damage and/or recover from stress (including target genes such as p21, GADD45, PDDB2, among others).

The p53-Mdm2 network regulation exerts responses driven by p53 activation and stabilization, genes transactivation, and function modulation mediated by post-translational modifications. Epigenetic modifications also take part, through proteins like histone and non-histone acetyltransferases (HATs and non-HATs, respectively). p300 function as HAT in chromatin remodeling and as a transcriptional co-factor for many proteins (p53, E2F, Rb, Smads, RUNX, and BRCA1), oncoproteins (such as myb, jun, fos) and also transforming viral proteins (such as E1A, E6, and large T antigen) [9]. HDAC1 is a deacetylase and acts as a co-repressor involved in differentiation and proliferation control. It is upregulated in malignant compared to benign tissue, and targets several transcription factors, including p53 [21].

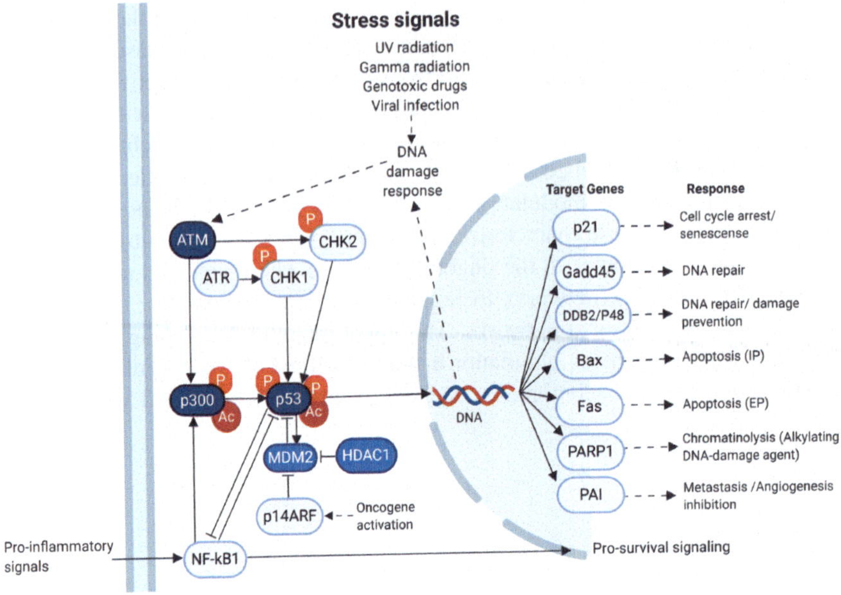


**Fig. 1** Tumour suppressor activities regulated by p53. Signal transduction is represented by edges (lines) and nodes (network element), where ⊣ represents signaling inhibition, and → represents signaling activation, Ac is acetylation, and p is phosphorylation

### 2.1.1 The p53-Mdm2 Network Regulated by p300 and HDAC1

The p53 activation in genotoxic stress is mediated through numerous protein interactions and gene transactivation, which induces cell cycle arrest, DNA repair, senescence, or cell death [55]. ATM (Ataxia Telangiectasia Mutated) protein has been related to p53 phosphorylation in response to ionizing radiation, stabilizing and activating p53 [10]. The p300 binds to p53 and acts as co-activator for p53-mediated transactivation and stabilization [20]. In response to DNA damage, p300-mediated acetylation of p53 has been reported, forming a complex with Mdm2 and p53 [33]. Besides, Mdm2 recruits HDAC1, inhibiting p300-mediated acetylation of p53, which promotes Mdm2-mediated p53 degradation [27]. The p53-Mdm2 interaction is related to different cell damage stimuli, such as DNA damage, alterations suffered by viral infections, among others [30, 34]. Finally, in non-genotoxic conditions, p53 is expressed at a very low level due to direct inhibition by negative regulator Mdm2, which promotes p53 ubiquitination that leads to proteasomal degradation. In this way, the p53-Mdm2 network regulates post-translational modifications as specific responses mediated by genotoxic stressors [27].

In Fig. 2, the stress response model of the p53-Mdm2 network regulated by p300 and HDAC1 is presented. Under stress-free conditions, p53 remains at low levels, due to Mdm2 inhibitory interaction ($K_8$). Mdm2 mediates p53 ubiquitylation and promotes degradation via proteasome ($K_7$). The Mdm2-p53 complex can also disso-

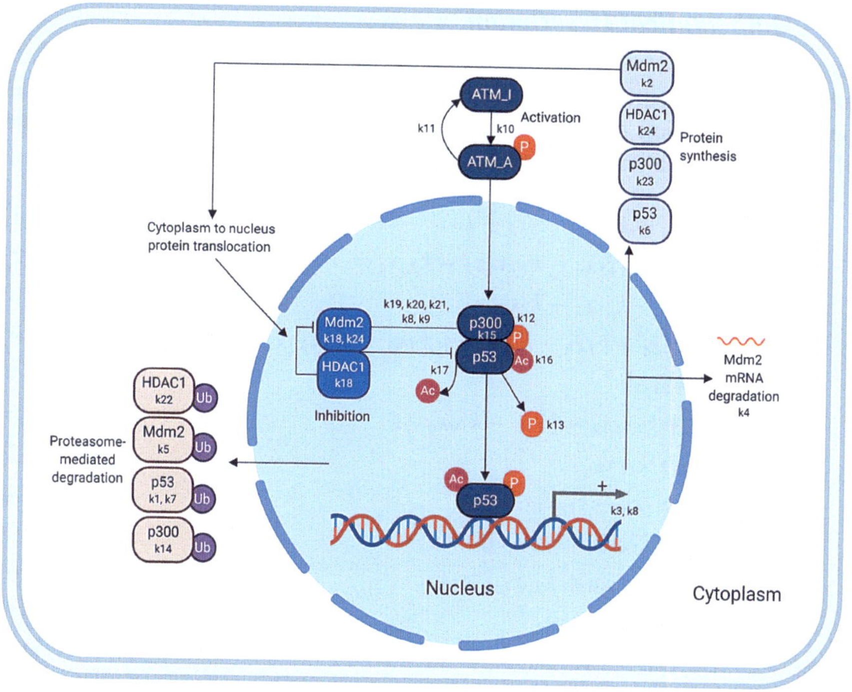


**Fig. 2** Biochemical model of p53-Mdm2-p300-HDAC1 regulatory network. Nodes represent chemical species and model parameters, edges establish functional interactions as activation or inhibition, and Ub stands for ubiquination

ciate to Mdm2 and p53 with a constant rate ($K_9$). Mdm2 and p300 have been shown to form Mdm2-p300 complex ($K_{20}$), which promotes p53 degradation mediated by polyubiquitination ($K_1$). Once genotoxic damage occurs, p53 stabilization is mediated by ATM kinase activity ($K_{10}$). ATM promotes p53 phosphorylation ($K_{12}$) and p53 reduces affinity for Mdm2, while p300 increase complex formation with p53 ($K_{15}$) and p300 acetylates and stabilizes p53 ($K_{16}$). Activated p53 has a half-life of 2 h and promotes the transcription of numerous genes ($K_3$), so they can be translated into protein ($K_2$), ($K_{24}$), ($K_{23}$), ($K_6$). At the same time, synthesis ($K_{23}$) and degradation ($K_{14}$) of p300 occur at constant rates. A complex of Mdm2-p53-p300 is formed at a particular time, where acetylation of p53 does not occur ($K_{19}$). This ternary complex can be dissociated ($K_{21}$) once p53 phosphorylation occurs, thus the affinity of contact with Mdm2 is reduced, and the dissociated p53-p300 complex activity is strengthened. Transcriptional activity of p53 promotes Mdm2 synthesis, and Mdm2 recruits deacetylase HDAC1. In the nucleus, HDAC1 will form an inhibition complex with Mdm2 ($K_{18}$), meanwhile HDAC1 deacetylates p53 ($K_{17}$), which promotes Mdm2-mediated p53 degradation ($K_7$), ($K_{13}$). Once the proteins finish their function, they will be sent to proteasome-mediated degradation for the case of HDAC1 ($K_{22}$), Mdm2 ($K_5$), p53 ($K_1$), and p300 ($K_{14}$), respectively.

## 2.2 Mathematical Description

Following [2], based on standard principles of mass-action law and biochemical transcription kinetics, the behavior of p53-Mdm2 network regulated by p300 and HDAC1 is mathematically described as follows:

$$\dot{x}_1 = -k_1x_1x_{14} + k_6 - k_8x_1x_4 + k_9x_6 - k_{12}x_1x_8 + k_{13}x_9 + k_{17}x_{11}x_{12}, \quad (1)$$
$$\dot{x}_2 = -k_{14}x_2 - k_{15}x_2x_9 - k_{19}x_6x_2 - k_{20}x_4x_2 + k_{23}, \quad (2)$$
$$\dot{x}_3 = -k_{18}x_4x_3 - k_{22}x_3 + k_{24}, \quad (3)$$
$$\dot{x}_4 = k_2x_5 - k_5x_4 + k_7x_6 - k_8x_1x_4 + k_9x_6 - k_{18}x_4x_3 - k_{20}x_4x_2 + k_{21}x_{13}, \quad (4)$$
$$\dot{x}_5 = k_3x_1 - k_4x_5, \quad (5)$$
$$\dot{x}_6 = -k_7x_6 + k_8x_1x_4 - k_9x_6 - k_{19}x_6x_2, \quad (6)$$
$$\dot{x}_7 = -k_{10}x_7 + k_{11}x_8, \quad (7)$$
$$\dot{x}_8 = k_{10}x_7 - k_{11}x_8 - k_{12}x_1x_8, \quad (8)$$
$$\dot{x}_9 = k_{12}x_1x_8 - k_{13}x_9 - k_{15}x_9x_2, \quad (9)$$
$$\dot{x}_{10} = k_{15}x_2x_9 - k_{16}x_{10} + k_{21}x_{13}, \quad (10)$$
$$\dot{x}_{11} = k_{16}x_{10} - k_{17}x_{11}x_{12}, \quad (11)$$
$$\dot{x}_{12} = -k_{17}x_{11}x_{12} + k_{18}x_4x_3, \quad (12)$$
$$\dot{x}_{13} = k_{19}x_6x_2 - k_{21}x_{13}, \quad (13)$$
$$\dot{x}_{14} = -k_1x_1x_{14} + k_{20}x_4x_2, \quad (14)$$

where $x_1$ is p53 protein, $x_2$ is p300 protein, $x_3$ is HDAC1 protein, $x_4$ is Mdm2 protein, $x_5$ is Mdm2 messenger RNA, $x_6$ is Mdm2-p53 complex, $x_7$ is inactivated ATM protein, $x_8$ is activated ATM protein, $x_9$ is phosphorylated p53 protein, $x_{10}$ is phosphorylated p53-p300 complex, $x_{11}$ acetylated p53 protein, $x_{12}$ is Mdm2-HDAC1 complex, $x_{13}$ is Mdm2-p53 with p300 complex, and $x_{14}$ is Mdm2 with p300 complex. The respective parameters are presented in Table 1.

### 2.2.1 Characteristics of p53-Mdm2 Network Regulated by p300 and HDAC1

- The model (1)–(14) describes the core components of the p53 network that are relevant to determining p53 dynamics in response to gamma radiation-induced DNA damage. The mathematical model considers several parameters with non-linear behaviors such as molecule production, degradation, translocation of network components, complex formation rate, translation rate, phosphorylation, and acetylation. Other genes/molecules regulations are ignored, gene mutations are not considered, and constant molecule concentrations in the cell are assumed.
- For developing more detailed models, the incorporation of other proteins that influence p53 activities is needed. The model here does not consider functional

**Table 1** Model parameters

| Parameter | Description | Value |
|---|---|---|
| $k_1$ | p53 degradation | $8.25 \times 10^{-4}$/s |
| $k_2$ | Mdm2 creation | $4.95 \times 10^{-4}$/s |
| $k_3$ | Mdm2-mRNA creation | $1.0 \times 10^{-4}$/s |
| $k_4$ | Mdm2-mRNA degradation | $1.0 \times 10^{-4}$/s |
| $k_5$ | Mdm2 degradation | $4.33 \times 10^{-4}$/s |
| $k_6$ | p53 synthesis | 0.078/s |
| $k_7$ | Mdm2-p53 degradation | $8.25 \times 10^{-4}$/s |
| $k_8$ | Mdm2-p53 synthesis | $11.55 \times 10^{-4}$/s |
| $k_9$ | Mdm2-p53 dissociation | $11.55 \times 10^{-6}$/s |
| $k_{10}$ | ATM activation | $1.0 \times 10^{-4}$/s |
| $k_{11}$ | ATM deactivation | $5.0 \times 10^{-4}$/s |
| $k_{12}$ | Phosphorylation of p53 | $5.0 \times 10^{-4}$/s |
| $k_{13}$ | Dephosphorylation of p53 | $5.0 \times 10^{-1}$/s |
| $k_{14}$ | p300 degradation | $1.0 \times 10^{-4}$/s |
| $k_{15}$ | p53-p300 formation | $1.0 \times 10^{-4}$/s |
| $k_{16}$ | Acetylation of p53 | $1.0 \times 10^{-4}$/s |
| $k_{17}$ | Deacetylation of p53 | $1.0 \times 10^{-5}$/s |
| $k_{18}$ | Creation of Mdm2-HDAC1 | $2.0 \times 10^{-4}$/s |
| $k_{19}$ | Creation of Mdm2-p53-p300 | $5.0 \times 10^{-4}$/s |
| $k_{20}$ | Formation of Mdm2-p300 | $5.0 \times 10^{-4}$/s |
| $k_{21}$ | Dissociation of Mdm2-p53-p300 | $1.0 \times 10^{-4}$/s |
| $k_{22}$ | Degradation of HDAC1 | $1.0 \times 10^{-4}$/s |
| $k_{23}$ | p300 synthesis | 0.08/s |
| $k_{24}$ | HDAC1 synthesis | $2.0 \times 10^{-4}$/s |

disruption by gene mutations, nor other p53 post-translational modifications such as sumoylation, external inhibitors, or altered ubiquitin-proteasome system.

- Likewise, this model does not include epigenetic regulation of gene expression mediated by histone modifications, which may alter the expression of genes dependent or independent of p53.

## 2.3 *Pinning Control Methodology*

A mathematical model for gene regulatory networks is proposed by using the framework of complex networks [5, 57]. Consider a general network consisting of $N$ non-identical nodes with nonlinear functional couplings, where each node is a scalar dynamical system. The proposed network is defined as

$$\dot{x}_i = f_i(x_i) + g_i(t, x_1, x_2, \ldots, x_N), \quad i = 1, 2, \ldots, N, \tag{15}$$

where $x_i \in \mathbb{R}$ is the species (genes, proteins, mRNAs and others) in the network, $i = 1, 2, \ldots, N$, $f_i : \mathbb{R} \mapsto \mathbb{R}$ represents the self-dynamics of node $i$ related to individual processes like proteins, the production or degradation process of RNA, and so on, and $g_i : \mathbb{R}^N \mapsto \mathbb{R}$ denotes the nonlinear coupling function between nodes, associated with changes of $x_i$ due to translation, transcription, activation, inhibition or other interaction processes.

To control (15) in a desirable manner for treatment or intervention purposes, the control goal is to force (15) to track a reference trajectory given as

$$y = y_r(t).$$

The above control objective is achieved by applying a small number of controllers to the network, according to the pinning control technique [11, 39, 58] as follows. Consider the first $l$ nodes selected to be pinned. Thus, the controlled network can be written as

$$\dot{x}_i = f_i(x_i) + g_i(t, x_1, x_2, \ldots, x_N) + h_i(x_i)u_i, \quad i = 1, 2, \ldots, l. \tag{16}$$

$$\dot{x}_i = f_i(x_i) + g_i(t, x_1, x_2, \ldots, x_N), \quad i = l+1, l+2, \ldots, N. \tag{17}$$

where $h_i : \mathbb{R} \mapsto \mathbb{R}$ is a nonlinear function of the node state $i$, for $i = 1, 2, \ldots, l$, and $u_i$ denotes a local discontinuous feedback control law on the node $i \in l$ described by

$$u_i = \begin{cases} 1 + K_i(1 - e_i), & \text{if } \ |\varphi_i| < 1, \\ 1 + K_i(1 - sign(e_i)), & \text{if } \ |\varphi_i| > 1, \end{cases} \tag{18}$$

where $K_i$ is a positive control gain selected by the designer, $e_i$ is the tracking error between the desired output trajectory ($y_r(t)$) and the controlled state ($x_i$), given by

$$e_i = (x_i - y_r(t)), \tag{19}$$

with $\varphi_i = \frac{e_i}{S_i}$ being a designed auxiliary variable to reject chattering effect caused by $sign(\cdot)$ (signum function extracts the sign of a real number) [56], and $S_i$ is a signal filter given by

$$\dot{S}_i = -\alpha_i S_i + \omega_i, \quad i = 1, 2, \ldots, l, \tag{20}$$

where $\alpha_i$ and $\omega_i$ are positive gains to be selected.

## 2.4 p53, p300, and HDAC1 as Pinned Nodes

To select the pinned nodes, the virtual leader methodology presented in [36, 50] is used. The methodology consists in analyzing the interactions between proteins presented in Fig. 2, using the mathematical model ((1)–(14)). The nodes that directly or indirectly affect every other node's dynamical behaviors are candidates as pinned nodes. For this purpose, the spanning tree of the p53-Mdm2 network regulated by p300 and HDAC1 is identified, as shown in Fig. 3. Based on this analysis, p53, p300, and HDAC1 are adequate biological selection as the pinned nodes.

From (1) to (3), the differential equations for p53, p300, and HDAC1 (pinned nodes) are defined by cellular processes, respectively, as follows:

$$\dot{x}_1 = \overbrace{-k_1 x_1 x_{14}}^{\text{p53 degradation}} \overbrace{+k_6}^{\text{p53 synthesis}} \overbrace{-k_8 x_1 x_4}^{\text{Mdm2-p53 synthesis}} \overbrace{+k_9 x_6}^{\text{Mdm2-p53 dissociation}} \overbrace{-k_{12} x_1 x_8}^{\text{p53 phosphorylation}} \overbrace{+k_{13} x_9}^{\text{p53 dephosphorylation}} \overbrace{+k_{17} x_{11} x_{12}}^{\text{p53 deacetylation}},$$

$$\dot{x}_2 = \overbrace{-k_{14} x_2}^{\text{p300 degradation}} \overbrace{-k_{15} x_2 x_9}^{\text{p53-p300 complex}} \overbrace{-k_{19} x_6 x_2}^{\text{Mdm2-p53-p300 complex}} \overbrace{-k_{20} x_4 x_2}^{\text{Mdm2-p300 complex}} \overbrace{+k_{23}}^{\text{p300 synthesis}},$$

$$\dot{x}_3 = \overbrace{-k_{18} x_4 x_3}^{\text{Mdm2-HDAC1 complex}} \overbrace{-k_{22} x_3}^{\text{HDAC1 degradation}} \overbrace{+k_{24}}^{\text{HDAC1 synthesis}}.$$

In order to control the p53-Mdm2 network, it is necessary to modulate the concentration of p53, p300, and HDAC1, to achieve an increased expression of p53 levels and induce p53 and Mdm2 downregulation. As can be seen, mathematically, p53 degradation, Mdm2-p53 synthesis, p53 phosphorylation, p300 degradation, p53-p300 complex, Mdm2-p53-p300 complex, Mdm2-p300 complex, Mdm2-HDAC1 complex, and HDAC1 degradation have a negative sign; while the synthesis process has a positive sign. Due to this fact, it is proposed here to modify p53 production ($K_6$), p300 production ($K_{23}$), and HDAC1 production ($K_{24}$) processes by adding the control law (18).

The equations of the pinned nodes p53 ($x_1$), p300 ($x_2$), and HDAC1 ($x_3$) are

$$\dot{x}_1 = -k_1 x_1 x_{14} + \boxed{k_6 u_1} - k_8 x_1 x_4 + k_9 x_6 - k_{12} x_1 x_8 + k_{13} x_9 + k_{17} x_{11} x_{12}, \quad (21)$$

$$\dot{x}_2 = -k_{14} x_2 - k_{15} x_2 x_9 - k_{19} x_6 x_2 - k_{20} x_4 x_2 + \boxed{k_{23} u_2}, \quad (22)$$

$$\dot{x}_3 = -k_{18} x_4 x_3 - k_{22} x_3 + \boxed{k_{24} u_3}, \quad (23)$$

$$\dot{x}_4 = k_2 x_5 - k_5 x_4 + k_7 x_6 - k_8 x_1 x_4 + k_9 x_6 - k_{18} x_4 x_3 - k_{20} x_4 x_2 + k_{21} x_{13}, \quad (24)$$

$$\dot{x}_5 = k_3 x_1 - k_4 x_5, \quad (25)$$

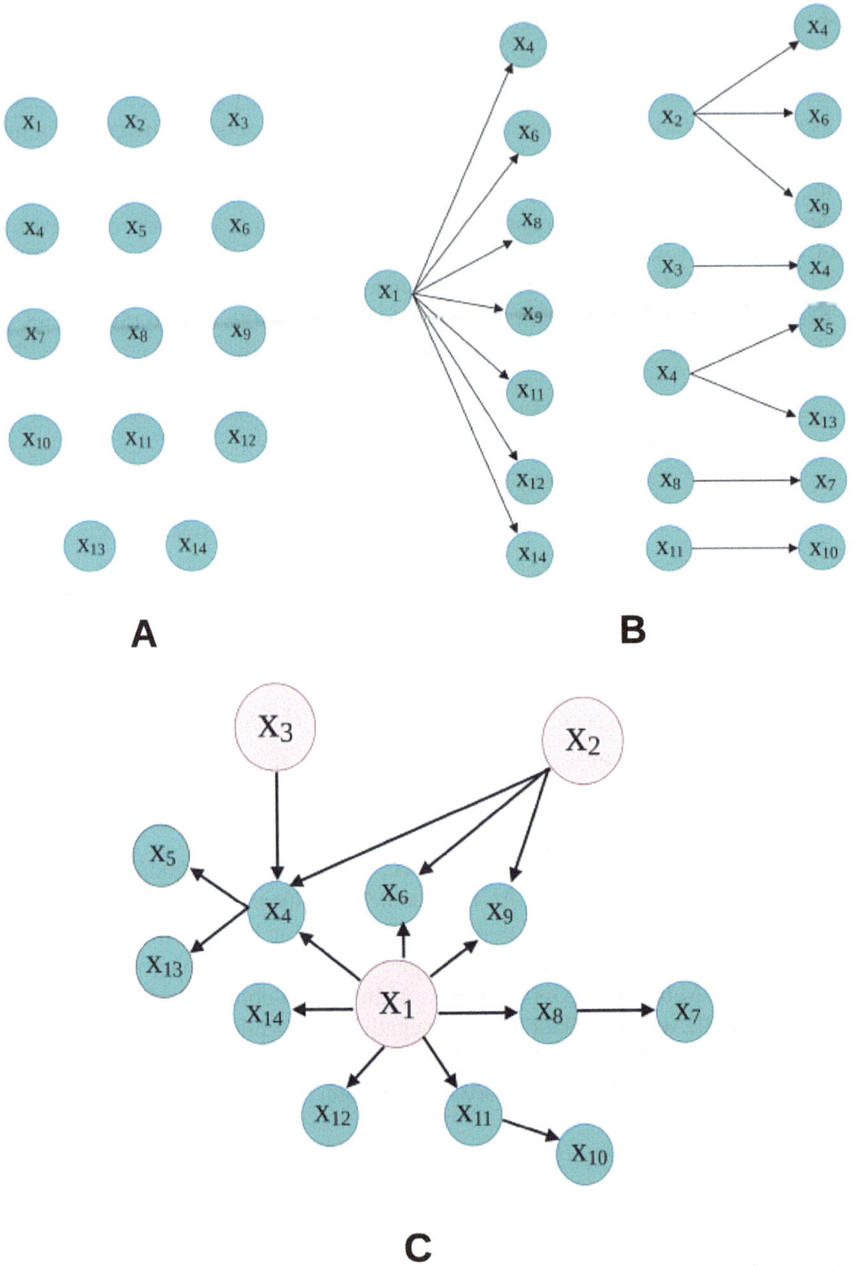


**Fig. 3** Spanning tree of the p53-Mdm2 network regulated by p300 and HDAC1. In **A** states, which are proteins and mRNA of the network, are presented. In **B**, based on the mathematical model (1)–(14) and a biological analysis, states are illustrated with their dependent molecules. Finally, in **C**, the spanning tree, where p53, p300, and HDAC1 do not have direct dependence (virtual leader) from other states, is presented

$$\dot{x}_6 = -k_7x_6 + k_8x_1x_4 - k_9x_6 - k_{19}x_6x_2, \tag{26}$$

$$\dot{x}_7 = -k_{10}x_7 + k_{11}x_8, \tag{27}$$

$$\dot{x}_8 = k_{10}x_7 - k_{11}x_8 - k_{12}x_1x_8, \tag{28}$$

$$\dot{x}_9 = k_{12}x_1x_8 - k_{13}x_9 - k_{15}x_9x_2, \tag{29}$$

$$\dot{x}_{10} = k_{15}x_2x_9 - k_{16}x_{10} + k_{21}x_{13}, \tag{30}$$

$$\dot{x}_{11} = k_{16}x_{10} - k_{17}x_{11}x_{12}, \tag{31}$$

$$\dot{x}_{12} = -k_{17}x_{11}x_{12} + k_{18}x_4x_3, \tag{32}$$

$$\dot{x}_{13} = k_{19}x_6x_2 - k_{21}x_{13}, \tag{33}$$

$$\dot{x}_{14} = -k_1x_1x_{14} + k_{20}x_4x_2. \tag{34}$$

# 3 Results

Simulations are performed using Matlab/Simulink with the fourth-order Runge-Kutta integration method and a fixed step-size of $1 \times 10^{-3}$.

## 3.1 *Behaviors of the p53-Mdm2 Genetic Regulatory Network Without Control Action*

### 3.1.1 The p53-Mdm2 Oscillation Response

Model (1)–(14) presents the oscillation response corresponding to an ionizing gamma-radiation dose, simulated over 48 h, using parameters in Table 1. This response is the normal behavior under radiation effect. Oscillation pattern is presented in a period of 6.2 h, as shown in Fig. 4. This response is due to post-translational modifications of p53 and the negative interactive loop of Mdm2-mediated ubiquitination, according to [13, 18].

### 3.1.2 Mdm2 Overexpression and p53 Downregulation

Model (1)–(14) presents Mdm2 overexpression and p53 downregulation when the following parameters are modified: $k_2 = 2.95 \times 10^{-3}$/s, $k_7 = 8.25 \times 10^{-5}$/s, $k_8 = 11.55 \times 10^{-5}$/s, $k_9 = 11.55 \times 10^{-4}$/s, and $k_{23} = 0.025$/s. This case, illustrated in Fig. 5, represents an oncogenic scenario, where Mdm2 overexpression suppresses functions of p53 through an increase rate of degradation, leading to failure to genotoxic damage responses, which is validated in [19, 32, 46].

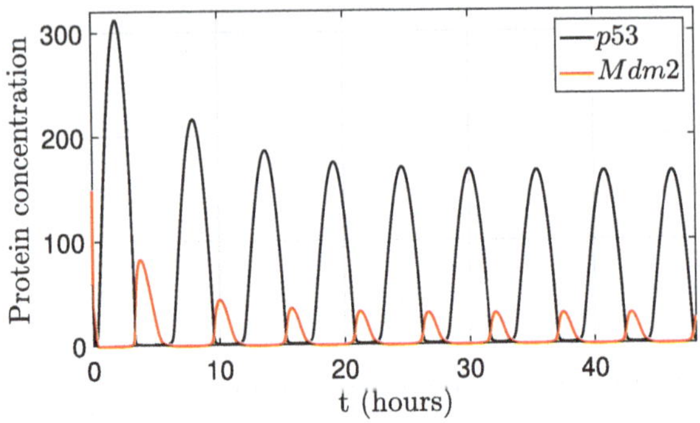


**Fig. 4** The p53 and Mdm2 oscillation response to ionizing radiation damage

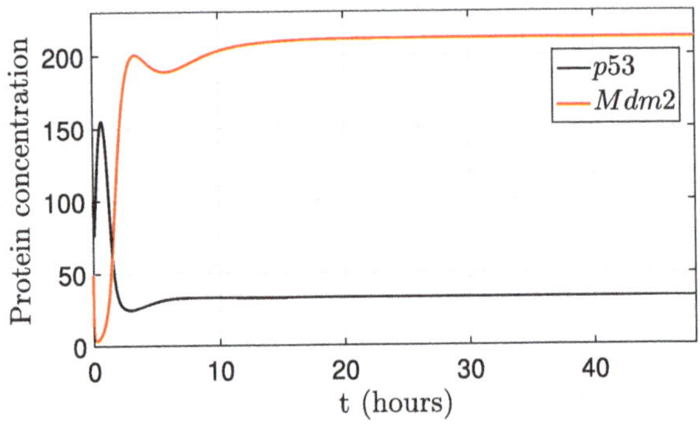


**Fig. 5** Mdm2 overexpression and p53 downregulation

### 3.1.3 Increased Expression of p53 Levels

Model (1)–(14) presents an increased expression of p53 levels when the following parameters are modified: $k_3 = 1 \times 10^{-5}$/s, $k_4 = 1 \times 10^{-3}$/s, and $k_{23} = 0.8$/s. The p53 dynamical behavior presented in Fig. 6 is related to downstream genes involved in signaling pathway, which can produce cell cycle arrest, DNA repair, senescence and/or apoptotic response. Experimental evidence indicates that p53 level can dictate the response of the cell, such that lower levels of p53 result in arrest, whereas higher level results in apoptosis [3, 15, 23, 31, 48].

### 3.1.4 The p53 and Mdm2 Downregulation

Model (1)–(14) presents p53 and Mdm2 downregulation when the following parameters are modified: $k_1 = 1.25 \times 10^{-6}$/s, $k_6 = 0.2$/s, and $k_8 = 11.55 \times 10^{-2}$/s. This case represents null activity for both p53 and Mdm2, as can be seen in Fig. 7, where a pattern shows the knockout cell lines and mice models [38]. Some models have been generated to explore the dependent and independent functional relationships and regulation of p53-Mdm2 signaling in development, in tissue remodeling, in aging, and in cancer [17].

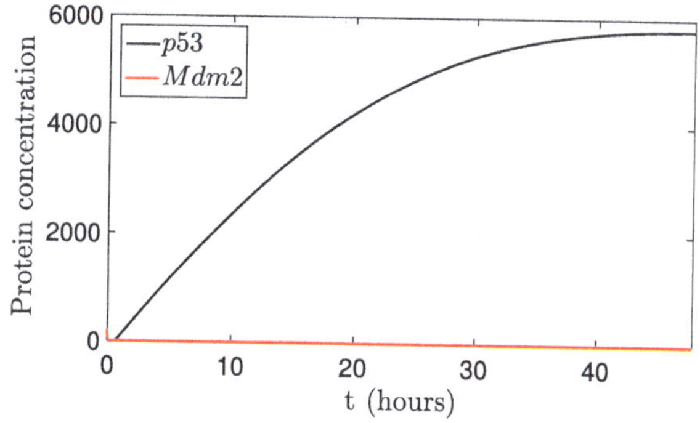


**Fig. 6** Increased expression of p53 levels response under gamma-radiation induction

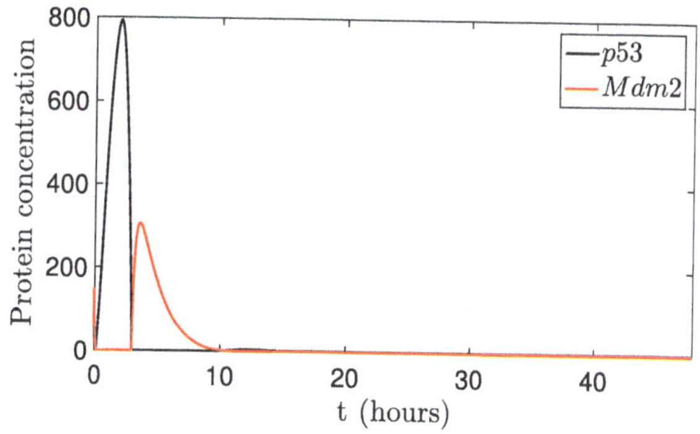


**Fig. 7** The p53 and Mdm2 downregulation

## 3.2 *Behaviors of the p53-Mdm2 Genetic Regulatory Network with Control Action*

This section illustrates the p53 and Mdm2 behavior under pinning control actions. Two cases are considered: (1) to achieve an increased expression of p53 level, and (2) to induce p53 and Mdm2 downregulation. For a 24 h lapse, the network runs without any control action, and presents Mdm2 overexpression and p53 downregulation as carcinogenic initial behavior for both cases.

### 3.2.1 Case 1: Achievement of a p53 Level Increased Expression

Starting at 0 h, the whole network runs without any control, and the system presents Mdm2 overexpression and p53 downregulation as initial behavior. At 24 h, the proposed control law is injected; consequently, the system gradually tracks the increased expression of p53 reference, as can be seen in Fig. 8A. Figure 8B–D illustrate the simulation results for the control input signals $u_1(t)$, $u_2(t)$, and $u_3(t)$ applied to the pinned nodes.

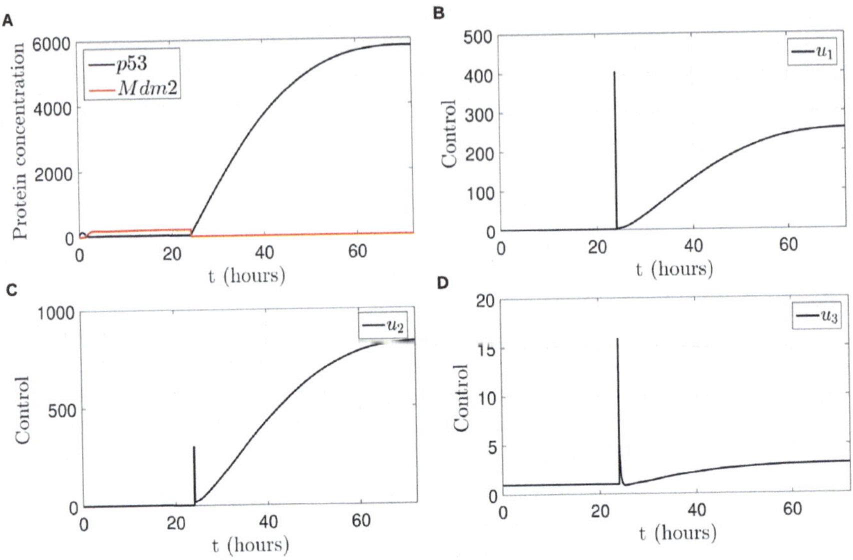


**Fig. 8** Proposed scenario simulations to achieve p53 level increased expression in the p53-Mdm2 network regulated by p300 and HDAC1. In **A**, increased expression of p53 response under pinning control (24–72 h) is achieved. In **B**, control signal $u_1(t)$ applied in p53. In **C**, control signal $u_2(t)$ applied in p300. Finally, in **D**, control signal $u_3(t)$ applied in HDAC1

### 3.2.2 Case 2: Induction of p53 and Mdm2 Downregulation

Starting at 0 h, the whole network runs without any control, and the system presents Mdm2 overexpression and p53 downregulation as initial behavior. At 24 h, the proposed control law is injected; consequently, the system gradually tracks the p53 and Mdm2 downregulation references as can be seen in Fig. 9A. Throughout, $u_1(t)$ is turn off as can be seen in Fig. 9B. Figure 9C, D illustrate the simulation results for the control input signals $u_2(t)$ and $u_3(t)$, applied to the pinned nodes.

## 4 Discussion

There are different strategies to combat cancer, for example inducing cytotoxicity, genotoxicity, and activity modification at specific targets in a signaling pathway, through either chemical agents, radiation, or genetic engineering. In cancer cells, the genomic instability and mutations do not allow the tumour suppression function to perform. This also implies that each type of cancer has a group of different abnormalities that may even be unique to cell clones within a single tumour.

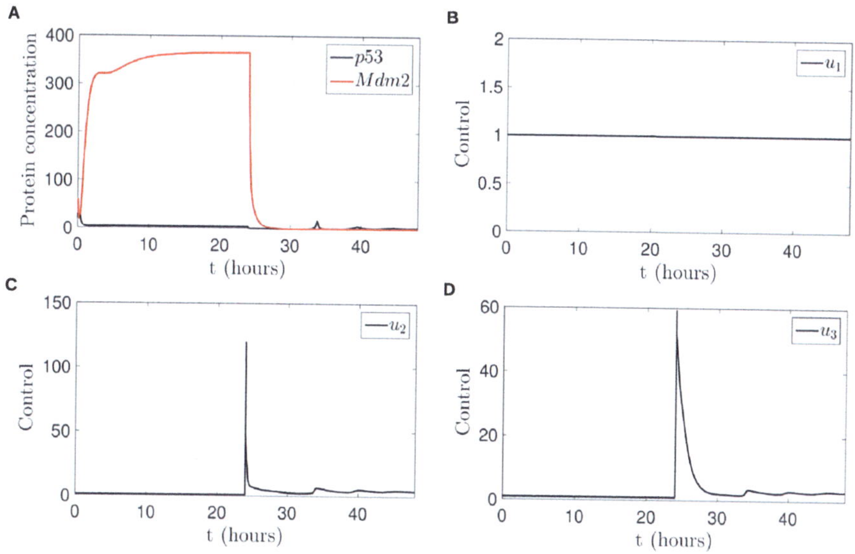


**Fig. 9** Proposed scenario simulations to induce p53 and Mdm2 downregulation in the p53-Mdm2 network regulated by p300 and HDAC1. In **A**, p53 and Mdm2 downregulation response under pinning control (24–72 h) is achieved. In **B**, control signal $u_1(t)$ applied in p53 is turn off. In **C**, control signal $u_2(t)$ applied in p300. Finally, in **D**, control signal $u_3(t)$ applied in HDAC1

Hence, to illustrate the applicability of the proposed pinning control, two cases are studied on a deterministic network model corresponding to tumour suppressor p53, Mdm2, HDAC1, and p300.

## 4.1 Behaviors Induced Without Control Action

### 4.1.1 The p53-Mdm2 Oscillation Response

In normal unstressed cells, p53 remains inactive and latent expressing low levels, due to the targeted degradation by the protein E3 ubiquitin-protein ligase Mdm2 (from *MDM2* proto-oncogene). Mdm2 binds to p53 and marks it for proteasome degradation, preventing p53 accumulation in the nucleus and its transcriptional activities. It is also known that in non-genotoxic conditions, p53 is rapidly degraded and is estimated having a short half-life [43]. However, under cell stress in exposition to gamma radiation, the genetic network induces a behavior compatible with an increase in the number of damped pulses of p53, as demonstrated in [35]. Gamma radiation causes many double-strand breaks (DSBs) in the DNA, which activates the p53-Mdm2 network. Cells detecting non-repaired DSBs along time, restarting a new pulse of p53 activity. While the DNA damage is still present, pulses in the oscillatory behavior continue until DNA is successfully repaired, and the cell goes into senescence or cell cycle arrest, or the cell could activate a program to die [35].

### 4.1.2 Mdm2 Overexpression and p53 Downregulation

Among the most important abnormalities that have been reported in cancer is the functional inactivation of tumour suppressor genes mediated by the over-activation of their negative regulators [32, 43, 46]. This is the case of Mdm2 overexpression, which mediates the accelerated degradation of p53; thus, p53 will show a decreased response to DNA damaging agents and, therefore, the loss of p53 tumour-suppressor signaling outcomes and acceleration of tumour evolution [46]. Overexpressed Mdm2 in human tumours has been reviewed in [49] for more than forty different types of malignancies, including solid tumours, sarcomas, and leukemias.

### 4.1.3 Increased Expression of p53 Levels

The p53 progressive accumulation is mediated by p53 post-translational modifications and increased stability and transactivation [8, 15, 29, 33, 43]. The p53 co-activation by p300 and ATM in a damage response could generate the activation of downstream proteins. Increased expression of p53 levels allows the induction of genes associated with cell fates, such as senescence or apoptosis [40]. According to [60], full phosphorylation of p53 residues corresponds to severe DNA damage. Along with p53 arrival and accumulation in mitochondria within 30 min after radiation stimuli, apoptosis is induced. In apoptotic response, mitochondrial p53 interacts with Bak/Bax and starts a process for the mitochondrial outer membrane permeabilization, cytochrome C release, and caspase activation to complete the cell death program.

### 4.1.4 The p53 and Mdm2 Downregulation

The level of p53 activity is autoregulated in cells due to the ability of p53 to induce the expression of its inhibitor *MDM2*. However, in some stressful circumstances, other proteins can regulate p53 and Mdm2 levels. This is the case for the activation of stress-activated protein c-jun N-terminal kinase 1 (JNK) that can promote apoptosis along or independent of p53 and decide cell fate in response to stress. JNK mediates phosphorylation of p53 in U4 cells (model of normal colonocytes) exposed to resembling digested fibers, induces p53 downregulation, not stabilization, and occurs in the absence of Mdm2. Stressing colon epithelial cells with resembling digested fibers induced activation of JNK and a rapid loss of p53. Targeted destruction of p53 in intestinal epithelial cells may be part of an apoptotic program that controls remodeling tissue of intestinal crypt column in vivo [62].

Also, in another study, p53 downregulation was observed in cells infected by some viruses that induce apoptosis, limiting viral replication. More evidence for the role of p53 pathways in viral defense is complex and is mediated by interplay with other regulators [41]. This evidence broadens the possibilities of regulation of the p53-Mdm2 network and its outcomes.

## *4.2 Behaviors Induced by the Pinning Control Technique*

The different types of cellular abnormalities in tumours require multiple strategies to combat carcinogenesis. A good scenario would be to know these abnormalities in advance before applying treatment to customize the strategy. Reversing carcinogenic behaviors can be accomplished theoretically, stirring a genetic network's dynamics, applying control techniques in a gene regulatory model network [59]. The more significant details are known about regulating a normal and altered cellular regulatory network, the more precision the mathematical models will acquire to simulate a cancer process's conditions. They will be a valuable support in the design of new anticancer therapies.

In the model (1)–(14), key regulators in the p53-Mdm2 network are presented as pinned nodes, p53, p300, and HDAC1, which accomplish control objectives on the network. In this sense, tumour suppressor activity of p53 is carried out primarily as a transcriptional factor, through post-translational modifications, which include phosphorylation and acetylation, allowing p53 stability and activating its transcriptional functions. Interestingly, the balance between acetylation and deacetylation is critical during gene transcription, and disruption of this balance can alter the cell cycle-promoting activities or its inhibition and other cellular functions [65].

### 4.2.1 Case 1: Achievement of a p53 Level Increased Expression

The control law (18) is added at the production of p53, p300, and HDAC1 nodes, which yields an increased p53 level expression and stabilization through post-translational modifications and increased transactivation. For the oncogenic scenario of overexpressed Mdm2, it is possible to generate p53 progressive accumulation, which is assumed to be activated by a mechanism linked to post-radiation activation of the ATM/ATR pathway, and p53-dependent and independent induction of downstream proteins [6]. ATM activation promotes p53 phosphorylation that reduces the affinity for Mdm2 [51] and allows p53 to interact with other co-activators such as p300, forming a complex. The p53-p300 complex causes p53 acetylation at specific lysine residues that are common for both acetylation and ubiquitination [26, 52]. Acetylation of lysine residues inhibits ubiquitination mediated by the interaction with Mdm2-HDAC1 complex [27, 37], allowing p53 levels to increase. Sustainment of elevated p53 protein allows the induction of genes associated with terminal fates, such as senescence or apoptosis [40]. With respect to sustained p53 levels, experimental evidence shows proliferation suppression of MCF-7 dividing cells under radiation stimuli, and also some p53 target genes correlate with p53 protein levels such as p21, GADD45A, Mdm2, PPM1D at first 24 h post-radiation stimuli. Post-translational modifications of p53 through different cofactors bind p53 to affect the choice of downstream gene programs [55, 61]. It seems that levels of p53 activation, time of stimulus, and damage type, are all important factors to cell fate decisions according to [48].

The p300 as a pinned node regulates a series of functions in the p53-Mdm2 genetic network model, among which it stands out the acetyltransferase activity in p53 to mediate transcription of target genes in response to DNA damage. After DNA damage, p300 participates in the activation of p53 mediated by acetylation [26], along with phosphorylation by the action of various kinases such as ATM [51]. ATM mediates phosphorylation at multiple p53 sites, including Ser46, in response to ionizing radiation. At the same time, p53 is activated by p300. The latter increases the stability of the formation of the p53-DNA-p300 complex. The p300 stimulates the transcriptional activity of p53 on p53-regulated promoters, and it enhances the responsiveness to a physiological upstream modulator of the p53 function, for example, under ionizing radiation stimuli [4]. Besides, p300 can modulate the p53 network activities at other levels, such as at the epigenetic level, by regulating the acetylation of target histone proteins (epigenetic modifications not included in this model). There is evidence that, if p300 were inactive, it would promote a failure in the activation of p53 and its responses, with the consequence of a failure in the execution of apoptosis due to DNA damage. The p300 functions as a versatile signal integrator of many transcription factors to facilitate transcriptional activation or repression, and multiple transduction pathways.

Finally, HDAC1 as a pinned node belongs to an enzyme family known as deacetylases that work as transcriptional co-activators to contribute to the balance between acetylation and deacetylation in regulating gene expression, which is thus linked to the control of cell fate. Further, it has been observed that p53 acetylation is a reversible process, and for it, Mdm2 recruits HDAC1 to form an Mdm2-HDAC1 complex, which deacetylates p53 [27]. In this sense, HDAC1 also has an important regulatory role in response to genotoxic damage. HDAC1 interacts with Mdm2, forming a complex and promoting p53 deacetylation and playing a role in DNA damage response through direct interaction with FUS (Fused-in-Sarcoma), a multi-functional ribonucleoprotein expressed in the nucleus. FUS participates in the maintenance of genome stability and DNA repair. FUS and HDAC1 interact and are required for DSB's repair, genome stability maintenance, and neurons survival [63].

### 4.2.2 Case 2: Induction of p53 and Mdm2 Downregulation

The p53-Mdm2 network has regulations mediated by other proteins to induce tissue remodeling or protection against viral infections. In these types of stressors, p53 has been seen to be downregulated or inactivated, as already been noted in

Sect. 4.1.4. This network executes different types of regulation in response to the differential choice of downstream gene programs. The p53 post-translational regulation contributes to cell program choice, especially phosphorylation, dephosphorylation, acetylation, and deacetylation.

As a regulator of a tumour-suppression mechanism, HDAC1 plays a role in senescence activation, while overexpressed in cervical cancer cells and contributes to decreased cellular proliferation. HDAC1 overexpression has been reported having increased levels of interaction in deacetylated Sp1 (Sp1 transcription factor) with p300, leading to the expression of *PP2AC* (phosphatase 2A subunit). Also, PP2Ac co-regulate senescence induction with active hypophosphorylated pRb (retinoblastoma protein) [1]. Inactive phosphorylated pRb is one of the participants in cell-cycle progression regulation through cyclin-dependent kinases (CDKs) under mitogenic stimuli [25].

Moreover, it was reported in [53] that the knock-down model of HDAC1 and HDAC2 accelerates the preleukemic state along with p53 deletion in transgenic preleukemic mice. In contrast, in the preleukemic phase of acute promyelocytic leukemia (APL), HDAC1 counteracts *PML-RAR* oncogenic fusion activities. In this scenario, HDAC1 blocks differentiation, impairs genomic stability and increases self-renewal in hematopoietic progenitors. Remarkably, short-term treatment of preleukemic mice with an HDAC inhibitor accelerates leukemogenesis. Nevertheless, in the knock-down of HDAC1 in APL mice, the leukemic animals enhance the survival duration. Thus, HDAC1 has a dual role in tumorigenesis: oncosuppressive in the early stages, and oncogenic in established tumour cells.

Additionally, the dual function of HDAC1 was also observed in young and old livers in association with different protein complexes. The in vitro and in vivo results suggest a model, where high levels of HDAC1, in association with chromatin remodelers and tissue-specific factors, promote an ever-changing chromatin structure, leading to failure of growth-promoting gene activation and aging [65].

Activation and stabilization of p53 depend on p300 activities. In this sense, p300 C/H1 region is a specific binding site for p53 and Mdm2. Moreover, Mdm2 also bound to p300 such that interactions of p53-Mdm2-p300 participate in p53 degradation mediated by Mdm2. Taken together, complex interactions between p53 and its co-regulators are involved in controlling p53 response [19].

The multiple effects of stirring the dynamics of the p53-Mdm2 network and other regulators must be carefully weighted to estimate the effects on downstream dependent pathways and its consequence on cell fates. Therefore, induced senescence, cell cycle arrest, or apoptosis by p300, HDAC1 and/or p53 modulation may be an effective approach to developing new alternatives for functional studies in tumours and the study of new therapies in cancer.

## 5 Conclusions

This chapter presents a control strategy to regulate cellular response in cancer in the p53-Mdm2 genetic regulatory network, based on pinning control techniques. To select pinned nodes, virtual leader methodology is applied. To illustrate the p53 and Mdm2 protein modulation under pinning control, two cases are considered. In the first case, to achieve a p53 level increased expression, the control action ($u_1(t)$, $u_2(t)$ and $u_3(t) \in \mathbb{R}$) is applied to p53, p300 and HDAC1 respectively, as show in Eqs. (21)–(23). In the second case, to induce p53 and Mdm2 downregulation, the control action $u_1(t)$ is turn off, $u_2(t)$ is applied in p300, and $u_3(t)$ is applied in HDAC1. From the above results, it can be clearly seen that the pinning controllers achieve regulation successfully for the p53-Mdm2 network regulated by p300 and HDAC1. The benefits from analysis of accurate gene regulatory network models for medicine, system biology, and biotechnology, provide a strong incentive for cooperation between experimentalists, control engineering, and computational scientists.

## References


1. Alberts, A., Thorburn, A., Shenolikar, S., Mumby, M., Feramisco, J.: Regulation of cell cycle progression and nuclear affinity of the retinoblastoma protein by protein phosphatases. Proc. Natl. Acad. Sci. **90**(2), 388–392 (1993)
2. Arora, A., Gera, S., Maheshwari, T., Raghav, D., Alam, J., Singh, B., Agarwal, S.: The dynamics of stress p53-mdm2 network regulated by p300 and HDAC1. PLoS ONE **8**(2), e52736 (2013)
3. Arya, A., El-Fert, A., Devling, T., Eccles, R., Aslam, M., Rubbi, C., Vlatković, N., Fenwick, J., Lloyd, B., Sibson, D., et al.: Nutlin-3, the small-molecule inhibitor of MDM2, promotes senescence and radiosensitises laryngeal carcinoma cells harbouring wild-type p53. Br. J. Cancer **103**(2), 186–195 (2010)
4. Avantaggiati, M., Ogryzko, V., Gardner, K., Giordano, A., Levine, A., Kelly, K.: Recruitment of p300/ CBP in p53-dependent signal pathways. Cell **89**(7), 1175–1184 (1997)
5. Barabási, A., Albert, R.: Emergence of scaling in random networks. Science **286**(5439), 509–512 (1999)
6. Batchelor, E., Loewer, A., Mock, C., Lahav, G.: Stimulus-dependent dynamics of p53 in single cells. Mol. Syst. Biol. **7**(1), 488 (2011)
7. Boccaletti, S., Latora, V., Moreno, Y., Chavez, M., Hwang, D.: Complex networks: structure and dynamics. Phys. Rep. **424**(4), 175–308 (2006)
8. Brown, C., Lain, S., Verma, C., Fersht, A., Lane, D.: Awakening guardian angels: drugging the p53 pathway. Nat. Rev. Cancer **9**(12), 862–873 (2009)
9. Chan, H., La Thangue, N.: p300/ CBP proteins: HATs for transcriptional bridges and scaffolds. J. Cell Sci. **114**(13), 2363–2373 (2001)
10. Chehab, N., Malikzay, A., Stavridi, E., Halazonetis, T.: Phosphorylation of Ser-20 mediates stabilization of human p53 in response to DNA damage. Proc. Natl. Acad. Sci. **96**(24), 13777–13782 (1999)
11. Chen, G.: Pinning control and controllability of complex dynamical networks. Int. J. Autom. Comput. **14**(1), 1–9 (2017)
12. Chen, G., Wang, X., Li, X.: Fundamentals of Complex Networks: Models, Structures and Dynamics. Wiley, New York (2014)
13. Ciliberto, A., Novák, B., Tyson, J.: Steady states and oscillations in the p53/ Mdm2 network. Cell Cycle **4**(3), 488–493 (2005)

14. Davidson, E.: The Regulatory Genome: Gene Regulatory Networks in Development and Evolution. Elsevier (2010)
15. El-Deiry, W.: Regulation of p53 downstream genes. Semin. Cancer Biol. **8**(5), 345–357 (1998)
16. Galluzzi, L., Vitale, I., Abrams, J., et al.: Molecular definitions of cell death subroutines: recommendations of the nomenclature committee on cell death 2012. Cell Death Differ. **19**(1), 107–120 (2012)
17. Gannon, H., Jones, S.: Using mouse models to explore MDM-p53 signaling in development, cell growth, and tumorigenesis. Genes & Cancer **3**(3–4), 209–218 (2012)
18. Geva-Zatorsky, N., Rosenfeld, N., Itzkovitz, S., Milo, R., Sigal, A., Dekel, E., Yarnitzky, T., Liron, Y., Polak, P., Lahav, G., et al.: Oscillations and variability in the p53 system. Mol. Syst. Biol. **2**(1), 1–13 (2006)
19. Grossman, S., Perez, M., Kung, A., Joseph, M., Mansur, C., Xiao, Z., Kumar, S., Howley, P., Livingston, D.: p300/ MDM2 complexes participate in MDM2-mediated p53 degradation. Mol. Cell **2**(4), 405–415 (1998)
20. Gu, W., Shi, X., Roeder, R.: Synergistic activation of transcription by CBP and p53. Nature **387**(6635), 819–823 (1997)
21. Halkidou, K., Gaughan, L., Cook, S., Leung, H., Neal, D., Robson, C.: Upregulation and nuclear recruitment of HDAC1 in hormone refractory prostate cancer. Prostate **59**(2), 177–189 (2004)
22. Haupt, Y., Maya, R., Kazaz, A., Oren, M.: Mdm2 promotes the rapid degradation of p53. Nature **387**(6630), 296–299 (1997)
23. Hsing, A., Faller, D., Vaziri, C.: DNA-damaging aryl hydrocarbons induce Mdm2 expression via p53-independent post-transcriptional mechanisms. J. Biol. Chem. **275**(34), 26024–26031 (2000)
24. Hu, B., Gilkes, D., Farooqi, B., Sebti, S., Chen, J.: MDMX overexpression prevents p53 activation by the MDM2 inhibitor Nutlin. J. Biol. Chem. **281**(44), 33030–33035 (2006)
25. Hume, A., Finkel, J., Kamil, J., Coen, D., Culbertson, M., Kalejta, R.: Phosphorylation of retinoblastoma protein by viral protein with cyclin-dependent kinase function. Science **320**(5877), 797–799 (2008)
26. Ito, A., Lai, C.H., Zhao, X., Saito, S., Hamilton, M., Appella, E., Yao, T.P.: p300/ CBP-mediated p53 acetylation is commonly induced by p53-activating agents and inhibited by MDM2. EMBO J. **20**(6), 1331–1340 (2001)
27. Ito, A., Kawaguchi, Y., Lai, C., Kovacs, J., Higashimoto, Y., Appella, E., Yao, T.: MDM2-HDAC1-mediated deacetylation of p53 is required for its degradation. EMBO J. **21**(22), 6236–6245 (2002)
28. Kamijo, T., Weber, J., Zambetti, G., Zindy, F., Roussel, M., Sherr, C.: Functional and physical interactions of the ARF tumor suppressor with p53 and Mdm2. Proc. Natl. Acad. Sci. **95**(14), 8292–8297 (1998)
29. Kaufmann, W., Kaufman, D.: Cell cycle control, DNA repair and initiation of carcinogenesis. FASEB J. **7**(12), 1188–1191 (1993)
30. Kessis, T., Slebos, R., Nelson, W., Kastan, M., Plunkett, B., Han, S., Lorincz, A., Hedrick, L., Cho, K.: Human papillomavirus 16 E6 expression disrupts the p53-mediated cellular response to DNA damage. Proc. Natl. Acad. Sci. **90**(9), 3988–3992 (1993)
31. Khan, S., Guevara, C., Fujii, G., Parry, D.: p14 ARF is a component of the p53 response following ionizing irradiation of normal human fibroblasts. Oncogene **23**(36), 6040–6046 (2004)
32. Knights, C., Liu, Y., Appella, E., Kulesz-Martin, M.: Defective p53 post-translational modification required for wild type p53 inactivation in malignant epithelial cells with mdm2 gene amplification. J. Biol. Chem. **278**(52), 52890–52900 (2003)
33. Kobet, E., Zeng, X., Zhu, Y., Keller, D., Lu, H.: Mdm2 inhibits p300-mediated p53 acetylation and activation by forming a ternary complex with the two proteins. Proc. Natl. Acad. Sci. **97**(23), 12547–12552 (2000)
34. Kruse, J., Gu, W.: Modes of p53 regulation. Cell **137**(4), 609–622 (2009)
35. Lahav, G., Rosenfeld, N., Sigal, A., Geva-Zatorsky, N., Levine, A., Elowitz, M., Alon, U.: Dynamics of the p53- Mdm2 feedback loop in individual cells. Nat. Genet. **36**(2), 147–150 (2004)

36. Lewis, F., Zhang, H., Hengster-Movric, K., Das, A.: Cooperative Control of Multi-agent Systems: Optimal and Adaptive Design Approaches. Springer, Berlin (2013)
37. Li, M., Luo, J., Brooks, C., Gu, W.: Acetylation of p53 inhibits its ubiquitination by mdm2. J. Biol. Chem. **277**(52), 50607–50611 (2002)
38. Li, M., Brooks, C., Wu-Baer, F., Chen, D., Baer, R., Gu, W.: Mono-versus polyubiquitination: differential control of p53 fate by Mdm2. Science **302**(5652), 1972–1975 (2003)
39. Li, X., Wang, X., Chen, G.: Pinning a complex dynamical network to its equilibrium. IEEE Trans. Circuits Syst. I Regul. Pap. **51**(10), 2074–2087 (2004)
40. Luo, Q., Beaver, J., Liu, Y., Zhang, Z.: Dynamics of p53: a master decider of cell fate. Genes **8**(2), 66 (2017)
41. Marques, J., Rebouillat, D., Ramana, C., et al.: Down-regulation of p53 by double-stranded RNA modulates the antiviral response. J. Virol. **79**(17), 11105–11114 (2005)
42. Moll, U., Petrenko, O.: The Mdm2-p53 interaction. Mol. Cancer Res. **1**(14), 1001–1008 (2003)
43. Momand, J., Zambetti, G., Olson, D., George, D., Levine, A.: The mdm-2 oncogene product forms a complex with the p53 protein and inhibits p53-mediated transactivation. Cell **69**(7), 1237–1245 (1992)
44. Momand, J., Wu, H., Dasgupta, G.: MDM2 master regulator of the p53 tumor suppressor protein. Gene **242**(1–2), 15–29 (2000)
45. Newman, M.: The structure and function of complex networks. SIAM Rev. **45**(2), 167–256 (2003)
46. Oliner, J., Kinzler, K., Meltzer, P., George, D., Vogelstein, B.: Amplification of a gene encoding a p53-associated protein in human sarcomas. Nature **358**(6381), 80–83 (1992)
47. Peter, I., Davidson, E.: Genomic Control Process: Development and Evolution. Academic (2015)
48. Purvis, J., Karhohs, K., Mock, C., Batchelor, E., Loewer, A., Lahav, G.: p53 dynamics control cell fate. Science **336**(6087), 1440–1444 (2012)
49. Rayburn, E., Zhang, R., He, J., Wang, H.: Mdm2 and human malignancies: expression, clinical pathology, prognostic markers, and implications for chemotherapy. Curr. Cancer Drug Targets **5**(1), 27–41 (2005)
50. Ren, W., Beard, R.: Distributed Consensus in Multi-vehicle Cooperative Control. Springer, Berlin (2008)
51. Saito, S., Goodarzi, A., Higashimoto, Y., Noda, Y., Lees-Miller, S., Appella, E., Anderson, C.: ATM mediates phosphorylation at multiple p53 sites, including Ser46, in response to ionizing radiation. J. Biol. Chem. **277**(15), 12491–12494 (2002)
52. Sakaguchi, K., Herrera, J., Saito, S., Miki, T., Bustin, M., Vassilev, A., Anderson, C., Appella, E.: DNA damage activates p53 through a phosphorylation-acetylation cascade. Genes & Dev. **12**(18), 2831–2841 (1998)
53. Santoro, F., Botrugno, O., Dal Zuffo, R., Pallavicini, I., Matthews, G., Cluse, L., et al.: A dual role for Hdac1: oncosuppressor in tumorigenesis, oncogene in tumor maintenance. Blood **121**(17), 3459–3468 (2013)
54. Secchiero, P., Bosco, R., Celeghini, C., Zauli, G.: Recent advances in the therapeutic perspectives of Nutlin-3. Curr. Pharm. Des. **17**(6), 569–577 (2011)
55. Sionov, R., Haupt, Y.: The cellular response to p53: the decision between life and death. Oncogene **18**(45), 6145–6157 (1999)
56. Slotine, J., Li, W.: Applied Nonlinear Control. Prentice Hall, Englewood Cliffs (1991)
57. Suarez, O., Vega, C., Sanchez, E., González-Santiago, A., Pardo-Garcia, A.: Pinning control applied to gene regulatory networks. In: IEEE Latin American Conference on Computational Intelligence (LA-CCI), pp. 1–6. IEEE (2018)
58. Suarez, O., Vega, C., Sanchez, E., González-Santiago, A., Rodríguez-Jorge, O., Pardo-Garcia, A.: Abnormal p53 degradation and apoptosis induction in p53- Mdm2 network using pinning control strategy. Revista Colombiana de Tecnologías de Avanzada **2**(32), 1–7 (2018)
59. Suarez, O., Vega, C., Sanchez, E., González-Santiago, A., Rodríguez-Jorge, O., Alanis, A., Chen, G., Hernandez-Vargas, E.: Pinning control for the p53- Mdm2 network dynamics regulated by p14 ARF. Front. Physiol. **11**, 976 (2020)

60. Tian, X., Liu, F., Zhang, X., Li, J., Wang, W.: A two-step mechanism for cell fate decision by coordination of nuclear and mitochondrial p53 activities. PLoS ONE **7**(6), e38164 (2012)
61. Vousden, K., Lu, X.: Live or let die: the cell's response to p53. Nat. Rev. Cancer **2**(8), 594–604 (2002)
62. Wang, J., Friedman, E.: Downregulation of p53 by sustained JNK activation during apoptosis. Mol. Carcinog. **29**(3), 179–188 (2000)
63. Wang, W., Pan, L., Su, S., Quinn, E., Sasaki, M., Jimenez, J., Mackenzie, I., Huang, E., Tsai, L.: Interaction of FUS and HDAC1 regulates DNA damage response and repair in neurons. Nature **16**(10), 1383–1391 (2013)
64. Wang, Z., Zhan, Y., Xu, J., Wang, Y., Sun, M., Chen, J., Liang, T., Wu, L., Xu, K.: $\beta$-sitosterol reverses multidrug resistance via BCRP suppression by inhibiting the p53-mdm2 interaction in colorectal cancer. J. Agric. Food Chem. **68**(12), 3850–3858 (2020)
65. Willis-Martinez, D., Richards, H., Timchenko, N., Medrano, E.: Role of HDAC1 in senescence, aging, and cancer. Exp. Gerontol. **45**(4), 279–285 (2010)
66. Zhu, X., Gerstein, M., Snyder, M.: Getting connected: analysis and principles of biological networks. Genes & Dev. **21**(9), 1010–1024 (2007)

# Heterogeneous Tumour Modeling Using PhysiCell and Its Implications in Precision Medicine

**Miloš Savić, Vladimir Kurbalija, Igor Balaz, and Mirjana Ivanović**

## 1 Introduction

### *1.1 Precision Medicine—General Introduction*

In modern society, with the constantly growing economy and demanding working and living conditions, a large portion of the population is facing stressful life that triggers chronic health problems like cancer, cardiovascular or neurological diseases. Such a situation requires a shift in healthcare towards more interdisciplinary, multi-disciplinary, and holistic initiatives [6, 38].

Accordingly, the importance of the development of appropriate services to improve the quality of life of persons who suffers from chronic diseases has been widely recognized. For that, the collection of huge amounts of patients' complex data (like clinical, environmental, nutritional, everyday activities…) is needed. It is necessary to properly aggregate such data, analyze it and present it to the clinical doctors and caregivers so they can devise better recommendations for adequate treatment and actions to improve patient's health.

This approach is the essence of the emerging research called personalized medicine or precision medicine. According to the US National Research Coun-

M. Savić (✉) · V. Kurbalija · M. Ivanović
Faculty of Sciences, University of Novi Sad, Trg Dositeja Obradovića 3, Novi Sad, Serbia
e-mail: svc@dmi.uns.ac.rs

V. Kurbalija
e-mail: kurba@dmi.uns.ac.rs

M. Ivanović
e-mail: mira@dmi.uns.ac.rs

I. Balaz
Faculty of Agriculture, University of Novi Sad, Trg Dositeja Obradovića 8, Novi Sad, Serbia
e-mail: igor.balaz@df.uns.ac.rs



I. Balaz and A. Adamatzky (eds.), *Cancer, Complexity, Computation*,
Emergence, Complexity and Computation 46,
https://doi.org/10.1007/978-3-031-04379-6_7

cil, "personalized medicine" is an older term with a similar meaning as "precision medicine." However, the term "personalized" could be misunderstood to imply that medical treatments are being developed uniquely for each individual.

According to the Precision Medicine Initiative [15], precision medicine is "an emerging approach for disease treatment and prevention that takes into account individual variability in genes, environment, and lifestyle for each person." Precision medicine, where an individual patient's molecular information is processed, also can be an adequate way to try to recognize different risk factors and help in the prevention of critical diseases like cancer [20]. Regardless of the scope of data included in consideration, the main characteristic of the approach is to obtain a more accurate prediction of which treatment and prevention strategies will have the best results for which groups of patients. It has the potential to improve the traditional symptom-driven and one-size-fits-all approach allowing earlier medical diagnostics, interventions, and tailoring better and economically personalized treatments.

Additionally in the last several decades, we are facing enormous production of biological data that causes a paradigm shift in medical research. Technological support to study molecular changes over the whole genome brings new dimensions to the concept of precision medicine. Such a comprehensive approach supported by powerful ICT tools and methods raised hope for the development of superior diagnostic and therapeutic instruments. This is especially relevant to cancer as its incidence is globally increasing.

Regardless of the promised theoretical benefits of precision medicine, its role in day-to-day healthcare is still relatively limited. However, given the enormous technological development and rapid research actions in a lot of medical-related disciplines, the expectations are that this approach will expand in medical and healthcare areas in the coming years.

## 1.2 Treatment Approaches in Precision Medicine

Cancer treatment options are numerous. From traditional chemotherapy where chemicals are used to kill fast-growing cells (both cancerous and non-cancerous) to more precise treatments such as hormone therapy, immunotherapy and targeted therapy. All of these more precise treatments rely on performing tumour analysis for patient-specific biomarkers (genes, proteins and other tumour markers). A closely related field is Pharmacogenomics. It is still in its infancy but its main focus is on discovering how genes affect a person's response to specific drugs. Pharmacogenomics combines the science of drugs and genomics to try to develop effective but safe medications and propose adequate doses that are tailored to variations in a person's genes. Expectations of this approach are very optimistic for the development of tailored drugs to treat serious health problems like cancer [20], cardiovascular disease [3], and so on.

In cancer research, there are several approaches to identifying appropriate molecular targets. One approach is to identify overexpressed proteins. The very first targeted therapy, tamoxifen, made in 1962, has been developed to inhibit the growth of estrogen receptor-positive (ER+) breast cancer cells [16]. The second breakthrough happened also by identifying overexpressed proteins - Human Epidermal Growth Factor receptor 2 (HER-2). HER-2 positive tumours are sensitive to the inhibition of Her2 receptor function and to date, a large number of therapies has been developed. They include monoclonal antibodies (trastuzumab, pertuzumab, margetuximab-cmkb), antibody-drug conjugates (different conjugates with trastuzumab), pan-her inhibitors (neratinib), signal transduction inhibitors (lapatinib) and tyrosine-kinase inhibitors (tucatinib).

Another approach is to determine whether cancer cells produce mutant proteins that drive cancer progression. An example is the cell growth signalling protein BRAF, which is in many melanomas presented in altered form, BRAFV600E. Inhibitor of B-Raf enzyme, Vemurafenib, has been approved in 2011 to treat patients with inoperable or metastatic melanoma that contains mutated BRAF protein. Currently, several clinical trials investigate possibilities of combined treatments to further improve the prognosis of patients with BRAFV600E mutation [24].

Yet another approach is to search for chromosome instabilities (CIN) that are present in cancer cells but not in normal cells [35]. Such instabilities are defined as an increase in the rate at which whole chromosomes or chromosomal fragments are gained or lost, typically resulting in aneuploidy or abnormal DNA content. Such changes have been observed in many cancers but are best understood in colorectal cancer [5, 21]. Two main strategies for targeting CIN in cancer are CIN-reducing and CIN-inducing approaches. The CIN-reducing approach aims to reduce the rate of CIN (by inhibiting abnormal processes), while the CIN-inducing approach aims to increase the level of chromosome missegregation and/or DNA damage to induce cell death [1].

## *1.3 The Role of Microenvironment in the Future Development of Precision Medicine*

All the approaches described above are focused on searching for single-cell-level targets. When successful, they enable precise targeting of specific cancer phenotype/genotype thus increasing treatment efficacy and making them less harmful for a patient. However, the main limiting factor of targeting dominant tumour cells only, is the high probability of developing drug resistance. It has been shown that the tumour microenvironment plays a major role in rapidly inducing drug resistance via a cascade of signalling events that transiently protect tumour cells from apoptosis induced by therapeutic chemicals [25]. The tumour microenvironment is a spatially and functionally complex network of the vasculature, stromal cells, Cancer Stem Cells (CSC), cancer-associated fibroblasts (CAF), and tumour-associated macrophages (TAM). In the following paragraphs, we will briefly outline each of the main microenvironment constituents and highlight their importance in determining the clinical outcome of treatment and the development of treatment resistance.

To have a lasting effect a drug should reach close to 100% of tumour cells [8]. To achieve that, once it leaves blood vessels a drug should penetrate as far as possible through a tumour tissue [33, 34]. Diffusion depth depends on both the chemical properties of a drug and microenvironment properties (density of extracellular matrix, tissue pressure, interstitial fluid pressure). Moreover, tissue oxygenation strongly affects the response of tumour cells to an applied drug [41]. The maximum diffusion distance of oxygen from blood vessels is about 100 microns and cell necrosis is observable at distances of 150 microns or more from the areas supplied by blood vessels [37]. Therefore, vasculature density is strongly associated with clinical outcomes in a number of cancers [14, 17, 30].

Cancer stem cells (CSC) have been increasingly recognized as the main reason for tumour relapse and metastasis [26]. This tumour cell subpopulation is more resistant than differentiated cancer cells to most of the conventional anticancer therapies, antimitotic agents, or radiation. They also can differentiate into tumour cells of various phenotypes and are regulated by a variety of processes including Notch, Hedgehog, NF-kB, Wnt and TGF-beta pathways [9].

CAFs maintain the structural framework of a tumour and, in contrast to normal fibroblasts, have increased proliferation, enhanced extracellular matrix production and unique cytokine secretion [27]. By secreting a variety of active factors, they modulate cancer metastasis through synthesis and remodelling of the extracellular matrix (ECM), and influence angiogenesis, tumour mechanics, drug access and therapy responses [28]. Like other elements of the tumour microenvironment, CAF number and function strongly influence treatment outcomes [2, 13].

TAM's promotion of resistance is mainly due to the secretion of a variety of cytokines that induce anti-apoptotic programs in cancer cells and stimulate tumour cell proliferation [4]. Also, activated TAMs promote metastasis by producing soluble factors [40] and can release the angiogenic molecules and express a series of enzymes involved in the regulation of angiogenesis [4].

## *1.4 Modelling in Precision Medicine*

Mathematical and computational modeling and simulations are getting more and more important in natural sciences and medical domains. The main aim is to develop and use different efficient algorithms, data structures, and visualization techniques for user-friendly and explainable communication tools for human-computer interaction. Being based on different mathematical theories and applications of ICT tools, approaches to models can take different routes, like dynamical, statistical, differential equations, game-theoretic, and so on. However, sometimes it is difficult to make a clear classification especially when modeling approaches overlap or combine producing a variety of models/systems/tools. Computer modeling and simulation of biological and medical systems encompass works with cellular subsystems, comprehensive data analyses, 2-D and 3-D visualization of the complex connections of cellular processes, and so on.

In this chapter, the focus is on tumour and drug design. Tumours are heterogeneous cellular entities, and their growth is influenced by changing microenvironment and dynamical interactions between cancerous and healthy cells. Such interactions influence changes in cell phenotypic behaviors, like proliferation, apoptosis, and migration. To obtain reliable experimental results (in spite of the fact that it enormously increases computational complexity) it is necessary to consider different spatial and temporal scales. Until recently most computational tumour models were focused on scale-specific models. As tumour growth span multiple scales, it is necessary to apply another computational approach and include modeling across different biological scales (like some of the biological spatial scales: atomic, molecular, microscopic (tissue/multicellular), and macroscopic (organ) scales). Usually, a model is considered multiscalar if considers at least two spatial scales and/or includes processes with at least two temporal scales. *"Simulating cancer behavior across multiple biological scales in space and time is increasingly being recognized as a powerful tool to refine hypotheses, focus experiments, and enable more accurate predictions."* [7].

There is still no clear and specific classification of multiscale models, but since in this chapter we are concentrated on tumour and drug design, we will briefly consider several types of models appropriate for cancer diseases according to [7].

Discrete (individual-based) modeling is based on defining a set of rules for each cell. Individual cells are represented in space and time and the simulation process tracks and updates their internal states. Discrete modeling is an adequate approach for studying carcinogenesis, genetic instability, and cell-cell and cell-matrix interactions.

For modeling larger-scale systems, a better solution is the approach where tumour tissue is presented as a continuum medium instead of individual cells. This approach is known as continuum modeling (population-based) because model variables (like cell volume fractions, nutrient, oxygen, growth factors) are described as continuous fields usually using PDE equations. However, such models cannot be used to examine individual cell dynamics and discrete events.

An approach that can mitigate shortcomings of both continuous and discrete approaches is so-called hybrid modeling. This approach assumes the coupling of a continuous model with a discrete one. Hybrid modeling assumes integration and the strengths of both continuum and discrete models i.e. it assumes interaction or coupling between at least two models not based on the same formalism. Such models can couple biological phenomena from the molecular and cellular scales to those at the tumour scale. There are different definitions of hybrid modeling but generally speaking, two categories can be distinguished: composite and adaptive hybrid modeling [7]. Applications of hybrid models are numerous like: biological networks modeling, cancer growth modeling, cancer immunology and so on [32].

A tumour is in dynamic communication with the microenvironment (that shapes the cell-level behavior) through biochemical and biophysical processes. Therefore, the growth of tumour cannot be observed as an isolated process. It must be considered as an integral part of a dynamical biochemical and biophysical environment i.e. as a 3-D multicellular system [23]. According to [23] to model and perform simulations for different aspects of cancer, a 3-D multicellular simulation platform should "simulate the birth, death, and motion of tumour cells; simulate biochemical microenvironments with multiple diffusing substrates; simulate the biomechanics of cells and the extracellular matrix; simulate the evolving blood vasculature; simulate interstitial and microvascular flow; integrate the above models, along with molecular-scale models to drive cell phenotype; integrate high-throughput experimental data to calibrate and validate models; do so reproducibly, using interoperable data formats".

A recent and very important trend in modeling and simulation in medical domains is oriented towards a specific field in precision medicine, pharmacometrics. To optimize the treatment regimens and the design of clinical trials, personalized pharmacokinetic/pharmacodynamic (PK/PD) modeling and simulations are utilized. In pharmacokinetic the goal is to find out the relation between doses and concentration while the goal of pharmacodynamic is to find out the relation between concentration and effects that produce for an individual patient. Multimodel approaches are also very welcomed in this area where several models are simultaneously fitted predominantly based on how well they fit the data [10].

Some researchers have gone a step ahead in pharmacometrics technology including Pharmacokinetic-Pharmacodynamic/Toxicodynamic (PK-PD/TD) modeling and simulation [19]. They also work on interactions between antiemetic and anticancer drugs in order to achieve an appropriate dosing scheme and a good balance between maximum drug efficacy and minimum toxicity for individual patients. Mentioned approaches are based on inter and multidisciplinary methods that, in addition to pharmacometrics also include biomathematics, pharmacology, and several ICT disciplines with expectations to evaluate the effect of individual patient factors on drug exposure and different doses of drugs.

In this chapter, we present the results of experiments performed using EvoNano PhysiCell that represents the extended version of PhysiCell system. Original PhysiCell system [12] is a physics-based 3-D multicellular systems simulator realized as a general-purpose toolkit. Multicellular systems are used in different domains including cancer (metastasis, growth, drug design for tumour cells). To achieve good results during simulation it is important to observe how individual cells grow, divide, interact, move, and die. Tissue-scale dynamics should be studied in the microenvironment as usually cells are affected by biochemical and biophysical signals. So, reliable simulation for multicellular systems should include: tissue microenvironments (the "stage") with multiple diffusing chemical signals (oxygen, drugs, and so on); and the dynamics tissue microenvironments i.e. many interacting cells (the "players" upon the stage). *It builds upon a multi-substrate biotransport solver to link cell phenotype to multiple diffusing substrates and signaling factors. It includes biologically-driven sub-models for cell cycling, apoptosis, necrosis, solid and fluid volume changes, mechanics, and motility "out of the box"* [12].

PhysiCell is a powerful framework realized in a modular manner allowing a wide range of users to extend, rewrite, or even replace its originally implemented functions. The system also has been parallelized and supports the dynamics and interactions of even millions of cells in 3-D microenvironments, with microenvironment-dependent phenotypes.

# 2 PhysiCell Based Simulator for Precision Medicine

This section will give an overview of new concepts introduced in the original PhysiCell which are required for a high-quality precision medicine simulation. These concepts include several types of cells involved in the simulation (cancer cells, cancer stem cells, healthy cells, vascular cells, and cancer-associated fibroblast (CAF) cells) and several substrates which are added into PhysiCell microenvironment (angiogenic factor, prostaglandin, cytokine, drug substrate). The first subsection will give a high-level overview of these concepts by providing explanations about processes that take place in the simulation and by explaining all necessary parameters. The second subsection will provide low-level overview with important implementation details.

## 2.1 Functionalities and Features

This section will provide an overview of introduced cell types and substrates in the order in which they are implemented. This is important since some later concepts require the presence of particular initial concepts.

*Cancer and cancer stem cells*

Cancer stem cells (CSC) are the first types of cells introduced in EvoNano PhysiCell extension. Together with Differentiated cancer cells (DCC) they are the main representatives of the potential tumour tissue. These two cell types share common behaviour, while some parameters (e.g. for apoptosis, necrosis, division) are different. CSCs are generally considered more resistant regarding death rate, division, resistance to drugs, etc.

DCC are part of the original PhysiCell simulator and they are always simulated in our simulator, while the simulation of CSC is controlled through parameter `cancer_stem_cells_enabled` of the Boolean type. The initial number of DCCs in the simulation is given in the parameter `random_cancer_cells`. The general behaviour of these cells is controlled through several simulation parameters:

- `random_cancer_cells` which defines the number of DCCs in the initial simulation set-up.
- `cancer_cell_persistence_time` and `cancer_stem_cell_persistence__time` define the persistence time for DCCs and CSCs respectively. For both cell types, this value is set to 15 minutes.
- `cancer_cell_migration_speed` and `cancer_stem_cell_migration_speed` define the migration velocity for DCCs and CSCs respectively. For both cell types, this value is set to 0.25 micrometers per minute.
- `cancer_cell_relative_adhesion` and `cancer_stem_cell_relative_adhesion` define the value of relative adhesion for DCCs and CSCs respesctively. This value is interpreted as the force for resistance to deformation and/or volume exclusion [29]. For both cell types, this parameter value is set to 0.05 (dimensionless).

- cancer_cell_apoptosis_rate and cancer_stem_cell_apoptosis_rate define the rate for entering the cell into apoptosis (programmed cell death). The probability of entering in the apoptosis death state is proportional to this parameter and to the delta time [12]. CSC dies very rarely at least via apoptosis death model, so the value of this parameter is set to zero (1/minute). The value for DCC is set to 4.065e-5.

The process of DCC and CSC division is more complicated and it is controlled through several parameters which will be described below. One DCC can be divided into two DCCs or into one DCC and one CSC (with some probability). One CSC can be divided into: two CSC, two DCC, and one CSC and one DCC (all outcomes with defined probability). This process of division is controlled through the following simulation parameters:

- cancer_cell_division_rate and cancer_stem_cell_division_rate define the division rate for DCCs and CSCs respectively. This value for both types of cells is set to 0.005 (dimensionless).
- CSC_probability_CC_division defines the probability for making one CSC in DCC division. This probability is set to value 0.05 (5%). Consequently, one DCC will be divided in two DCCs in 95% of cases.
- symmetric_division_probability defines the probability of splitting one CSC into two cells of the same type (two DCC or two CSC). This value is set to to 0.01 (1%) so in 99% of divisions one CSC will be divided in one DCC and one CSC.
- symmetric_division_double_CC_probability defines the probability of dividing one CSC into two DCCs when symmetric division is selected (previous item). This value is set to 0.01 (1%) so in 99% of symmetric divisions one CSC will be divided into two CSCs. Since the probability of symmetric division is 1%, and that under these circumstances the probability of creating two DCCs from one CSC is 1%, overall the probability of creating two DCCs from one CSC is just 0.01%.

The parameters cancer_stem_cell_o2_necrosis_threshold and cancer_stem_cell_o2_necrosis_max are defined only for the CSCs, and they are used to simulate higher resistance of CSCs to the lack of oxygen and for the entering necrosis process. When the level of oxygen is lower than o2_necrosis_max (PhysiCell parameter) the cell immediately becomes necrotic. When the level of oxigen is between o2_necrosis_max and o2_necrosis_threshold (also PhysiCell parameter) the necrotic process is started according to stochastic necrosis model [12]. For the values of oxygen higher than o2_necrosis_threshold, there is no necrosis. The default PhysiCell values which are also applied for DCCs are o2_necrosis_threshold=5.0 and o2_necrosis_max=2.5. For the CSCs, these values are significantly lower which simulate higher resistance to low oxygen values: cancer_stem_cell_o2_necrosis_threshold=0.1 and cancer_stem_cell_o2_necrosis_max=0.01.

There are still some more parameters that refer to DCCs and CSCs that are connected to the other concepts (vascularity, drug diffusion etc.) and they will be described in corresponding sub-sections.

*Healthy cells*
Healthy cells are introduced in EvoNano PhysiCell extension to simulate the impact of different drugs on healthy tissue. These cells are not so important for the mechanism of tumour growth and dynamics so many aspects of these cells are turned off. Healthy cells are immobile and they do not divide nor die, either via apoptosis or necrosis.

The presence of healthy cells in the simulation is controlled via parameter `healthy_cells_active` of type Boolean. Healthy cells are randomly distributed throughout the initial space. The number of healthy cells in the initial space is controlled via parameter `random_healthy_cells` (dimensionless). When the initial space needs expansion, due to the tumour growth, the number of healthy cells is controlled with parameter `random_healthy_cells_expansion_factor` (dimensionless).

*Vascular cells and an angiogenic factor*
This mechanism simulates the growth of vascular networks within and around tumour tissue which ensures an additional supply of oxygen to the growing tumour. An angiogenic factor and vascular cells mechanism can be turned on/off with parameter `vascularity_active` of type Boolean. When turned on, this mechanism introduces a new substrate into simulation: an angiogenic factor (AF). In that case, all cancer cells secrete AF which represents an activation signal to vascular cells. Vascular cells then try to improve the creation of a vascular network in the direction of a higher concentration of AF. With such a mechanism, a growing tumour indicates the higher production of vascular cells, which results in a richer vascular network, which as a consequence provides a higher supply of oxygen to tumour tissue.

The main parameters (diffusion coefficient and decay rate) for AF are taken from [11]. The amount of AF which is secreted from one cancer cell is given in the parameter `angiogen_in_cancer_cell`.

Vascular cells are initially placed randomly in the simulation space. The number of the vascular cells in the initial space is specified by a parameter `vascular_seed_points` (current value is 1). Afterward, they are dividing in the direction of high AF density if the concentration of AF is above the defined threshold value. This threshold is given by parameter `critical_angiogen_to_expand_vascularity` (current value is set to 0.055 according to [39]). In such a way, we simulate the growth of capillaries and blood supply of tumour. However, not all vascular cells in whose environment the concentration of AF is higher than the threshold value divide because that would cause the exponential growth of the vascular network. To prevent this exponential growth we introduced the probability of dividing vascular cells in the presence of high AF concentration. This value is specified in the parameter `vascular_cell_expansion_probability`, the current value is 0.01 which is determined empirically.

When the concentration of AF is too high (higher than parameter critical_angiogen_to_stop_o2_secretion) the secretion of oxygen from vascular cells stops. With this mechanisms, we simulate the death of vascular cells which happen in the deep interior of the tumour as a result of tumour necrosis (necrotic core).

In the situation when the simulation space should be expanded, the expansion of vascular network is controlled with parameter expand_vascularity_when__expanding_space (Boolean parameter, default value is true). The quantity of this expansion is controlled through expansion_scale_factor parameter (current value is 1.75, empirically adjusted).

When the functionality of vascular cells is defined, the process of metastasis can be also introduced. In metastasis, cancer cells break away from the original (primary) tumour, travel through the vascular system, and form a new tumour in other organs or tissues of the body. Therefore, we introduce two more parameters which control the process of detachment of CSC: CSC_detachment_probability and CSC_detachment_critical_proximity. The first parameter defines the probability of CSC which is near some vascular cell to detach from the original tumour and enter the vascular system. The threshold distance which defines what is near is defined with the second parameter. Until now, we have not experimented with the metastasis process yet, so these parameters are set to the following values: 0.0 and 1000000, respectively.

*A prostaglandin*

Prostaglandins are a family of signaling molecules that regulate the invasive behaviour of cancer cells. Instead of implementing all members of the family as separate substrates, we implement an abstract prostaglandin substrate that regulates DCC-CSC conversion. In our implementation, a prostaglandin is secreted from cells when a cell dies either by apoptosis or necrosis. The simulation of this substrate is turned off/on with Boolean parameter prostaglandin_active. The parameters for diffusion and decay rate are taken as for substrate cytokine (described below) [18] since there are no reliable experimental data for prostaglandins.

The amount of a prostaglandin that is secreted from a cell is controlled via parameter prostaglandin_in_cancer_cell. This value is set to 8.8e-7.

High values of prostaglandin trigger the conversion of some DCCs to CSCs. The DCC->CSC conversion is triggered when the concentration of prostaglandin in substrate exceeds the value of parameter cancer_cell_prostaglandin_thr. Of course, only a small amount of DCCs is converted into CSC in the presence of high prostaglandin values. The quantity of conversions is controlled with parameter prostaglandin_conversion_probability which represents the probability of DCC->CSC conversion in the presence of prostaglandin. This parameter is empirically determined to value 0.00008, in order to obtain expected tumour growth and expected number of CSCs. Also, the number of prostaglandin induced DCC->CSC conversion is tracked within the simulation as an important indicator, and it is reported in the log file.

*CAF cells and cytokine*
Cancer-Associated Fibroblasts (CAFs) are special types of tumour cells that are responsible for providing physical support for other tumour cells and for secreting cytokines [22]. These cells appear near vascular cells and it is important to preserve the expected ratio of vascular and CAF cells. This ratio is defined with parameter `CAF_vascular_ratio` (value set to 0.6, determined empirically). The simulation of these cells and accompanying substrate cytokine is turned off/on with Boolean parameter `CAF_active`.

The division of CAF cells is turned off since it is observed that they appear in a random manner near the vascular network. We simulate the same behaviour in our model. Also, the apoptosis is turned off since CAF cells are unable to undergo apoptosis. The necrosis still exists although these cells are more resistant to low oxygen levels than DCCs (while CSCs are more resistant than CAF cells). The oxygen values which indicate necrosis are defined with parameters `caf_o2_necrosis_threshold` and `caf_o2_necrosis_max` in the same fashion as with CSCs. These values are set to values 1 and 0.1 respectively, to simulate oxygen resistance between DCCs and CSCs.

As already mentioned, CAF cells secrete cytokine, another substrate that is simulated. Te cytokine diffusion coefficient and decay rate parameters are taken from [18]. The amount of cytokine which is secreted from one CAF cell can be adjusted through parameter `cytokine_in_CAF`. This value is set up on 8.8e-7 according to [18].

The cytokine affects neighboring cells in a similar manner as prostaglandin. In the presence of high cytokine levels, some DCCs convert to CSCs. The critical cytokine concentration which triggers DCC->CSC conversion is defined in parameter `cancer_cell_cytokine_resistance`. The probability of DCC->CSC conversion in the presence of high cytokine values is given in parameter `cytokine_conversion_probability`. The values of these two parameters are set up on values 1e-12 and 0.01, after extensive experiments whose goal was to replicate the desired tumour growth and desired cell type ratios.

CAF cells have one more parameter: `o2_stop_cytokine_threshold`. This parameter defines the lowest value of oxygen for which the CAF cell is capable to secrete cytokine. Bellow this oxygen value (current value is 10) CAF cell does not secrete cytokine, although it does not die (oxygen levels for necrosis are much lower: 0.1 and 1). If oxygen values later return to values higher than this threshold CAF cell "wakes up" and secretes cytokine again.

*Drug diffusion mechanism*
The mechanism of drug administration is supported on the level of the substrate. We simulate nanoparticle-based cancer therapies by simulating functionalized nanoparticles (NP), i.e. the NPs with an attached particular drug. That drug can affect tumour in different ways: CSC and DCC death, stopping the division of cancer cells, or destroying tumour vascular network. In the current simulator, we support only drugs that cause cancer cell death, but the other mechanisms will be also supported in the future.

**Table 1** NP resistance values for different cell types

| | |
|---|---|
| healthy_cell_NP_resistance | 0.7 |
| cancer_cell_NP_resistance_lower_bound | 0.3 |
| cancer_cell_NP_resistance_upper_bound | 0.5 |
| cancer_stem_cell_NP_resistance_lower_bound | 0.5 |
| cancer_stem_cell_NP_resistance_upper_bound | 0.7 |

The simulation of the drug mechanism can be turned off/on via the Boolean parameter `NP_active`. NP substrate is secreted from vascular cells similarly to oxygen. The amount of secreted NPs from one vascular cell can be controlled through parameter `NP_in_vascular_cell`. Since the administration of the drug does not start at the same time as the tumour creation but only after a tumour is detected and analyzed, we simulate this behaviour with parameter `cancer_cells_to_activate_NP`, which specifies the number of cancer cells when the drug administration starts.

Secreted NP has an influence on all present cells. In the current version, if the concentration of NP exceeds some threshold value, the cell dies by apoptosis. The threshold values are different for all cells since different types of cells have different resistance to a particular drug. These threshold values are controlled by several parameters. These parameters and their current values are given in Table 1. All healthy cells have the same value of NP resistance. On the other hand, each DCC and CSC acquires a unique value from the interval `[lower_bound..upper_bound]`.

Figure 1 illustrates all previously introduced concepts. It shows the state of the tumour after 15 simulation days. This condition is obtained using previously described parameters and their values which are mentioned in the text. The cell types have the following colors:

- grey: DCC
- blue: CSC
- maroon: Vascular cell
- yellow: CAF cell
- orange: Necrotic cell

Moreover, the empty space in the middle of tumour represents necrotic core, but the old necrotic cells are removed from the simulation due to efficiency issues.

## 2.2 *EvoNano PhysiCell Implementation*

The implementation of the EvoNano PhysiCell simulator for precision medicine is organized into several custom C/C++ modules (modules put in `custom_modules` folder of PhysiCell) and the main module realizing necessary initialization and the main loop of the simulator. The custom modules are:

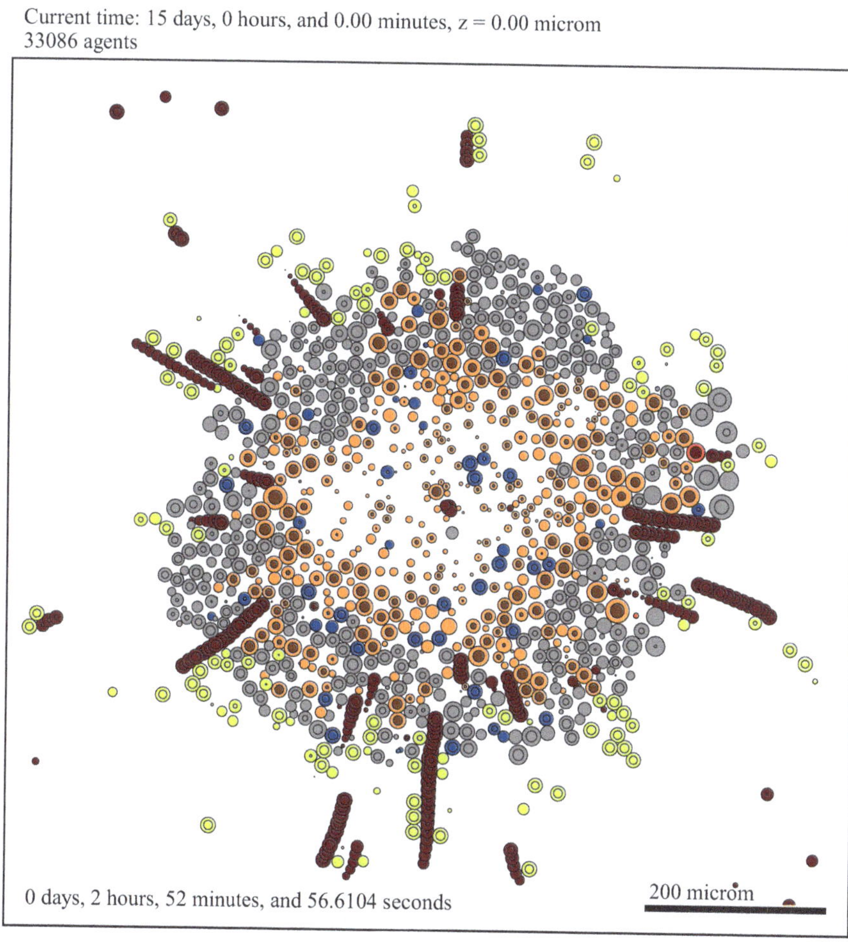


**Fig. 1** An example of tumour after 15 simulation days

1. `cancc` that contains functionalities related to cancer cells supported by the simulator,
2. `csvlog` that defines functions for emitting log messages in CSV (comma separated values) log files,
3. `evo_nano_3D` that defines functions for creating cell types, initializing cell definitions, initializing the microenvironment and tissue and realizing specific rules for cancer cell creation and division,
4. `healthyc` that implements normal (non-cancer, healthy) cells,
5. `spexp` that implements the microenviroment with dynamic space expansion capabilities,
6. `utils` that contains various utility functions for sampling random numbers,

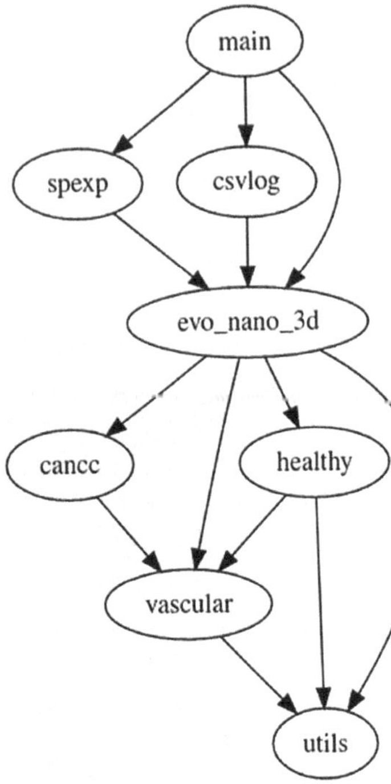


**Fig. 2** Dependencies between modules of the EvoNano PhysiCell simulator

7. `vascular` that implements functionalities related to vascular cells and vascular network growth.

As usually for C/C++ modules, each module of our simulator is realized in two files, i.e. one header file exposing the interface of the module to other modules and one cpp file containing the full implementation of the specified functionalities. Figure 2 shows dependencies between the modules of the simulator including also the main module.

*Utils module*
The `utils` module is located at the bottom of the module dependency graph and it is the only module that does not depend on other EvoNano PhysiCell modules. This module implements two routines for sampling real-valued random numbers from uniform distributions: one for sampling a real number from the interval $[A, B]$ and one for sampling a real number from $[A, B]$ with a given probability $p$ or from the interval $[C, D]$ with 1 - $p$ probability ($A$, $B$, $C$, $D$ and $p$ are input parameters). This module also defines two methods returning the borders of the simulation space inside which are sampled coordinates of:

1. initially created cancer and health cells (cells that are formed when the simulation starts), and
2. healthy and vascular cells after extending the simulation space.

*Vascular module*

In EvoNano PhysiCell, blood vessels are implemented as chains of vascular cells. This means that one capillary is composed of adjoined vascular cells, while the complete vascular network is a set of capillaries (which are not necessarily mutually inter-connected). The `vascular` module defines various routines related to vascular cells and their functions. This module defines 4 important global variables:

1. `bool vascularity_active`—a boolean variable indicating whether the vascular network is simulated,
2. `bool NP_simul`—a boolean variable indicating whether NP (nano-particles or drugs) is simulated,
3. `bool NP_active`—a boolean variable indicating whether NP diffusion is active when NP is simulated (vascular cells may not start diffusing NP immediately when the simulation starts, but after some time specified in the configuration file; NP diffusion may also stop and later resume in more complex simulation scenarios), and
4. `std::vector<Cell*>* vascular_cells`—a pointer to vector containing pointers to all vascular cells (each cell PhysiCell is an instance of the `Cell` class).

The module exposes the following functions to other EvoNano PhysiCell modules:

1. `void init_vascular_network()`—creates the initial vasculature when the simulation starts,
2. `void expand_vascular_network(double sm)`—creates new vascular cells in a new space region after the simulation space dynamically expands (`sm` is the seed multiplier),
3. `void init_vascular_cells()`—initializes vascularity configuration parameters and creates a PhysiCell cell definition object for vascular cells,
4. `Cell* create_vascular_cell(double x, double y, double z)`—creates a new vascular cell at the given position $(x, y, z)$ and returns the pointer to it, and
5. `void divide_vascular_cell(Cell* c)`—"divides" the vascular cell $c$ by creating a new vascular cell in the immediate neighborhood of $c$.

The initial vascular network is formed by creating $k$ vascular seed cells at the beginning of the simulation, where $k$ is one of the EvoNano PhysiCell configuration parameters. The coordinates of each seed cell are randomly sampled. The total amount of $k\dot{s}$ vascular seed cells are also created when the simulation space is expanded, where $s$ is the seed multiplier. Their coordinates are also randomly sampled but considering only the newly created regions of space.

Vascular cells are defined as non-divisible, immortal, and non-motile cells of constant volume as shown by the program code fragment given below. Two custom variables are also associated to each vascular cell: `num_expansions` and `flagged_for_division`. The value of the first variable reflects how many times a vascular cell was "divided". The second variable indicates whether the cell is designated to be "divided" in the next simulation cycle. The vascular cell can be "divided" exactly once, i.e. if num_expansions is equal to 0 then the cell can be flagged for division and after division num_expansions is increased by 1.

**Program code fragment: initialization of vascular cell definition**

```
Cell_Definition vascular_cell_def;

void create_vascular_cell_definition() {
    ...
    vascular_cell_def = cell_defaults;

    // turn off motility
    vascular_cell_def.phenotype.motility.is_motile = false;

    // turn off cell division
    int s = live.find_phase_index(PhysiCell_constants::live);
    int e = live.find_phase_index(PhysiCell_constants::live);
    vascular_cell_def.phenotype.
        cycle.data.transition_rate(s, e) = 0.0;

    // turn off cell death
    int apoptosis = cell_defaults.phenotype.
        death.find_death_model_index("Apoptosis");
    vascular_cell_def.phenotype.death.rates[apoptosis] = 0.0;
    int necrosis = cell_defaults.phenotype.
        death.find_death_model_index("Necrosis");
    vascular_cell_def.phenotype.death.rates[necrosis] = 0.0;

    // turn off volume updates
    vascular_cell_def.functions.volume_update_function = NULL;

    // custom variables
    vascular_cell_def.custom_data.add_variable
        ("num_expansions", "dimensionless", 0);
    vascular_cell_def.custom_data.add_variable
        ("flagged_for_division", "dimensionless", 0);

    // set function performing phenotype updates
    vascular_cell_def.functions.update_phenotype =
        vascular_cell_update_phenotype;
    ...
}
```

The function for creating vascular cells reduces to the PhysiCell function for creating cells (also called PhysiCell agents) with the vascular cell definition defined in the `vascular` module. This function also adds the newly created vascular cell to the vector of all vascular cells.

The function for updating the phenotype of vascular cells defines how vascular cells react to the angiogen. Let $a$ denotes the concentration of the angiogen near a vascular cell $v$. The impact of $a$ to $v$ is defined by the following rules implemented in the function for handling the angiogen:

- If $a$ is higher than the angiogen threshold to stop oxygen secretion then $c$ immediately stops releasing oxygen. When $a$ drops below this threshold then $c$ reactivates its oxygen release.
- If $a$ is higher than the critical angiogen threshold to expand the vascular network and $c$ was not previously expanded then it is considered as the candidate for "division". The cell is marked for division with a certain probability that is one of EvoNano PhysiCell parameters controlling vascular network growth.

The function for phenotype updates of vascular cells also handles NP. If NP is simulated and active then vascular cells release a certain concentration of NP into the environment.

As already emphasized, the division of a vascular cell is actually the creation of a new vascular cell. The new cell is created to touch the cell marked for division in the opposite direction of the angiogen gradient. The function for dividing vascular cells also increases `num_expansions` variable to prevent further divisions and ensure that vascular cells form a chain-like structure.

*Healthy module*

Healthy (normal, non-cancer) cells are implemented in the `healthy` module. In reality, normal cells are divisible and mortal. However, their division rates are negligible compared to cancer cells. Thus, normal cells are in EvoNano PhysiCell implemented as non-divisible cells. Additionally, a normal cell can die by entering apoptosis only if it is exposed to a high concentration of NP.

The module defines the following two global variables:

- `bool healthy_cells_active`—a boolean flag indicating whether healthy cells are simulated, and
- `int NP_HC_deaths`—the total number of normal cell deaths due to critical exposure to NP.

The public functionalities of the module are given by the following functions:

- `void init_healthy_cells()`—initializes configuration parameters and creates the definition object for healthy cells.
- `Cell* create_healthy_cell(double x, double y, double z)`—creates and returns a new healthy cell at given coordinates.
- `void expand_healthy_cells(int m)`—creates new healthy cells in the new region of space after the simulation space is dynamically expanded ($m$ is a

multiplier factor, i.e. the total number of newly created normal cells is $k\dot{m}$, where $k$ is the number of normal cells created at the start of the simulation).

The function defining how healthy cells react to NP is called by the function performing phenotype updates at regular time intervals. This function checks whether the concentration of the NP near a healthy cell is higher than critical. If it is then the apoptotic death of the cell is activated (see the program code fragment given below).

**Program code fragment: function for handling NP by healthy cells**

```
void healthy_cell_handle_NP(Cell* c) {
    // is NP simulated?
    if (!NP_simul)
        return;

    // is NP active?
    if (!NP_active)
        return;

    int NP = microenvironment.find_density_index("NP");
    double NP_near = microenvironment.
        nearest_density_vector(c->position)[NP];

    if (NP_near > HC_NP_resistance) {
        int apoptosis =  cell_defaults.phenotype.
            death.find_death_model_index("Apoptosis");
        c->start_death(apoptosis);
        ++NP_HC_deaths;
    }
}
```

*Cancc module*

The functionalities of three different types of cancer cells, DCC, CSC, and CAF, are implemented in the `cancc` module. This module defines 6 global variables. The first three global variables are boolean flags that can be set to enable or disable certain functionalities related to cancer cells:

1. `bool cancer_stem_cells_enabled`—CSC cells are simulated if this flag is set to true, otherwise CSC cells are not created during the simulation.
2. `bool prostaglandin_simul`—if this flag is set to true then DCC and CSC cells release prostaglandin during their death cycle.
3. `bool cytokine_caf_simul`—this boolean flag indicates whether CAF cancer cells releasing cytokine are simulated or not.

The next three global variables are counters:

1. `NP_CC_deaths`—the number of cancer cell deaths caused by NP.
2. `prostaglandin_conversions`—the number of DCC converted to CSC due to critical exposure to prostaglandin.
3. `cytokine_conversions`—the number of DCC converted to CSC due to critical exposure to cytokine.

The module exports the following functions:

- `void init_cancer_cells()`—initializes configuration parameters related to cancer cells and creates definition objects for all three cancer cell types.
- `Cell* create_cancer_cell(double x, double y, double z, bool s)`—creates a new DCC cell (if $s$ is false) or a new CSC cell (if $s$ is true) at the given position.
- `Cell* create_CAF_cell(double x, double y, double z)`—creates a new CAF cell at the given position.
- `void divide_cancer_cell(Cell* c)`—divides a DCC cell into two cancer cells.
- `void divide_cancer_stem_cell(Cell* c)`—divides a CSC cell into two cancer cells.

The function `init_cancer_cells` creates three different instances of the `Cell_Definition` class, one object per cancer cell type. These objects reflect the main differences between different cancer cell types, i.e. each cell type defines its cell division rate, motility parameters, adhesion strengths, apoptosis rates, oxygen thresholds to enter necrosis, secretion, and uptake parameters for relevant substrates (all those parameters are specified in the configuration file). Two custom data variables are associated with each cancer cell:

- `NP_resistance`—the critical concentration of NP to start apoptotic cell death.
- `NP_prev`—the concentration of NP near the cell in the previous iteration of the simulation. This variable enables quantifying temporal change of NP near the cell and to adjust cell division rate accordingly.

All cancer cells update their base phenotype parameters according to the oxygen phenotype update model defined by PhysiCell. However, each cancer cell type has its own phenotype update function that additionally defines how cancer cells react with substrates and release them. The `cancc` module implements customized phenotype update functions relying on the standard oxygen phenotype update model that additionally defines how cancer cells release and interact with substrates. The phenotype update function for DCC is shown in the code inset given below.

**Program code fragment: phenotype update function for DCC.**

```
void cancer_cell_phenotype_update(
    Cell* c, Phenotype& phenotype, double dt)
{
    // dying cancer cells release prostaglandin
    if (c->phenotype.death.dead && prostaglandin_simul) {
        int p = microenvironment.
            find_density_index("prostaglandin");
        microenvironment.
            nearest_density_vector(c->position)[p] =
            PROSTAGLANDIN_IN_CANCER;

        return;
    }

    // standard oxygen-based phenotype update model
    update_cell_and_death_parameters_O2_based(
        c, phenotype, dt);

    // phenotype updates based on NP
    cancer_cell_handle_NP(c);

    if (prostaglandin_simul) {
        // check for prostaglandin-based DCC->CSC conversion
        cancer_cell_handle_prostaglandin(c);
    }

    if (cytokine_caf_simul) {
        // check for cytokine-based DCC->CSC conversion
        cancer_cell_handle_cytokine(c);
    }
}
```

The function for handling NP operates according to the following rules. If the concentration of NP near the cancer cell is higher than the NP resistance threshold for that cell then the cell enters apoptotic death. The concentration of NP near the cancer cell also affect the cell division rate: if it is close to zero then the cell division rate is returned to the original value from the configuration file, otherwise, the cell division rate is divided by the increase of NP concentration (i.e., more NP implies slower division of cancer cells).

Functions for handling prostaglandin and cytokine are structurally similar. If the concentration of prostaglandin (resp. cytokine) is higher than critical then the DCC cell is converted into a CSC cell with a given probability that controls the conversion rate.

**Program code fragment: function for handling prostaglandin.**

```
void cancer_cell_handle_prostaglandin(Cell* c) {
    int p = microenvironment.
        find_density_index("prostaglandin");
    double pnear = microenvironment.
        nearest_density_vector(c->position)[p];

    double prob = UniformRandom();
    if (prostaglandin_near > cc_prostaglandin_thr &&
        prob <= prost_conv_prob)
    {
        // DCC -> CSC conversion
        c->convert_to_cell_definition(cancer_stem_cell_def);
        prostaglandin_conversions++;
    }
}
```

Cancer cells have heterogeneous resistances to NP. This means the resistance to NP is randomly sampled from a uniform distribution where each cell type has its own upper and lower bounds of the distribution. The NP threshold above which a cell enters apoptotic death is determined upon cell creation as it can be seen in the following program code fragment.

**Program code fragment: function for creating DCC and CSC cells**

```
Cell* create_cancer_cell(
    double x, double y, double z, bool stem)
{
    Cell* c;

    if (stem)
        c = create_cell(cancer_stem_cell_def);
    else
        c = create_cell(cancer_cell_def);

    c->assign_position(x, y, z);
```

```
        // determine the NP resistance level
        if (NP_simul) {
            int ind = c->custom_data.
                find_variable_index("NP_resistance");

            double u = UniformRandom();
            if (stem)
                c->custom_data[ind] = CSC_NP_LOWER +
                    u * (CSC_NP_UPPER - CSC_NP_LOWER);
            else
                c->custom_data[ind] = CC_NP_LOWER +
                    u * (CC_NP_UPPER - CC_NP_LOWER);
        }

        return c;
    }
```

EvoNano PhysiCell functions for dividing cancer cells rely on the `divide()` method implemented in the PhysiCell `Cell` class. This method called on an arbitrary cell *c* returns a new cell *n* that is placed near *c*. The new cell *n* has the same characteristic as *c* and the volume of both cells is equal to half of the volume of *c* prior to division. The function for dividing DCC always creates a new DCC if CSC cells are not simulated. If CSC cells are simulated then a new DCC is converted to a CSC with a certain probability (see the program code fragment given below). The division of CSC cells is realized similarly by following appropriate rules for symmetric and asymmetric CSC division.

**Program code fragment: function for dividing DCC cells**

```
void divide_cancer_cell(Cell* c) {
    if (!cancer_stem_cells_enabled) {
        Cell* new_cell = c->divide();
        return;
    }

    double pCSC = UniformRandom();
    Cell* new_cell = c->divide();

    // CSCP: the probability to convert to a CSC
    if (pCSC <= CSCP) {
        new_cell->convert_to_cell_definition(
```

```
                cancer_stem_cell_def);
        }
    }
```

*Evo_nano_3d module*

The evo_nano_3d module implements the main rules of the simulator (cell division and detachment rules) and various setup functions that are called from the main module. This module also implements a function providing a periodic report about all cells and a custom cell coloring function for visualizations produced by PhysiCell. This module initializes the default cell definition object that is reused to build definitions of all supported cell types. The default cell definition object specifies the default cell cycle model and oxygen uptake and secretion parameters.

The module defines the following two setup functions:

- void setup_microenvironment() to initialize the BioFVM microenvironment (please recall that BioFVM is a library used by PhysiCell to simulate diffusion)
- void setup_tissue() to setup the vascular network and create initial DCC and healthy cells that are randomly placed in the environment. The number of initially created DCC and healthy cells is specified in the configuration file.

The function for making cell report iterates through the vector of all cells (this vector is maintained by PhysiCell) and makes the distribution of cells per cell type. This function also counts how many cells entered the death cycle.

The function defining EvoNano PhysiCell cell division rules is an umbrella function: it iterates through the vector of cells marked for division and depending on the cell type calls the appropriate cell division function defined in other EvoNano PhysiCell modules. Additionally, this function takes care of CAF cells by creating them near vascular cells as the vascular network grows such that the ratio of CAF to vascular cells never exceeds the bound specified in the configuration file.

**Program code fragment: function implementing cell division rules.**

```
void evonano_cell_division(
    Cell_Container* cell_container, double t)
{
    for (int i = 0;
          i < cell_container->cells_ready_to_divide.size();
          i++)
    {
        Cell* c = cell_container->cells_ready_to_divide[i];
        int type = c->type;
```

```
        if (type == CANCER_CELL_TYPE)
            divide_cancer_cell(c);
        else if (type == CANCER_STEM_CELL_TYPE)
            divide_cancer_stem_cell(c);
        else if (type == VASCULAR_CELL_TYPE) {
            divide_vascular_cell(c);
            if (cytokine_caf_simul)
                create_CAF_near_vascular(c);
        } else
            c->divide();
    }

    cell_container->cells_ready_to_divide.clear();
}
```

The module also defines the CSC detachment rule. Namely, if a CSC cell is close to some vascular cell then it can go to the vascular network with a certain probability. The number of detached CSC cells is a proxy indicator of the degree of tumour metastasis. Both the critical distance and the detachment probability are specified in the configuration file. Also, the user can disable CSC detachments.

*Spexp module*
The BioFVM environment in PhysiCell has constant dimensions. For a large tumour it is necessary to allocate large dimensions of the simulation space prior to starting the simulation. If the tumour evolves slowly then BioFVM most of the time simulates diffusion through empty space regions located far from actual cancer cells, which in turn may drastically slow down the simulation. Thus, we decided to make a general-purpose extension of PhysiCell that enables dynamic space expansion in run time. With this capability, it is possible to start the simulation in a relatively small space that is appropriately expanded as the tumour grows. Consequently, BioFVM does not simulate diffusion through empty space regions which significantly improves the execution time for large tumours.

Dynamic space expansion functionalities are implemented in the `spexp` module. This module defines two public functions called from the `main` module:

1. `bool check_conditions_to_expand()`
2. `void expand_microenvironment(double voxel_size)`

The first function checks if there is a cell close to the space borders. If there is a cell such that its distance to the closest border is smaller than some threshold then the first function returns true indicating that the space should be expanded by calling the second function. The distance threshold is one of the simulation parameters and it is specified in the configuration file.

The second function for expanding the microenvironment first makes its copy, i.e. the density of each substrate in each voxel is saved in a temporary 3D matrix. Then, a new bigger microenvironment is created and substrate densities are restored from the temporary matrix. The increase along all space axes is specified in the configuration file. The new microenvironment is registered to each cell by calling the appropriate method. Finally, all substrate gradient vectors and the BioFVM diffusion solver are reinitialized.

*Csvlog module*
The `csvlog` module is the simplest EvoNano PhysiCell module. This module logs the state of the simulation at regular time intervals by exporting relevant information for all cells (cell type, position, and volume) into a csv file. The concentration of oxygen near each active cell is also recorded.

*Main module*
The `main` module defines the main function which starts the simulation and implements the main simulation loop that executes for the specified maximum simulation time. Prior to entering the main simulation loop, all necessary initialization steps are performed:

1. the configuration file is parsed and loaded into an internal data structures,
2. the random number generator is initialized,
3. cell type definition objects are instantiated, and
4. the BioFVM microenvironment is created and populated with initially created cells.

The main loop takes care of events occurring at different time scales: diffusion of substrates (0.1 min scale), cell mechanics updates (1 min scale) and cell processes (10–100 min scale). Each iteration of the main loop corresponds to one event at the lowest time scale (substrate diffusion), whereas events at higher time scales are triggered by checking appropriate time step sizes. In each iteration of the main loop the following actions are performed:

1. BioFVM functions are executed to simulate the diffusion of the specified substrates and update the microenvrionment accordingly.
2. It is checked whether the simulation space should be expanded (by calling the function from the `spexp` module). If the criterion for expanding space is satisfied then the microenviroment, the vascular network, and the set of healthy (normal) cells are expanded.
3. It is checked whether NP should be activated in the case that it is simulated. The diffusion of NP activates either after the specified number of days or after the tumour reaches some critical mass.
4. If it is time for cell events then the position, volume, and phenotype of each cell is updated and the routines from the `evo_nano_3d` module implementing custom EvoNano cell division and detachment rules are called.
5. It is checked whether the simulation state should be logged and appropriate log operations are performed.

## 3 Results

The configuration parameters of EvoNano PhysiCell could be calibrated such that the growth of a simulated tumour is aligned with some theoretical model. To set parameters of the simulator corresponding to normal physiological conditions we use the Gompertzian growth model [31]. The Gompertzian growth is given by the following equation

$$V(t) = V_0 e^{\frac{\alpha}{\beta}(1-e^{-\beta t})},$$

where $V(t)$ is the total tumour volume at time $t$, $V_0$ is the initial tumour volume, and $\alpha$ and $\beta$ are constants regulating the growth rate. In our experiments we use $\alpha = 0.58$ and $\beta = 0.072$ according to the recent study by Vaghi et al. [36]. In our simulations, we always start from 100 regular cancer cells (DCCs) that are randomly placed within the initial simulation space of a small dimension ($x = y = z = 200$μm). Then, we simulate 30 days of tumour growth with dynamic space expansion capabilities described in the previous section. Since the volume of a single cell when crated by the simulator is 0.000002494 mm$^3$, the volume $V_0 = 0.0002494$ mm$^3$. Figure 3 shows the growth of a simulated tumour after calibrating the simulator's parameters according to the previously described Gompertzian growth model. The volume of the tumour is computed as the total volume of all cancer cells and it is averaged over 10 simulations. It can be seen that only in the first 10 days there are small deviations from the theoretical model and after that initial period the volume of the simulated tumour is almost identical to the predictions by the theoretical model.

The evolution of the total number of live cancer cells, dying cancer cells and vascular cells is shown in Fig. 4. One randomly place vascular cell is created at the beginning of each simulation in order to initialize the vascular network. This network

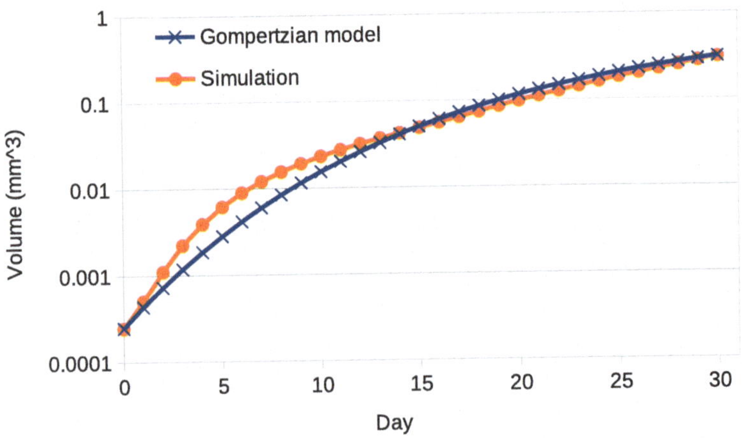


**Fig. 3** The daily evolution of the volume of a tumour simulated by EvoNano PhysiCell compared to the volume determined by the Gompertzian model

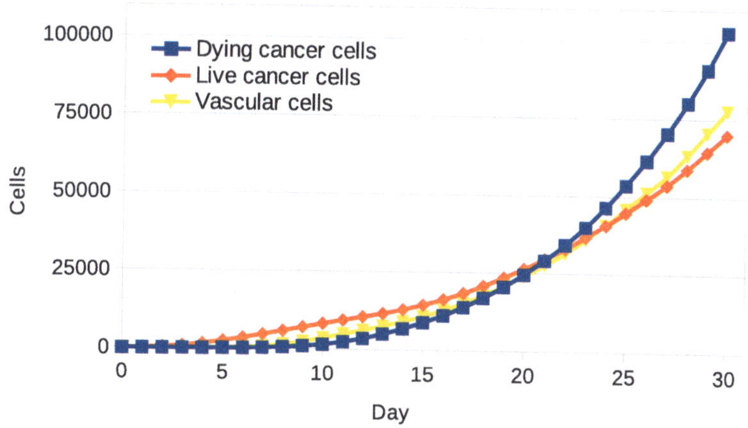


**Fig. 4** The daily evolution of the total number of live cancer cells, dying cancer cells and vascular cells in the simulated tumour

is then expanded according to nearby concentrations of the angiogen produced by cancer cells. It can be seen that the total number of vascular cells closely follows the total number of live cancer cells. On the other side, the total number of dying cancer cells is close to 0 up to day 10. On day 10, the necrotic core of the tumour emerges and the number of dying cancer cells grows at a faster rate than live cancer cells and vascular cancer cells.

Simulated cancer cells can enter either the necrotic or the apoptotic death cycle. Figure 5 shows the evolution of the percentages of necrotic and apoptotic cancer cells. It can be observed that there is a phase transition that corresponds to the emergence of the necrotic core happening at day 10: in the first days cancer cells mostly die by apoptosis and later, due to low oxygen conditions, they die dominantly by necrosis.

As already mentioned, three different types of cancer cells are currently supported by the EvoNano PhysiCell simulator: DCC, CSC, and CAF cancer cells. The percentages of those three types of cancer cells considering the whole tumour (both live and dying cancer cells) are shown in Fig. 6. It can be seen that CSC cells constitute less than 6% of the tumour. More specifically, their percentage slowly grows from 2.94% on day 1 to 5.65% on day 30. On the other hand, DCC and CAF cells have a more complex dynamic with a phase transition. It can be seen that the percentage of CAF cells increases in the first 12 days and then starts to decline. The dynamic of DCC cells is exactly the opposite: the percentage of DCC cells first decline and after day 12 this percentage starts to grow. This phase transition can be also explained by the emergence of the necrotic core. After the emergence of the necrotic core, the number of dying cancer cells, which are mostly DCC cells, grows at a faster rate than the number of vascular cells. Consequently, DCC cells grow faster than CAF cells whose growth rate is always lower than the growth rate of vascular cells.

One of the important aspects of cancer simulation with the previously described setup is the count and origin of CSCs. These cells do not die in apoptosis, and they

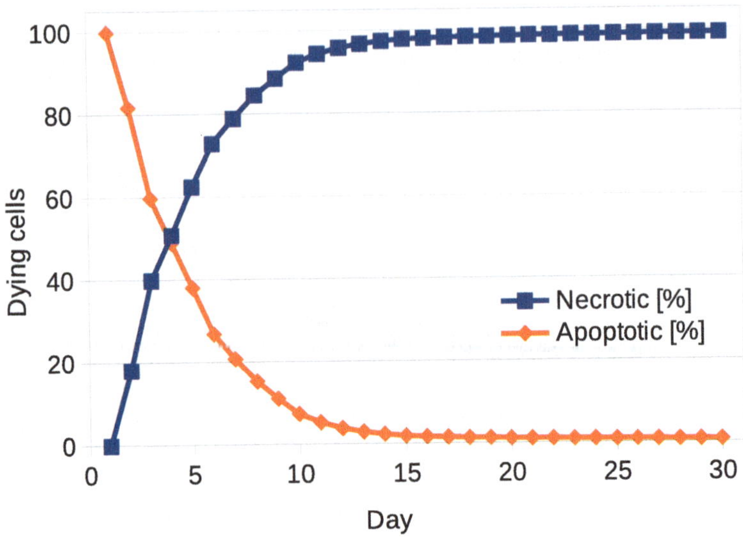


**Fig. 5** The percentage of necrotic and apoptotic cells in the total number of dying cancer cells

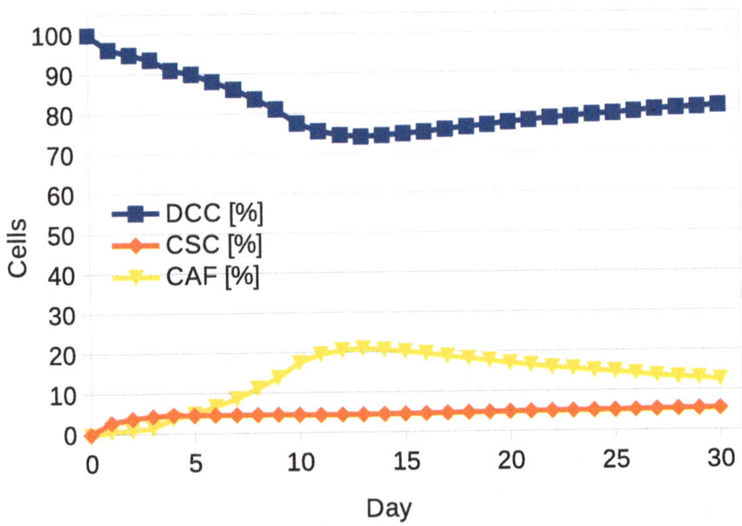


**Fig. 6** The percentage of DCC, CSC and CAF cancer cells in the simulated tumour

are more resistant to low oxygen levels. Therefore, they have an important influence on cancer growth. Furthermore, the emergence of these cells is three-folded:

- some CSCs appear from DCCs in the DCC->CSC conversion process under the influence of high levels of cytokine

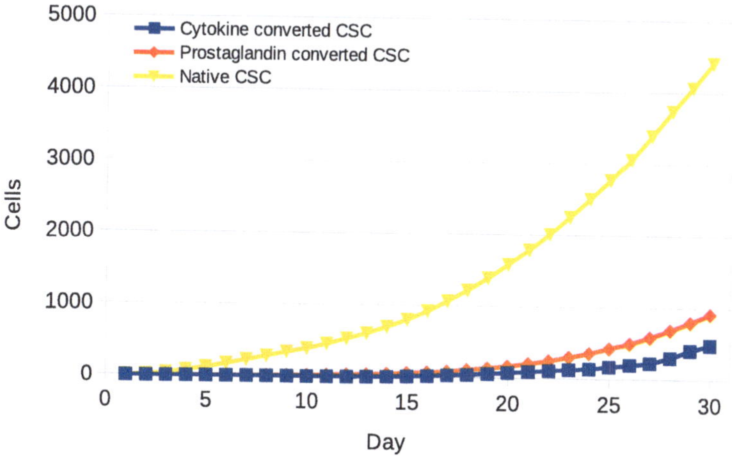


**Fig. 7** The daily count of CSCs and their origin

- some CSCs appear from DCCs in the DCC->CSC conversion process under the influence of high levels of prostaglandine
- all other CSCs are created as a result of DCCs or other CSCs division.

The number of created CSCs according to their origin is shown in Fig. 7. The distribution of prostaglandin converted and cytokine converted CSCs is not empirically known (from clinical research) but it is assumed that the number of converted CSCs is significantly smaller. The number of CSCs and their origin is controlled through numerous parameters as already described. The situation shown in Fig. 7 is a result of many experiments and parameter tuning, all for the purpose of obtaining Gompertzian growth.

In order to demonstrate the influence of all introduced concepts, we performed several simulations with different setups. We compared the simulations where all parameter values are the same (as described earlier) and where:

- prostaglandin and CAF/cytokine mechanisms are turned off
- only prostaglandin mechanism is turned off
- only CAF/cytokine mechanism is turned off
- all mechanisms are turned on.

The daily volumes of tumours with these setups are given in Fig. 8. It is evident that the simulation without prostaglandin behaves almost exactly as with all mechanisms turned on. Also, the simulation without cytokine/CAF behaves similarly as without both mechanisms. However, as seen in Fig. 7 the amount of prostaglandin conversions is higher than cytokine conversions. That may imply that prostaglandin will have a higher influence on cancer growth. However, this is not the case (as seen in Fig. 8) because cytokine/CAF mechanisms also influence the propagation of CAF cells, and

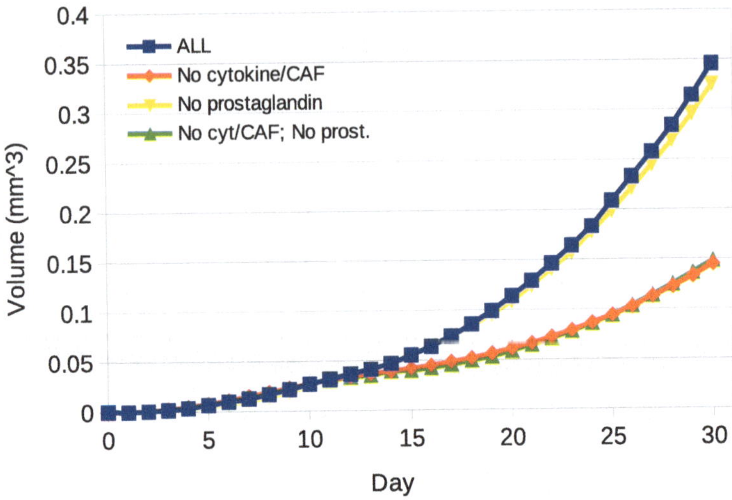


**Fig. 8** The daily evolution of the volume of a tumour simulated with different simulation setups

the amount of them is not negligible (see Fig. 6). All in all, the simulation with all mechanisms nicely follows Gompertzian growth with all specified parameters. If any of the mechanisms are turned off, the re-tuning of parameters is inevitable.

## 4 Conclusions

By intensive experiments, we simulate nanoparticle-based cancer therapies and the existence of NPs with an attached particular drug. That drug can affect cancer in different ways but at the moment we support only drugs that affect cancer cell death.

As we mentioned, for the experimental part we used the EvoNano cancer simulator. It is implemented in C++ relying on the PhysiCell library. The simulator realizes various types of cancer cells (differentiated cancer cells, cancer stem cells, and CAFs), healthy cells, and vascular cells as building blocks for the vascular network. Various substrates are supported (oxygen, angiogen, cytokine, prostaglandin) including also special nano-particle substrates simulating drug diffusion. The main technical feature of the simulator is dynamic space expansion that improves runtime efficiency and enables simulations of larger tumours compared to regular PhysiCell-based simulators without this feature.

The results of conducted experiments show that the large space of simulator parameters can be calibrated to be in line with the Gompertzian tumour growth model. In our simulations, we have observed the emergence of tumour necrotic cores and related phase transitions. We also empirically examined the dynamics of cancer cell conversions based on cytokine and prostaglandin, showing that re-tuning of simulator parameters is required if any of cancer cell conversion mechanisms are turned off or changed. In short, our results demonstrate high sensitivity of tumour development on the interplay of microenvironment signalling factors. Since this aspect of tumour physiology is often neglected in cancer modeling, we believe that our results will help in highlighting the importance of cell-communication signalling in analyzing tumour growth and treatment response.

## References

1. Bakhoum, S.F., Compton, D.A.: Chromosomal instability and cancer: a complex relationship with therapeutic potential. J. Clin. Investig. **122**, 1138–1143 (2012). https://doi.org/10.1172/JCI59954
2. Calon, A., et al.: Stromal gene expression defines poor-prognosis subtypes in colorectal cancer. Nat. Genet. **47**, 320–329 (2015)
3. Cavallari, L.H., Mason, D.L.: Cardiovascular pharmacogenomics-implications for patients with CKD. Adv. Chronic Kidney Dis. **23**(2), 82–90 (2016)
4. Chen, Y., Song, Y., Du, W., et al.: Tumor-associated macrophages: an accomplice in solid tumor progression. J. Biomed. Sci. **26**, 78 (2019). https://doi.org/10.1186/s12929-019-0568-z
5. Cisyk, A.L., Penner-Goeke, S., Lichtensztejn, Z., Nugent, Z., Wightman, R.H., Singh, H., McManus, K.J.: Characterizing the prevalence of chromosome instability in interval colorectal cancer. Neoplasia **17**, 306–316 (2015). https://doi.org/10.1016/j.neo.2015.02.001
6. Claeys, A., Vialatte, J.S.: Advances in genetics: towards a Precision Medicine? Technological, social and ethical scientific issues of personalised medicine [Les progrès de la génétique: vers une médecine de précision? Les enjeux scientifiques, technologiques, sociaux et éthiques de la médecine personnalisée] (2014)
7. Deisboeck, T.S., Wang, Z., Macklin, P., Cristini, V.: Multiscale cancer modeling. Annu. Rev. Biomed. Eng. **15**(13), 127–155 (2011)
8. Dewhirst, M.W., Secomb, T.W.: Transport of drugs from blood vessels to tumour tissue. Nat. Rev. Cancer **17**(12), 738–750 (2017). https://doi.org/10.1038/nrc.2017.93
9. Dreesen, O., Brivanlou, A.H.: Signaling pathways in cancer and embryonic stem cells. Stem Cell Rev. **3**(1), 7–17 (2007). https://doi.org/10.1007/s12015-007-0004-8. PMID: 17873377
10. Feng, K., Leary, R.H.: Toward personalized medicine with physiologically based pharmacokinetic modeling. Int. J. Pharmacokinet. **2**(1), 1–4 (2017)
11. Ghaffarizadeh, A., Friedman, S.H., Macklin, P.: BioFVM: an efficient, parallelized diffusive transport solver for 3-D biological simulations. Bioinformatics **32**(8), 1256–1258 (2016). https://doi.org/10.1093/bioinformatics/btv730. Epub (2015), PMID: 26656933; PMCID: PMC4824128
12. Ghaffarizadeh, A., Heiland, R., Friedman, S.H., Mumenthaler, S.M., Macklin, P.: PhysiCell: an open source physics-based cell simulator for 3-D multicellular systems. PLoS Comput. Biol. **14**(2), e1005991 (2018). https://doi.org/10.1371/journal.pcbi.1005991
13. Guinney, J., et al.: The consensus molecular subtypes of colorectal cancer. Nat. Med. **21**, 1350–1356 (2015)
14. Gujam, F.J., Going, J.J., Edwards, J., et al.: The role of lymphatic and blood vessel invasion in predicting survival and methods of detection in patients with primary operable breast cancer. Crit. Rev. Oncol. Hematol. **89**, 231–241 (2014)
15. https://medlineplus.gov/genetics/understanding/precisionmedicine/definition/
16. Jordan, V.C.: Tamoxifen: catalyst for the change to targeted therapy. Eur. J. Cancer **44**(1), 30–38 (2008)
17. Kahn, et al.: J. Clin. Invest. **131**(2), e136655 (2021). https://doi.org/10.1172/JCI136655

18. Kim, B.J., Hannanta-anan, P., Chau, M., Kim, Y.S., Swartz, M.A., et al.: Cooperative roles of SDF-1a and EGF gradients on tumor cell migration revealed by a robust 3D microfluidic model. PLoS ONE **8**(7), e68422 (2013). https://doi.org/10.1371/journal.pone.0068422
19. Kobuchi, S., Shimizu, R., Ito, Y.: Semi-mechanism-based pharmacokinetic-toxicodynamic model of oxaliplatin-induced acute and chronic neuropathy. Pharmaceutics **12**(2), 125 (2020)
20. Lahiri, C., Pawar, S., Mishra, R.: Precision medicine and future of cancer treatment. Precis. Cancer Med. **2**, 33 (2019). AME Publishing
21. Lengauer, C., Kinzler, K.W., Vogelstein, B.: Genetic instability in colorectal cancers. Nature **386**, 623–627 (1997). https://doi.org/10.1038/386623a0
22. Liu, T., Han, C., Wang, S., et al.: Cancer-associated fibroblasts: an emerging target of anti-cancer immunotherapy. J. Hematol. Oncol. **12**, 86 (2019). https://doi.org/10.1186/s13045-019-0770-1
23. Macklin, P., Frieboes, H.B., Sparks, J.L., Ghaffarizadeh, A., Friedman, S.H., Juarez, E.F., Jonckheere, E., Mumenthaler, S.M.: Progress towards computational 3-D multicellular systems biology. Adv. Exp. Med. Biol. **936**, 225–246 (2016). https://doi.org/10.1007/978-3-319-42023-3-12
24. Mauri, G., Bonazzina, E., Amatu, A., Tosi, F., Bencardino, K., Gori, V., Massihnia, D., Cipani, T., Spina, F., Ghezzi, S., Siena, S., Sartore-Bianchi, A.: The evolutionary landscape of treatment for BRAFV600E mutant metastatic colorectal cancer. Cancers (Basel) **13**(1), 137 (2021). https://doi.org/10.3390/cancers13010137. PMID: 33406649; PMCID: PMC7795863
25. Meads, M.B., Gatenby, R.A., Dalton, W.S.: Environment-mediated drug resistance: a major contributor to minimal residual disease. Nat. Rev. Cancer **9**(9), 665–674 (2009). https://doi.org/10.1038/nrc2714
26. Pattabiraman, D.R., Weinberg, R.A.: Tackling the cancer stem cells - what challenges do they pose? Nat. Rev. Drug Discov. **13**(7), 497–512 (2014). https://doi.org/10.1038/nrd4253. PMID: 24981363; PMCID: PMC4234172
27. Ping, Q., Yan, R., Cheng, X., et al.: Cancer-associated fibroblasts: overview, progress, challenges, and directions. Cancer Gene Ther. **28**, 984–999 (2021). https://doi.org/10.1038/s41417-021-00318-4
28. Sahai, E., Astsaturov, I., Cukierman, E., et al.: A framework for advancing our understanding of cancer-associated fibroblasts. Nat. Rev. Cancer **20**, 174–186 (2020). https://doi.org/10.1038/s41568-019-0238-1
29. Simpson, M.J., Towne, C., McElwain, D.L.S., Upton, Z.: Migration of breast cancer cells: understanding the roles of volume exclusion and cell-to-cell adhesion. Phys. Rev. E. **82**, 041901 (2010). https://doi.org/10.1103/PhysRevE.82.041901
30. Sleeman, J.P., Thiele, W.: Tumor metastasis and the lymphatic vasculature. Int. J. Cancer **125**, 2747–2756 (2009)
31. Speer, J.F., Petrosky, V.E., Retsky, M.W., Wardwell, R.H.: A stochastic numerical model of breast cancer growth that simulates clinical data. Cancer Res. **44**(9), 4124–4130 (1984)
32. Stéphanou, A., Volpert, V.: Hybrid modelling in biology: a classification review. Math. Model. Nat. Phenomena **11**(1), 37–48 (2016)
33. Stillman, N.R., Kovacevic, M., Balaz, I., Hauert, S.: In silico modelling of cancer nanomedicine, across scales and transport barriers. NPJ Comput. Mater. **6**, 92 (2020). https://doi.org/10.1038/s41524-020-00366-8
34. Stillman, N.R., Balaz, I., Tsompanas, M.A., et al.: Evolutionary computational platform for the automatic discovery of nanocarriers for cancer treatment. NPJ Comput. Mater. **7**, 150 (2021). https://doi.org/10.1038/s41524-021-00614-5
35. Thompson, L.L., Jeusset, L.M., Lepage, C.C., McManus, K.J.: Evolving therapeutic strategies to exploit chromosome instability in cancer. Cancers (Basel) **9**(11), 151 (2017). https://doi.org/10.3390/cancers9110151. PMID: 29104272; PMCID: PMC5704169
36. Vaghi, C., Rodallec, A., Fanciullino, R., Ciccolini, J., Mochel, J.P., et al.: Population modeling of tumor growth curves and the reduced Gompertz model improve prediction of the age of experimental tumors. PLoS Comput. Biol. **16**(2), e1007178 (2020). https://doi.org/10.1371/journal.pcbi.1007178

37. Vaupel, P., Mayer, A.: A hypoxia in cancer significance and impact on clinical outcome. Cancer Metastasis Rev. **26**, 225–239 (2007)
38. Venne, J., Busshoff, U., Poschadel, S., Menschel, R., Evangelatos, N., Vysyaraju, K., Brand, A.: International consortium for personalized medicine: an international survey about the future of personalized medicine. Pers. Med. **17**(2), 89–100 (2020)
39. Vilanova, G., Colominas, I., Gomez, H.: Capillary networks in tumor angiogenesis: from discrete endothelial cells to phase-field averaged descriptions via isogeometric analysis. Int. J. Numer. Method Biomed. Eng. **29**(10), 1015–1037 (2013). https://doi.org/10.1002/cnm.2552. Epub (2013). PMID: 23653256
40. Wei, Y., Zhao, Q., Gao, Z., Lao, X.M., Lin, W.M., Chen, D.P., et al.: The local immune landscape determines tumor PD-L1 heterogeneity and sensitivity to therapy. J. Clin. Invest. **129**(8), 3347–3360 (2019). https://doi.org/10.1172/JCI127726
41. Wilson, W.R., Hay, M.P.: Targeting hypoxia in cancer therapy. Nat. Rev. Cancer **11**, 393–410 (2011)

# Local Quantitative and Qualitative Sensitivity Analysis of CSC Dynamical Simulation

**Branislava Lalic and Igor Balaz**

**Abstract** Modeling tumour development is a challenging goal that opens a window of opportunity for broad spectra of applications—from different feedback regulations in tumour system to timing and dosage of new treatments testing. When properly calibrated and validated, models can significantly improve our knowledge and contribute to narrowing clinical trials. In this study, we used the ODE model that includes only stem cells (wild and mutated) and differentiated cells to investigate inhibition of the differentiation feedback and the uncertainty in signalling parameters (self-renewal probability and division rate) parameterization of feedback regulation affects tumour growth. Local quantitative and qualitative sensitivity analysis is performed using data for breast tumour tissue. Obtained results indicate that even slight variations in initial values of signaling parameters can lead to considerable differences in tumour size over the course of 250 days. The model exerted the highest sensitivity to self-renewal probability for initial values above 0.5.

## 1 Introduction

According to the Cancer Stem Cell (CSC) hypothesis, only a fraction of tumour cells are tumourigenic [1]. CSC supposedly duplicates without limit and differentiate into multiple types of cancer cells [2, 3]. Regulation of both duplication and differentiation is a tightly controlled process. They depend on the interplay of internal cellular and external microenvironment factors. Essential internal features such as symmetric/asymmetric division, metabolic state, cellular quiescence and movements, apoptosis, and the existence of oxygen and its consumption play a significant role in tumour growth and malignancy [4]. At the same time, tumour cells and CSC resides in a complex tumour microenvironment. Tumour microenvironment is

B. Lalic (✉) · I. Balaz
Faculty of Agriculture, University of Novi Sad, Novi Sad, Serbia
e-mail: branislava.lalic@polj.edu.rs

I. Balaz
e-mail: igor.balaz@df.uns.ac.rs



I. Balaz and A. Adamatzky (eds.), *Cancer, Complexity, Computation*,
Emergence, Complexity and Computation 46,
https://doi.org/10.1007/978-3-031-04379-6_8

composed of tumour cells, CSC, Tumour-associated fibroblasts (TAF), endothelial cells, immune cells (e.g. tumour associated macrophages (TAM) and lymphocytes), and non-cellular components (collagen, hyaluronan, laminin) [5–8]. The interaction of tumour cells with their microenvironment is dynamic and bidirectional mediated by a large number of soluble factors (chemokines, cytokines, growth factors, etc.), or factors that enable horizontal genetic/biomaterial transfer including cfDNA, apoptotic bodies, and exosomes (for a detailed overview see: [4, 9, 10].

Attempting to model such complicated systems put us immediately in front of a crucial choice: should we aim for a descriptive or a conceptual model [11]. In a descriptive model, the goal is to create a digital twin of the biological system of interest by explicitly including as many processes as possible to reach a maximally detailed system description. If adequately calibrated and validated, such models could have very precise simulation outputs, suitable for clinical research. However, on the negative side, descriptive models require tracking many state variables that are impossible to analyze. Experimental values of many of these variables are hard or even impossible to obtain, leading to arbitrary parameter fitting. In such cases, despite an impressive number of used parameters, and close-fitting model outputs to experimental data it is questionable whether the model correctly represents the underlying biology [11].

On the other hand, conceptual models rely on a limited number of processes and reduced system descriptions. It leads to robust models of limited applicability but stable over the wide range of model parameters. Also, developing a conceptual model requires a deep understanding of the system in question so that the key mechanisms are highlighted and modeled.

In order to choose the overall approach for the specific modeling task, the obvious first question is "why". Why are we modeling a specific system, and what are the goals we want to achieve? In our case, do we want to understand the high-level mechanics of feedback regulation in tumour systems, do we want to decipher the relative importance of specific signals, or do we aim for a precise prediction of drug targets? Whatever is the goal, we should take equal care of defining the translation of experimental data to a model, performance metrics, and integration of obtained results into a multiscale pipeline [12].

Once the descriptive versus conceptual model choice is made, the next step in the deliberation is the correct mathematical approach for a given goal. Approaches in mathematical modeling of cancer development are numerous (for a comprehensive overview, see, for example, [13–18]. Roughly, they can be divided to deterministic (mechanistic) (see, for example, [19–22]), stochastic (see, for example, [23–26]) or combined [27]. Both stochastic and deterministic approaches have positive and negative aspects. Deterministic models highly rely on detailed knowledge of the biological and physicochemical background of processes that are modeled. Such an approach implies using a high number of parameters and computationally very demanding spatial and temporal integrations. On the other hand, stochastic models are far less computational demanding, firmly connected to well-developed computational procedures, but entirely dependent on input data. While deterministic models allow out-of-the-box view and design of new experiments (or even to find an error in current ones), it is not possible with stochastic models. Therefore, their combination looks like a promising concept. Stochastic models should identify patterns and make short-term predictions, while deterministic models should incorporate processes and system characteristics primarily affecting identified patterns.

As indicated at the beginning of this chapter, we will narrowly focus on the role of CSC and feedback regulation in tumour growth. Our goal is to investigate how the uncertainty of signaling parameters affects growth. Therefore, we opt for a conceptual, deterministic modeling approach. In this chapter, we use the model developed by Rodriguez-Brenesa et al. [21]. The model is accompanied by the excellent numerical methodology description, allowing full reproducibility of presented results and validation using available experimental data. The significant feature of this model is that it considers only the basic processes (division and differentiation), cell mutations, and feedback between cells, so it can be considered as a ground-based model. Since one of the landmark tumour properties is growth dynamics that escaped the organism's control, we focus on mutations (denoted as Sdiff) that inhibit stem-cell response to signals coming from cell differentiation and, therefore, lead to uninhibited growth. The first step towards calibrating and validating the model for various tumour types is to test model sensitivity. Therefore, we performed a local quantitative and qualitative sensitivity analysis of Sdiff mutation using an ensemble of characteristics and parameter values for different feedback parameterization functions.

## 2 Methodology

### 2.1 Model Description

The model developed by Rodriguez-Brenesa et al. [21] is the ordinary differential equations (ODE) model, which takes into account two populations of cells: wild stem cells ($S$) with unlimited reproductive potential, and differentiated cells ($D$). Only differentiated cells die, with rate $d$. Stem cells can divide with rate $\nu$. The result of division can be either two stem cells (self-renewal process) or two differentiated cells (differentiation process). If we denote the probability of self-renewal as p, the stem cell pool will then increase by $p\nu S$. Conversely, in the differentiation process, the stem-cell pool increases by $(1-p)\nu S$. Described processes lead to changes in the number of wild stem cells over time what can be expressed in the form

$$\dot{S} = (2p-1)\nu S \tag{1}$$

$$\dot{D} = 2(1-p)\nu S - dD. \tag{2}$$

The RHS of Eq. (2) accounts for the increase of differentiated cells population due to the differentiation of wild stem cells and decreased population due to cell death.

Two types of feedback loops are introduced in the model. Both of them are related to signaling molecules secreted by differentiated cells. The first feedback loop inhibits cell division, while the second one suppresses stem cell self-renewal and thus promotes terminal differentiation and death. Both, the intensity of division and probability of self-renewal do not assume specific mathematical terms but depend on the number of differentiated cells ($\nu(D)$ and $p(D)$, respectively. Therefore, the mathematical formulation of the described system and related processes can be written in the form

$$\dot{S} = (2p(D) - 1)\nu(D)S \tag{3}$$

$$\dot{D} = 2(1 - p(D))\nu(D)S - dD. \tag{4}$$

Feedbacks affecting division rate and self-renewal probability are parameterized using Hill's functions, which are commonly used to describe ligand–receptor interactions in the form:

$$\nu(D) = \frac{\nu_0}{1 + hD^m} \tag{5}$$

and

$$p(D) = \frac{p_0}{1 + gD^n}, \tag{6}$$

where $\nu_0$ and $p_0$ are initial values of $\nu$ and $p$, $m$ and $n$ are Hill's coefficients, while $h$ is division feedback and $g$ is the relative magnitude of the differentiation.

Analysis of equilibrium of the system described by Eqs. (3) and (4) and its unique, non-trivial stationary point led to conditions for differentiation probability in equilibrium point $\left(\hat{S}, \hat{D}\right)$ in the form

$$p\left(\hat{D}\right) = \frac{1}{2};\ -p'\left(\hat{D}\right) < \frac{1}{\left(2\hat{D}\right)};\ \hat{S} = \frac{d\hat{D}}{\nu\left(\hat{D}\right)}. \tag{7}$$

The further we are from the equilibrium point, the more likely the mathematical model is to enter numerical instability.

Rodriguez-Brenesa et al. [21] showed that introduced feedbacks have a fundamentally different effect on cell growth control. The feedback on the self-renewal probability is sufficient to control cell growth, while the division rate feedback alone cannot control cell growth. To assess the role of the differentiated feedback in tumour growth, they introduced Sdiff mutation. Stem cells that acquire this mutation are denoted as Sm. Then, the mathematical model (3)–(4) includes an additional equation for the time evolution of mutated stem cells

$$\dot{S}_m = (2p_0 - 1)\nu(D)S_m \tag{8}$$

and redesigned Eq. (4) for the evolution of differentiated cells pool in the form

$$\dot{D} = 2(1 - p(D))\nu(D)S - 2(1 - p_0)\nu(D)S_m - dD, \tag{9}$$

which incorporates a decrease in population due to stem cell mutation and an increase due to differentiation of mutated cells.

## 2.2 *Sensitivity Tests Design*

To test how the inhibition of the differentiation feedback affects $S$, $S_m$, and $D$ cells population over time, with added Sdiff mutations, we measured the model's sensitivity using a calculated number of all three cell types over two integration periods: 250 and 2500 days. The time scale (and unit for division rate) is chosen according to a typical time scale of division rate. Since the main goal is to assess how parameter uncertainty affects model performance, we varied model parameters around one reference value for each parameter. Therefore, this sensitivity analysis (SA) is local [28]. Qualitative SA is performed by visual inspection of model outputs using 3D scatter diagrams where the number of the stem ($S$, $S_m$) and differentiated ($D$) cells are analyzed for different values of division rate ($\nu$) and the probability of self-renewal ($p$) for different values of feedback factors ($g$ and $h$) and different functional parameterizations of introduced feedbacks. The sensitivity index used in quantitative SA to evaluate uncertainties of ensemble-based model outputs is ensemble spread (SPRD) [29]

$$\text{SPRD} = \sqrt{\frac{1}{m}\sum_{i=1}^{m}\frac{1}{(n-1)}\sum_{j=1}^{n}\left(A_i^{EA} - A_j\right)^2}, \tag{10}$$

where $A_i^{EA}$ is the ensemble average, $A_j$ is the ensemble member, $m$ is the sample size (250 or 2500 days), and $n$ is the ensemble size (21). Values of system characteristics ($\nu$, $p$) and parameters ($g$, $h$) used for quantitative and qualitative SA are presented in Table 1. Initial values are adopted from Rodriguez-Brenesa et al. [21].

Impact of feedback loop parameterisation (Eqs. (5) and (6)) on tumour growth is tested using different values for coefficients $m$ and $n$ (Table 2) in the Hill's functions for $\nu(D)$ and $p(D)$.

**Table 1** System characteristics and parameters used

| System characteristic/parameter | Initial value | Range |
|---|---|---|
| $\nu_0$ | 6.93 | 4.93–8.93 |
| $p_0$ | 0.6 | 0.3–0.7 |
| $g$ | $10^{-4}$ | $(5–15) \times 10^{-5}$ |
| $h$ | $4.3 \times 10^{-3}$ | $(2.3–6.3) \times 10^{-3}$ |

In all simulations, the death rate is fixed at $d = 0.0693$

**Table 2** Feedback loop parameterization

| Denotation | $m$ | $n$ |
|---|---|---|
| Par 1 | 1 | 1 |
| Par 2 | 1/2 | 1/2 |
| Par 3 | 2 | 2 |

### 2.3 *Model Calibration and Tumour Growth Simulation*

Model calibration is performed using the results of laboratory experiments with two groups of animals. At the beginning of the experiment, both groups are injected with 1000 cells of a very aggressive MDA-MB-231 cell line of breast tumour. Starting sixty days after injection tumour growth is regularly measured every 3–4 days during 39 days in group G12, and 29 days in group G1-4.

During calibration of the model, i.e., selection of parameterization procedure and parameter values that best describe tumour growth dynamics, the following criteria are applied [30]:

- Root mean square error (RMSE) is less than the standard deviation of measured values;
- the standard deviation of simulated values (std_s) is close to the standard deviation of measured values (std_m);
- the selected set of parameters has RMSE up to 10% above a minimum value.

Root mean square error is used to measure model performance since it provides a good overview of the data set, with significant errors weighted more than many minor errors [31].

## 3 Results and Discussion

### 3.1 Sensitivity Tests

To study model sensitivity to feedback parameterization and system characteristics/parameter values, we assume the healthy cell population is near equilibrium, i.e., initial values of $S$ ($S(0) = 173$) and $D$ ($D(0) = 1904$) satisfy conditions described by Eq. (7). Starting from near equilibrium, we investigate mutated cells' growth from low numbers ($S_{\mathrm{m}}(0) = 2$).

Results of SA performed and presented on Figs. 1, 2, 3, 4, 5, 6, 7, 8, 9, 10, 11, 12, and Tables 1 and 2 can be summarized as follows:

- The model is the most sensitive to stem cell self-renewal probability ($p$) values if its initial value is larger than 0.5. An increase in $p_0$ from 0.5 to 0.7 can increase $S_{\mathrm{m}}$ for 109 cells over just 250 days integration period (Fig. 3).

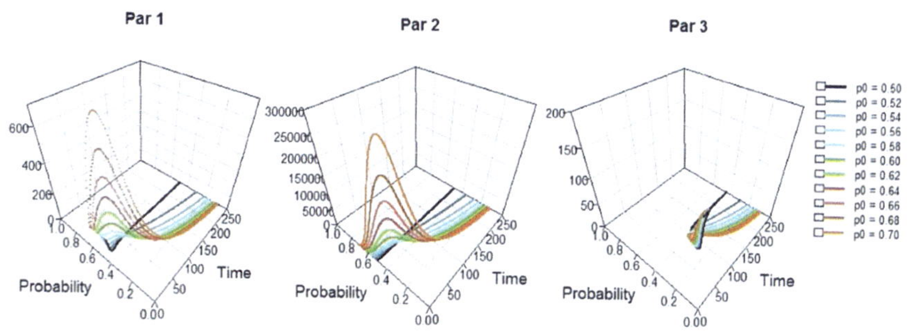


**Fig. 1** Wild stem cells' sensitivity on the initial value of the self-renewal probability of stem cells ($p_0$) and the parameterization of feedback loops in the ODE model

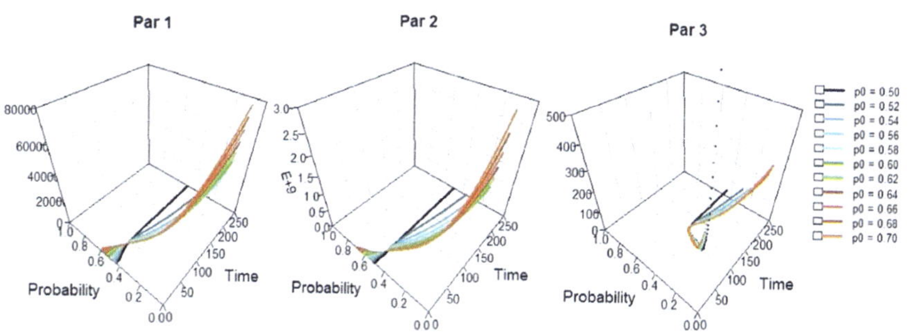


**Fig. 2** Differentiated cells' sensitivity on the initial value of self-renewal probability of stem cell ($p_0$) and parameterization of feedback loops in ODE model

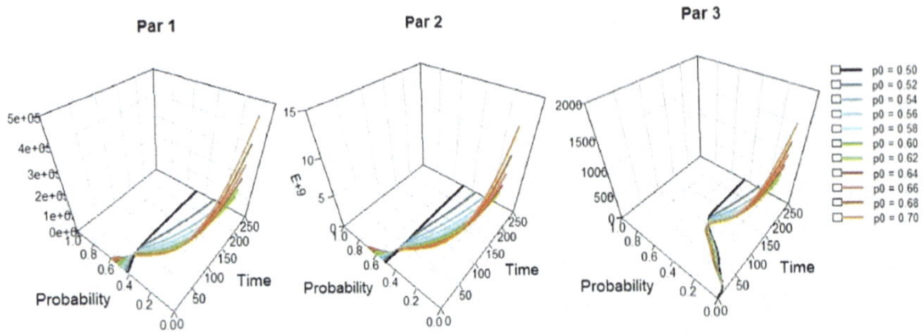


**Fig. 3** Mutated stem cells' sensitivity on the initial value of self-renewal probability of stem cell ($p_0$) and parameterization of feedback loops in ODE model

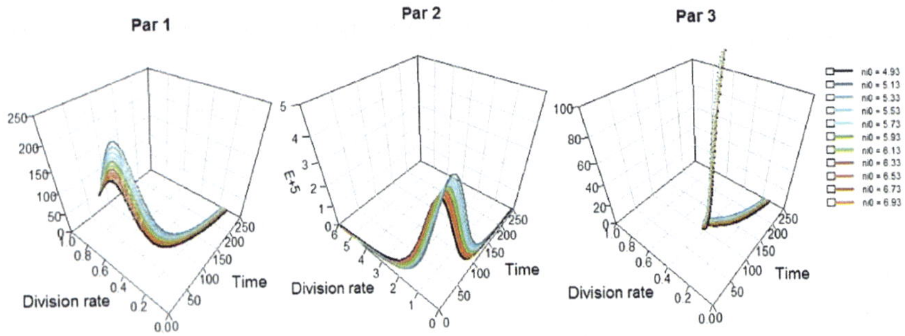


**Fig. 4** Wild stem cells sensitivity on the initial value of division rate ($\nu_0$) and parameterization of feedback loops in ODE model

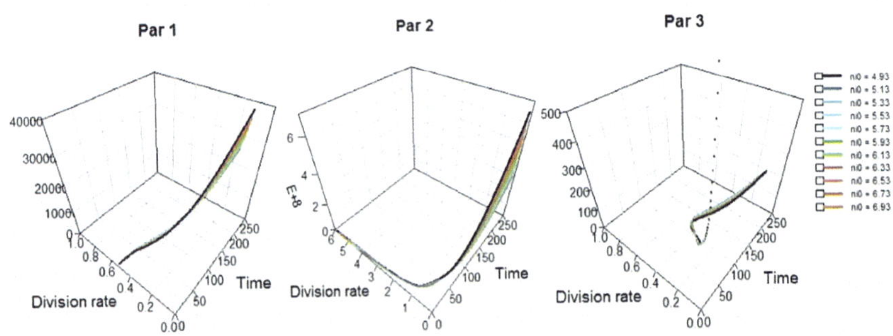


**Fig. 5** Differentiated cells' sensitivity on the initial value of division rate ($\nu_0$) and parameterization of feedback loops in ODE model

- Among all performed sensitivity tests, the highest spread is obtained for $D$ and $S_m$ cells (depending on parameterization applied) during the first 250 days, but for longer run (2500 days) $S_m$ spread is expected to be highest (Tables 1 and 2).

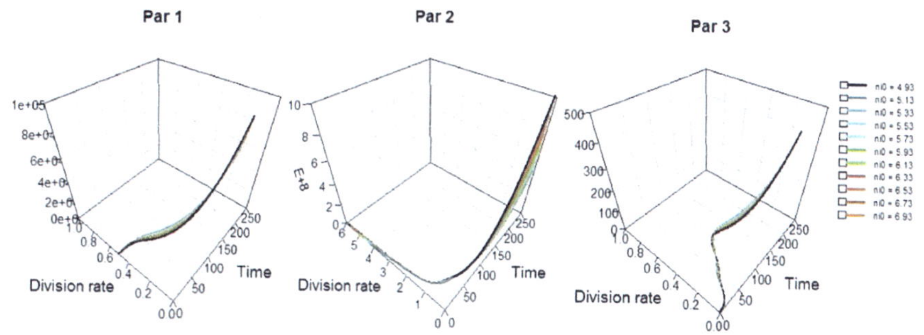


**Fig. 6** Mutated stem cells' sensitivity on the initial value of division rate ($\nu_0$) and parameterization of feedback loops in ODE model

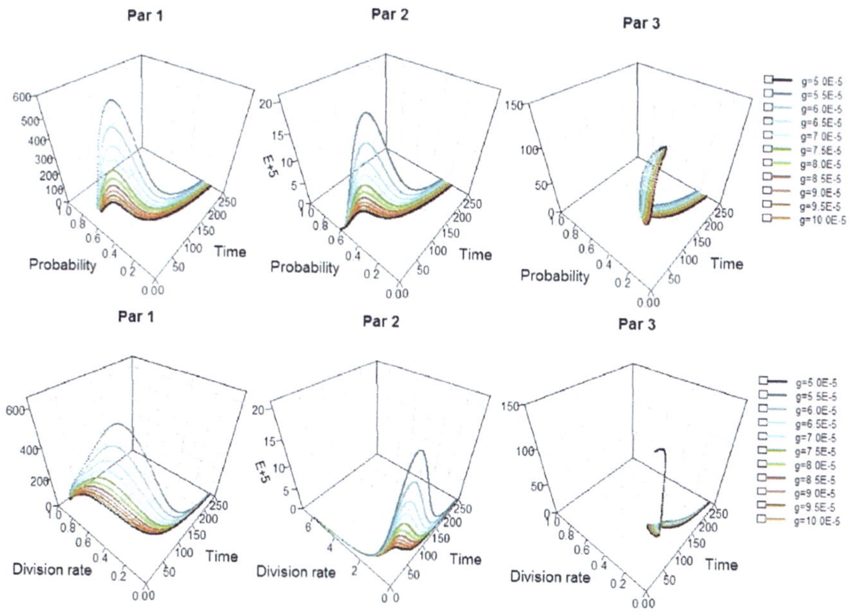


**Fig. 7** Wild stem cells' sensitivity to the relative magnitude of the differentiation ($g$) values and parameterization of feedback loops in the ODE model

- After less than 50 days, both $D$ and $S_{\mathrm{m}}$ start uninhibited growth even if the $p$ value drops below 0.5 due to differentiation-related feedback (Figs. 2 and 3).
- Model sensitivity on division rate ($\nu$) is significant but, in comparison to $p_0$, produces a smaller spread for $D$ and $S_{\mathrm{m}}$'s simulated values for all feedback parameterizations, while wild stem cells sensitivity is less pronounced for Par 1 and Par 3 (Figs. 4, 5, 6).
- The impact of model parameters g and h on model outputs are somewhat different (Figs. 7, 8, 9, 10, 11, 12). While $g$ parameter affects the spread of simulated cells

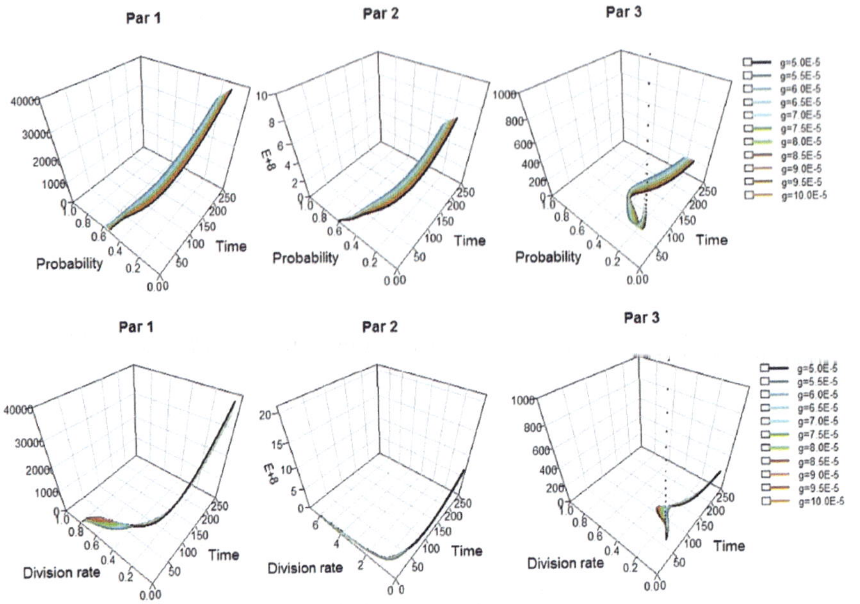


**Fig. 8** Differentiated cells' sensitivity on the relative magnitude of the differentiation ($g$) values and parameterization of feedback loops in ODE model

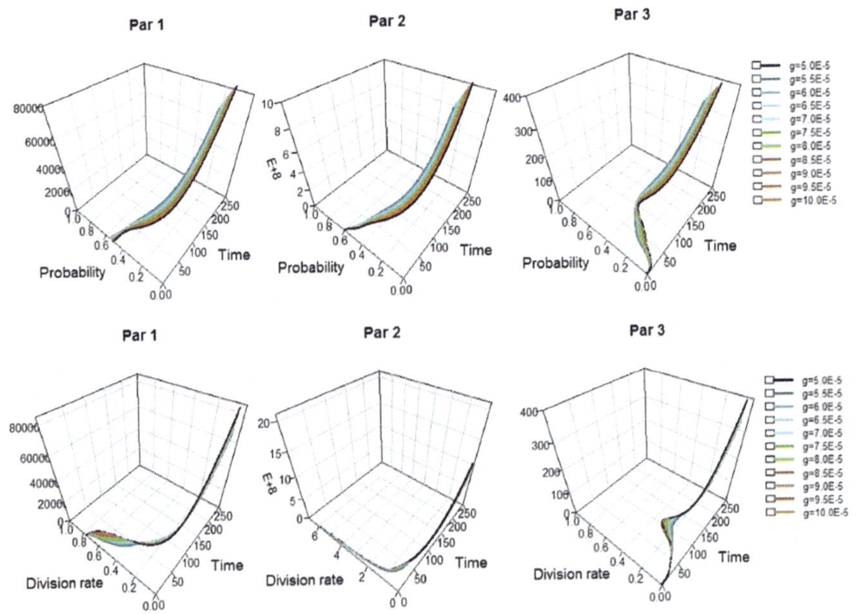


**Fig. 9** Mutated stem cells' sensitivity to the relative magnitude of the differentiation ($g$) values and parameterization of feedback loops in ODE model

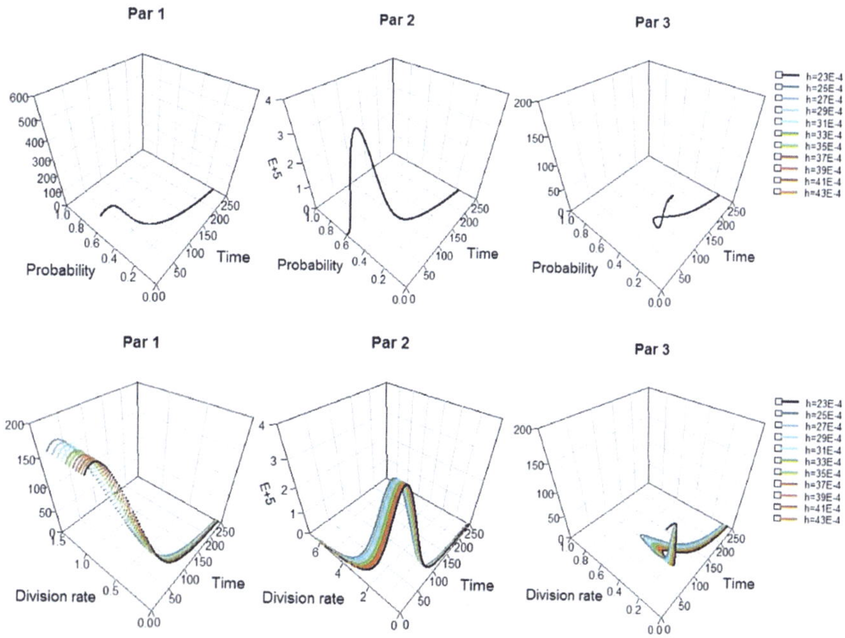


**Fig. 10** Wild stem cells' sensitivity on division feedback ($h$) values and parameterization of feedback loops in ODE model

with almost the same magnitude as the division rate ($\nu$), h parameter variations produce a practically negligible spread (app. 10–4) over both integration periods (Tables 1 and 2), except for $D$ and $S_m$'s for Par 2.

- For all feedback loop parameterizations and tested initial values of $p_0$ and $\nu_0$, the spread for both $D$ and $S_m$ is highest for Par 2 and lowest for Par 3.
- For both $p$ and $\nu$ system characteristics, and $g$ and $h$ parameters, the spread for all simulated cell types increases over time (Tables 3 and 4).

## 3.2 Tumour Growth Simulation

Our calibration procedure includes three parameterizations tested in Sect. 3.1. As a first step, we explored a wide range of possible initial values for self-renewal probability ($p_0$) and division rate ($\nu_0$) of stem cells, as well as relative magnitude of the differentiation ($g$) and division feedback ($h$). Since measurements start sixty days after injection, we assumed it as an initial day for calibration.

For both animal groups (Figs. 13 and 14), all calibration criteria are met for $m = 0.5$ and $n = 0.5$ (Par 2), $g = 0.0001$, $h = 0.0043$ and similar values for $p_0$ (0.71–0.72) and $\nu_0$ (16.5–19.0) (Table 5). The significant standard deviation of simulated values and RMSE can be found only for $p_0 = 0.75$ and $\nu_0 = 19.0$, but primarily as

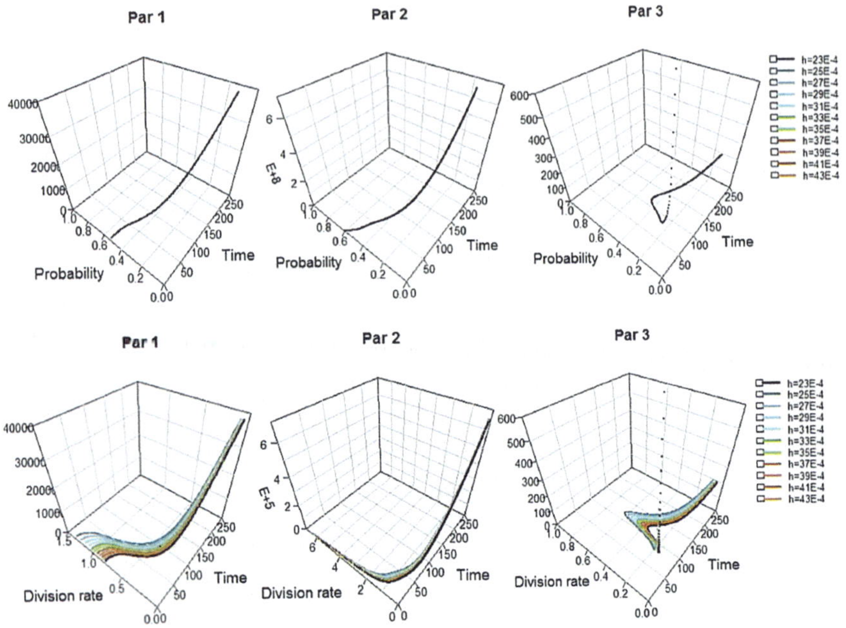


**Fig. 11** Differentiated cells' sensitivity on division feedback ($h$) values and parameterization of feedback loops in ODE model

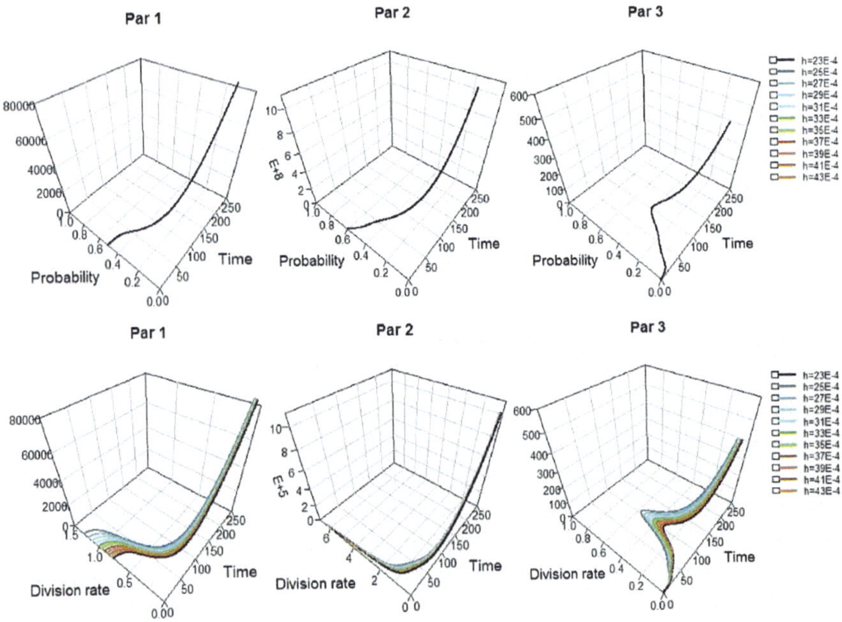


**Fig. 12** Mutated stem cells' sensitivity on division feedback ($h$) values and parameterization of feedback loops in ODE model

**Table 3** Spread of $S$, $D$ and $S_m$ simulated values during 250 days integration period

| Characteristic/parameter | Parameterization | $S$ | $D$ | $S_m$ |
|---|---|---|---|---|
| $p$ | 1 | 6850.00 | 0.2E+09 | 3.06E+09 |
| $p$ | 2 | 0.24E+12 | 0.37E+13 | 0.16E+11 |
| $p$ | 3 | 5.00 | 0.35E+04 | 0.33E+05 |
| $\nu$ | 1 | 492.00 | 0.20E+08 | 0.71E+08 |
| $\nu$ | 2 | 0.12E+10 | 0.18E+12 | 0.46E+09 |
| $\nu$ | 3 | 14.00 | 0.18E+03 | 0.72E+03 |
| $g$ | 1 | 7780.00 | 0.18E+07 | 0.13E+08 |
| $g$ | 2 | 0.90E+10 | 0.82E+11 | 0.58E+08 |
| $g$ | 3 | 12.00 | 0.23E+02 | 0.21E+03 |
| $h$ | 1 | 0.10E-09 | 0.83E-05 | 0.31E-04 |
| $h$ | 2 | 0.45E-03 | 0.16E+04 | 0.29E+04 |
| $h$ | 3 | 96E-10 | 0.27E-08 | 0.44E-09 |

**Table 4** Spread of $S$, $D$ and $S_m$ simulated values during 2500 days integration period

| Characteristic/parameter | Parameterization | $S$ | $D$ | $S_m$ |
|---|---|---|---|---|
| $p$ | 1 | 0.68E+03 | 0.26E+11 | 0.36E+14 |
| $p$ | 2 | 0.41E+04 | 0.17E+08 | 0.13E+25 |
| $p$ | 3 | 0.52E+00 | 0.78E+05 | 0.13E+09 |
| $\nu$ | 1 | 0.49E+02 | 0.18E+10 | 0.53E+12 |
| $\nu$ | 2 | 0.15E+03 | 0.54E+06 | 0.22E+23 |
| $\nu$ | 3 | 0.15E+01 | 0.22E+04 | 0.92E+06 |
| $g$ | 1 | 0.78E+03 | 0.23E+07 | 0.15E+10 |
| $g$ | 2 | 0.86E+04 | 0.26E+05 | 0.19E+20 |
| $g$ | 3 | 0.12E+01 | 0.57E+01 | 0.44E+04 |
| $h$ | 1 | 0.10E-10 | 0.11E-02 | 0.33E+00 |
| $h$ | 2 | 0.42E-10 | 0.13E-05 | 0.33E+10 |
| $h$ | 3 | 0.96E-11 | 0.58E-08 | 0.26E-05 |

a consequence of rather uncertain tumour growth in the G1-4 animal group before and after the 80th day of the experiment. Namely, it is quite difficult to conclude if it is an experimental error or unexpected growth. We intentionally selected this group in order to stress the challenges of biological processes parameterization and simulation.

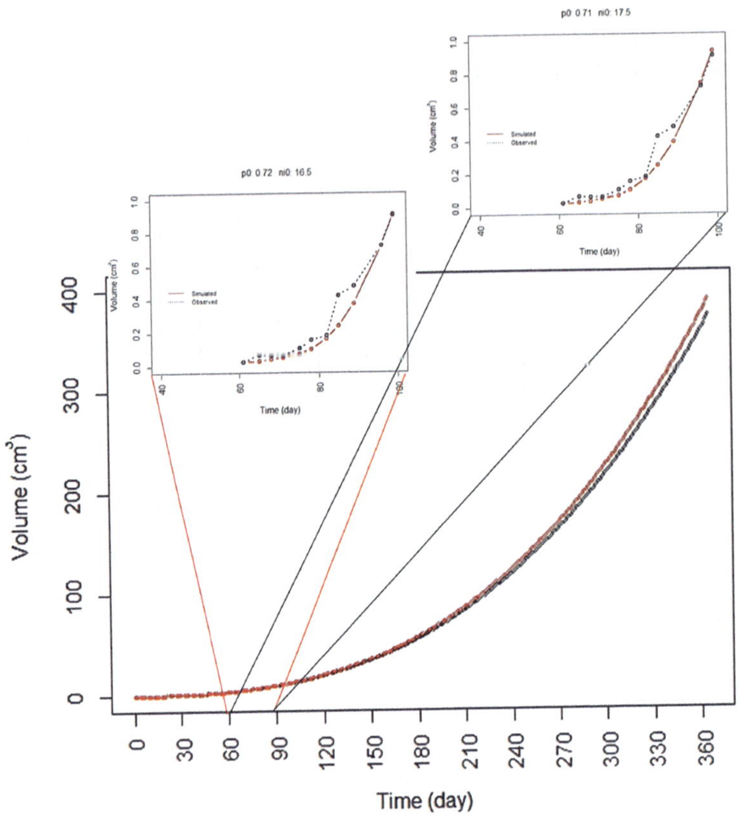


**Fig. 13** Calibration and simulation of tumour growth for G1-2 animal group using $m = n = 0.5$ and two sets of values for $p_0$ and $\nu_0$

## 4 Conclusion

The chosen model is as elementary as it could be. It includes just basic processes, cell types, and feedbacks. It cannot be simplified further to enhance robustness without eliminating basic processes. The model is the most sensitive to self-renewal probability for initial values above 0.5. The highest variation over time is obtained for different initial values of mutated stem cells and differentiated cells. After 250 days, differences in tumour size ($S + D + S_m$) can reach ~1 $mm^3$ of tumour tissue ($10^6$ cells) for only 0.02 difference in $p_0$. In other words, even slight variations in initial parameters can lead to huge differences in tumour size.

After the first model calibration using two groups of animals with different tumour sizes at the beginning of the simulation (24.3 $mm^3$ and 31.4 $mm^3$, respectively), we obtained the best calibration results for the same Hill's coefficients ($m$ and $n$), the same values of the relative magnitude of differentiation and division feedback ($g$ and $h$) and narrow range of initial values for self-renewal probability and division rate ($p_0$

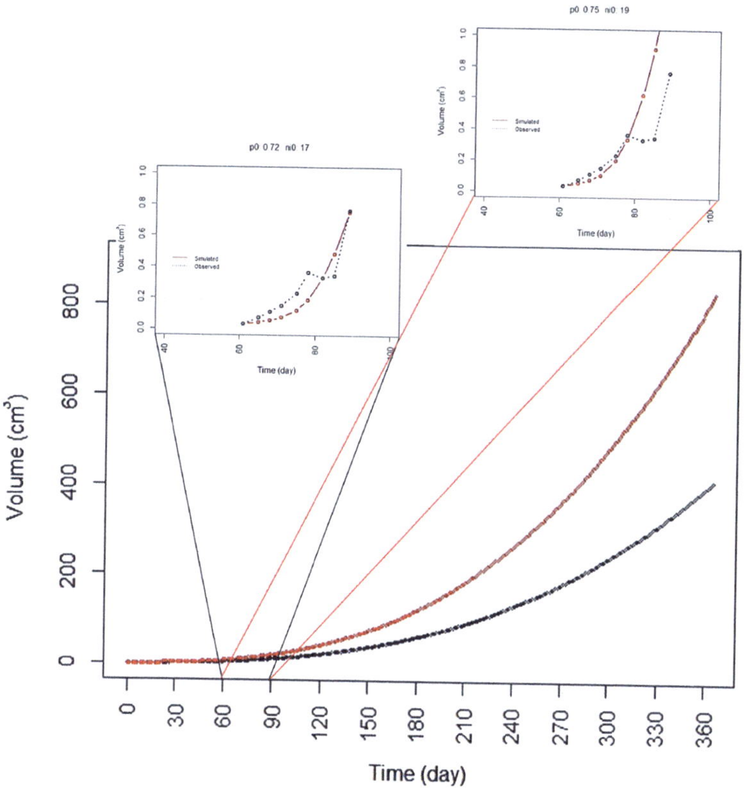


**Fig. 14** Calibration and simulation of tumour growth for G1-4 animal group using $m = n = 0.5$ and two sets of values for $p_0$ and $\nu_0$

**Table 5** Standard deviation of measured (*std_m*) and simulated (*std_s*) number of cancer cells ($S + D + S_m$) and their RMSE

| Animal group | $p_0$ | $\nu_0$ | *std_m* | *std_s* | RMSE |
|---|---|---|---|---|---|
| G1-2 | 0.72 | 16.5 | 3.02 | 3.04 | 0.67 |
| G1-2 | 0.71 | 17.5 | 3.02 | 3.12 | 0.63 |
| G1-4 | 0.72 | 17.0 | 2.23 | 1.31 | 1.63 |
| G1-4 | 0.75 | *19.0* | 2.23 | *4.84* | *3.10* |

and $\nu_0$)—all very promising. Even it should be confirmed using data from additional clinical trials with different cell lines, these simulations are already a good indication of model efficacy.

Our further research will be focused on:

- determination of $p$ and $\nu$ for different cancer tissues and development stages before and after treatment using available clinical trials;
- explore new feedback loops between different cell types and possible parameterizations;
- assess the effect of $p$ and $\nu$ changes during cancer development and identify the most favourable pattern.

Finally, the open question is to what extent obtained results are transferrable to the biological and medical domain. Suppose both physical and mathematical models are realistic and produce outputs that nicely correlate with experimental results. In that case, conditions like these described with Eq. (7) can offer some more insight into limits related to biological characteristics and parameter values, which are challenging to determine. However, the idea that physical and mathematical models can suggest biological limitations should always be taken with utmost precautions.

**Acknowledgements** This work was supported by the European Union's Horizon 2020 research and innovation programme under grant agreement No. 800983.

## References

1. Medema, J.P.: Cancer stem cells: the challenges ahead. Nat. Cell Biol. (2013). https://doi.org/10.1038/ncb2717
2. Beck, B., Blanpain, C.: Unravelling cancer stem cell potential. Nat. Rev. Cancer (2013). https://doi.org/10.1038/nrc3597
3. Reya, T., Morrison, S.J., Clarke, M.F., Weissman, I.L.: Stem cells, cancer, and cancer stem cells. Nature (2001). https://doi.org/10.1038/35102167
4. Azimzade, Y., Saberi, A.A., Sahimi, M.: Role of the interplay between the internal and external conditions in invasive behavior of tumors. Sci. Rep. (2018). https://doi.org/10.1038/s41598-018-24418-8
5. Arneth, B.: Tumor microenvironment. Medicina (Kaunas). (2020). https://doi.org/10.3390/medicina56010015
6. Jin, M.Z., Jin, W.L.: The updated landscape of tumor microenvironment and drug repurposing. Signal Transduct. Target. Ther. (2020). https://doi.org/10.1038/s41392-020-00280-x
7. Labani-Motlagh, A., Ashja-Mahdavi, M., Loskog, A.: The tumor microenvironment: a milieu hindering and obstructing antitumor immune responses. Front. Immunol. (2020). https://doi.org/10.3389/fimmu.2020.00940
8. Lau, A.N., Vander Heiden, M.G.: Metabolism in the tumor microenvironment. Annu. Rev. Cancer Biol. (2020). https://doi.org/10.1002/mco2.6
9. Baghban, R., Roshangar, L., Jahanban-Esfahlan, R.: Tumor microenvironment complexity and therapeutic implications at a glance. Cell Commun. Signal. (2020). https://doi.org/10.1186/s12964-020-0530-4
10. Whiteside, T.L.: The tumor microenvironment and its role in promoting tumor growth. Oncogene (2008). https://doi.org/10.1038/onc.2008.271
11. Beckman, R.A., Kareva, I., Adler, F.R.: How should cancer models be constructed? Cancer Conrol. (2020). https://doi.org/10.1177/2F1073274820962008
12. Balaz, I., Hauert, S., Adamatzky, A.: Editorial: computational approaches in cancer modelling. BioSystems (2021). https://doi.org/10.1016/j.biosystems.2021.104385

13. Altrock, P.M., Liu, L.L., Michor, F.: The mathematics of cancer: integrating quantitative models. Nat. Rev. Cancer (2015). https://doi.org/10.1038/nrc4029
14. Brady, R., Enderling, H.: Mathematical models of cancer: when to predict novel therapies, and when not to. Bull. Math. Biol. (2019). https://doi.org/10.1007/s11538-019-00640-x
15. Kareva, I.: Understanding Cancer from a Systems Biology Point of View. From Observation to Theory and Back. Elsevier, Amsterdam (2018). https://doi.org/10.1016/C2017-0-00190-8
16. Otto, S., Day, T.: A Biologist's Guide to Mathematical Modeling in Ecology and Evolution. Princeton University Press, Princeton (2007). https://doi.org/10.5860/choice.44-6894
17. Weiss, L.D., Komarova, N.L., Rodriguez-Brenes, I.A.: Mathematical modeling of normal and cancer stem cells. Curr. Stem Cell Rep. (2017). https://doi.org/10.1007/s40778-017-0094-4
18. Yin, A., Moes, D.J.A.R., van Hasselt, J.G.C., Swen, J.J., Guchelaar, H.J.: A review of mathematical models for tumor dynamics and treatment resistance evolution of solid tumors. CPT Pharmacomet. Syst. Pharmacol. (2019). https://doi.org/10.1002/psp4.12450
19. Bessonov, N., Pinna, G., Minarsky, A., Harel-Bellan, A., Morozova, N.: Mathematical modeling reveals the factors involved in the phenomena of cancer stem cells stabilization. PLoS ONE (2019). https://doi.org/10.1371/journal.pone.0224787
20. Meacci, L., Primicerio, M.: Mathematical models for tumors with cancer stem cells. Comp. Appl. Math. (2018). https://doi.org/10.1007/s40314-018-0707-2
21. Rodriguez-Brenesa, I.A., Komarova, N.L., Wodarz, N.: Evolutionary dynamics of feedback escape and the development of stem-cell-driven cancers. PNAS (2011). https://doi.org/10.1073/pnas.1107621108
22. Weeks, S.L., Barker, B., Bober, S., Cisneros, K., Cline, J., Thompson, A., Hlatky, L., Hahnfeldt, P., Enderling, H.: A multi-compartment mathematical model of cancer stem cell driven tumor growth dynamics. Bull. Math. Biol. (2014). https://doi.org/10.1007/s11538-014-9976-0
23. Komarova, N.: Stochastic modeling of drug resistance in cancer. J. Theor. Biol. (2006). https://doi.org/10.1016/j.jtbi.2005.08.003
24. López, A.G., Iarosz, K.C., Batista, A.M., Seoane, J.M., Viana, R.L., Sanjuán, M.A.F.: Nonlinear cancer chemotherapy: modelling the Norton-Simon hypothesis. Commun. Nonlinear Sci. Numer. Simul. (2019). https://doi.org/10.1016/j.cnsns.2018.11.006
25. Odoux, C., Fohrer, H., Hoppo, T., Guzik, L., Stolz, D.B., Lewis, D.W., Gollin, S.M., Gamblin, T.C., Geller, D.A., Lagasse, E.: A stochastic model for cancer stem cell origin in metastatic colon cancer. Cancer Res. (2008). https://doi.org/10.1158/0008-5472.CAN-07-5779
26. Yang, J., Sun, Z., Komarova, N.L.: Analysis of stochastic stem cell models with control. Math. Biosci. (2015). https://doi.org/10.1016/j.mbs.2015.06.001
27. Rodriguez-Brenes, I.A., Komarova, N.L., Wodarz, D.: The role of telomere shortening in carcinogenesis: a hybrid stochastic-deterministic approach. J. Theor. Biol. (2019). https://doi.org/10.1016/j.jtbi.2018.09.003
28. Pianosi, F., Beven, K., Freer, J., Hall, J.W., Rougier, J., Stephenson, D.B., Wagener, T.: Sensitivity analysis of environmental models: a systematic review with practical workflow. Environ. Model. Softw. (2016). https://doi.org/10.1016/j.envsoft.2016.02.008
29. Toth, Z., Talagrand, O., Candille, G., Zhu, Y.: Probability and ensemble forecasts. In: Jolliffe, I., Stephenson D.B. (eds.) Forecast Verification: A Practitioner's Guide in Atmospheric Science. Wiley, New York (2003). https://doi.org/10.1002/9781119960003
30. Pielke, P.A.: Mesoscale Meteorological Modeling. Academic Press, Cambridge (1984). https://doi.org/10.1016/C2009-0-02981-X
31. Mahfouf, J.F.: A numerical simulation of the surface water budget during HAPEX-MOBILHY. Bound. Lay. Meteorol. (1990). https://doi.org/10.1007/BF00154442

# The Role of Molecular Dynamics Simulations in Multiscale Modeling of Nanocarriers for Cancer Treatment

**Marina Kovacevic and Igor Balaz**

**Abstract** Nanoparticles hold great potential for improving the drug delivery of anticancer drugs. However, this potential is not fully utilized, evident from the small number of clinically approved nanoparticles. Nanoparticle design is evolving in complexity, yet most experimental methods cannot keep up since they lack the proper resolution for accurate characterization and testing necessary for clinical approval. The computational approach can advance research from the laboratory to clinical applications by offering insights into various phenomena with precision inaccessible to the experimental methods. It can also significantly reduce the time for new design testing and the costs associated with the experimental approach. To fully assess nanoparticles' efficacy, we need to consider a wide range of length and time scales. These scales include single atom resolution (for precise characterization of their physico-chemical properties), single cell scale (to assess nanoparticle-cell interactions and movement across the tissue), and whole tumour scale to evaluate their influence on the tumour. In this chapter, we present a Multiscale approach utilizing those scales with the focus on the role of the Molecular Dynamics Simulations.

## 1 Introduction

### *1.1 Role of Nanocarriers in Anticancer Treatment*

Traditional anticancer treatment faces serious challenges. Anticancer drugs are often poorly soluble or completely insoluble under physiological conditions. Once a drug enters the bloodstream, it encounters numerous barriers preventing it from reaching

M. Kovacevic (✉)
Faculty of Sciences, University of Novi Sad, Trg Dositeja Obradovica 3, Novi Sad, Serbia
e-mail: marinak@dh.uns.ac.rs

I. Balaz
Faculty of Agriculture, University of Novi Sad, Trg Dositeja Obradovića 8, Novi Sad, Serbia
e-mail: igor.balaz@df.uns.ac.rs



I. Balaz and A. Adamatzky (eds.), *Cancer, Complexity, Computation*,
Emergence, Complexity and Computation 46,
https://doi.org/10.1007/978-3-031-04379-6_9

its designated target [11]. Consequently, by ending up off-target, anticancer drugs cause serious side effects. Despite all the obstacles, some amount of a drug does end up where it should be, and the anticancer drug does what it was designed to do—inhibit tumour growth. The exact amount varies depending on the used drug, condition of patients, tumour type, etc., but it is estimated to be in the range of 0.01–0.1% of the injected dose [56, 61]. So in order to achieve the required efficacy, large doses and more frequent administration are necessary. However, this is far from the perfect solution as this approach can increase side-effects that arise from the non-specificity and non-selectivity of these drugs.

These issues can be solved by improving drug delivery [13, 74, 75]. In vivo fate of the molecules is determined by their physico-chemical and biochemical properties [11, 72]. Thus, drug can be attached to or encapsulated in a drug delivery vehicle specifically designed to carry it to its target.

Nanoparticles have proven to be an excellent tool for the cause. It has been shown that the use of nanoparticles improves solubility and stability of drugs in the organism, shields drugs from biodegradation, prolongs blood circulation time, induces preferential interactions between ligands, adjusts the charge density, etc. [23, 37, 49, 64]. Moreover, they can be functionalized with targeting moieties, tailored to release payload on demand, carry large quantities of drugs or carry different drugs as means of combinatorial therapy etc. [42, 76].

Their small size and unique characteristics offer possibilities to overcome barriers that traditional treatment encounters [16, 62].

They are incredibly versatile. They can be organic such as liposomes, dendrimers, polymer-based, etc., or inorganic—gold, silver, silica, etc. They come in various shapes, sizes and have a high surface area. These properties affect their reactivity, toughness, optical and other properties [31]. Additionally, they can be easily functionalized with various molecules. This provides the means to fine-tune their properties and tailor them to successfully control biocompatibility and biodistribution.

## 1.2 Issues with a Nanocarrier Design

Innovative, multifunctional, and complex nanoparticles have been designed, and synthesized [4, 32, 33, 36, 50, 53, 73]. However, this is not reflected in the design nor the numbers of clinically approved nanocarriers [43].

There are many possible explanations covering different stages in their development—from the initial approach to design to complex, expensive, and time-consuming in vitro and in vivo testing.

Although we cannot provide a definite solution to the problem mentioned above, in this chapter, we will give a brief overview of some issues we can encounter when designing novel nanocarriers for anticancer treatment and how to overcome at least some of them.

As mentioned in the introduction, in vivo fate of the molecules is determined by their physico-chemical and biochemical properties. In the case of the nanocarriers, these properties depend on the nanoparticle type, molecules chosen for the functionalization (from now on referred to as ligands), and the selected drug. Thus, a thorough understanding of each element's molecular structure is critical since it directly affects their interactions and hence the structure and stability of the final complex.

When we inject a nanocarrier into the bloodstream, we need to ensure that it will be stable in this environment so it stays in circulation long enough. However, stability has to be fine-tuned so the drug can be released from the complex once it reaches its target [60]. If the complex is not stable enough, the drug can be released prematurely and cause off-site toxicity. On the other hand, if the complex is too stable, the drug will not be released at the site of action.

A drug should also be shielded from biodegradation, immune system, and interactions with other molecules during the journey since these can alter both the structure and the intended function. To address these issues, we need to know precisely how does a nanocarrier with specific physico-chemical properties interacts with the physiological environment; how do changes in its physico-chemical properties relate to the changes in the physiological response; which properties or combination of properties ligands need to have to obtain specific structure, so our nanocarrier interacts with the environment in the way we predicted. Since we will inject more than one, we need to consider their collective behavior as well.

If we are successful, our nanocarriers will reach the tumour cells. Now they need to penetrate cells, go across the cell membrane, and deliver the drug to the site of action [43]. To achieve this, we need to understand how the nanocarriers interact with cells. For example, how many of them can penetrate the tumour? How deep into the tumour do they go? We also need to consider that tumours are heterogeneous structures containing more than one type of cell.

If we successfully overcome these hurdles, we need to know how our nanocarriers affect the tumour. Are they stopping the further growth of the tumour, are they successfully killing the tumour cells, how many cells? These are essential questions for adjusting the dosage and frequency of the administration.

Once we understand these relations, we can go back to the beginning and "program" nanocarriers' behavior by modifying their size, shape, and surface chemistry.

Sounds pretty straightforward. So what are the issues?

Let's consider the first step in the process—an adequate characterization of nanocarriers' attributes: size, composition, stability, and surface properties such as coating density, surface area, etc. [60].

This type of information is extremely challenging to obtain experimentally. Moreover, most experimental methods are not precise enough since they lack both the temporal and the spatial resolution necessary to investigate the dynamic changes and behavior of such small and complex systems [6].

Even the determination of nanocarrier size, which is one of the critical properties, is difficult to achieve precisely. There are numerous methods that can be used, such as dynamic light scattering (DLS), transmission electron microscopy (TEM), atomic force microscopy (AFM), etc. However, some methods are not suitable for all nanoparticle types and will not provide the accurate data we need. For example, TEM is not suitable for liposomes and proteins because they don't sufficiently deflect an electron beam, making them susceptible to imaging artifacts [60].

An additional issue is that nanocarriers are polydispersed. So after the synthesis, instead of having a solution of nanocarriers of a particular size, we end up with some size distribution. To consider them monodispersed, 90% of the distribution should lie within 5% of the median size [71]. However, this is extremely difficult or even impossible to achieve since most experimental methods used for the assessment of nanoparticle size lack the appropriate level of resolution [69].

Other critical points are the composition and surface properties of the nanocarriers [12]. As mentioned above, one of the advantages of using nanoparticles is the ease of their functionalization. As we gain a deeper understanding of the overall process, we can adjust the nanocarrier design more precisely to ensure that it ends up with the desired properties. This can be done, for example, by attaching multiple ligand types (e.g., one ligand type can be chosen to decrease non-specific interactions with proteins, one to increase the solubility, one can serve as a targeting ligand). As a result, the nanocarrier composition is getting more and more complex. Composition affects the structure and surface properties, and these parameters define the behavior of nanocarriers in vivo. If we want to have control over nanocarriers' in vivo mechanism of action, we need an accurate description of these properties. But their characterization and validation of our structure become challenging since there are no universal techniques for quantifying, for example, targeting ligands [20].

Although there are numerous additional examples we could mention about this stage, it would fall out of the scope of this chapter. So we will move forward to the cellular uptake of the nanocarriers.

There are several methods used to assess the cellular uptake, e.g., confocal microscopy, flow cytometry, etc. However, the issue with these methods is that it is difficult to distinguish between internalized material from the material bound to the outer plasma membrane [51]. As in the case of physico-chemical characterization, the underlying issue is the lack of the proper level of resolution.

Efficacy of the nanocarriers is assessed in vivo. However, there are obstacles, too, since the structure and size of the tumour itself can also affect the nanocarrier's efficacy. For example, small tumours, up to 2 mm in size, usually don't yet have developed vasculature, which can affect nanocarrier tumour penetration. Uptake can also vary with tumour growth rate [2]. Since properties we want to assess depend on multiple factors, we need to be careful when designing the experiments and interpreting results.

We can conclude that to establish a successful pipeline for designing and testing novel nanocarriers, we need accurate methods for their physico-chemical characterization, as well as a thorough understanding of the relation between their physico-chemical properties and tumour biology [60].

This requires knowledge from different fields, from chemistry, physics, biology, and medicine. In addition, methods with varying levels of resolution are also a requirement: from the atomistic level for the precise physico-chemical characterization, the cell level for accurately describing nanocarrier-cell interactions, and the whole tumour level for assessing overall nanoparticle efficacy.

Although this is possible to achieve experimentally, this approach is time-consuming, expensive, and not precise enough for establishing a successful and effective pipeline from the bench to the clinic.

## 2 Multiscale Problem Requires Multiscale Solution

As evident from the previous sections, we need to consider a wide range of length and time scales: molecular, cellular, tissue, etc. The accurate description on each level provides information about different aspects necessary for successful nanocarrier design. However, a common drawback among experimental methods is an inadequate resolution. In some cases, to measure one property with the appropriate level of accuracy, we need to perform multiple analyses. This is time-consuming and expensive.

Moreover, the clinical impact is not an independent phenomenon but is a result of the interactions of elements across multiple scales [17]. Thus focusing on the isolated elements cannot be sufficient for a complete understanding of the nanocarriers' in vivo behavior. Due to the versatility of nanoparticles and the wide range of their physico-chemical characteristics, we could use no standardized methods. As a result, there is no standardized pipeline that connects different resolutions.

These issues can be resolved with the computational approach. There are many available methods and models describing the relevant length and time scales with the precision inaccessible to the experiments. By combining them, we are presented with the opportunity to investigate realistic biological scenarios while significantly reducing the time for the new design testing and the costs associated with the experimental approach.

However, most computational models still focus on a single scale. Multiscale modeling presents a new set of challenges since we need to consider many components and parameters from different scales. We also need to consider the complex relationship between them [15]. But advances in model development and in computational power over the years are making this possible.

The first question that arises in Multiscale modeling is how to choose appropriate methods and scales?

Although in an "idealistic" scenario, we would consider everything from the quantum level to the entire organism, this would not be the best approach. Apart from this being technically extremely difficult or even impossible to achieve (at least from the perspective of the computational cost), we would also end up with an enormous amount of information, most of which would probably be redundant. Thus, it would become challenging to extract valuable details or patterns.

Instead, we need to focus on elements that contribute the most to the phenomena or process we want to investigate and choose the model accordingly.

When choosing the models (i.e., length and time scales), it is necessary to find the balance between the accuracy and the computational cost.

In this chapter, we will describe the Multiscale model for evaluating the efficacy of nanocarriers in cancer treatment. Thus, we focused on three different scales—Molecular scale, Cell scale, and Tissue scale.

1. **Molecular scale—Single-atom resolution**. Molecular level description of the nanocarriers is a crucial aspect that has a significant impact on the clinical outcome. If we understand, on this level, which properties determine the nanocarrier structure and dynamics, we can have more control over the final product. This step can lead to a more informative and efficient design of the nanocarriers.

Molecular dynamics (MD) simulations give us an atomistic level description of the simulated systems. This level of resolution is mostly experimentally inaccessible. Since physico-chemical characterization is the fundamental step when designing a nanocarrier, MD can provide detailed insight into its structure and dynamics.

This can be a standalone method used before synthesis to save time and resources. We can run a preliminary evaluation of a large number of different systems. Based on this assessment, we can choose the most promising designs and start the experimental work from there.

As a part of the Multiscale model for testing treatment efficacy, Molecular dynamics simulations can be used to extract realistic parameters for the higher scale simulations.

2. **Cell scale—Single cell resolution**. The efficacy of cancer treatment depends to a great extent on the extravasation to the tumour site, and tissue penetration within a tumour [40, 58]. To realistically evaluate the nanocarrier's efficacy, we need to understand nanocarrier-cell interactions and the movement of the nanocarriers through the tissue. This resolution can be accessed with stochastic reaction-diffusion simulations.

Input parameters for this scale can be obtained from both the Molecular scale and the Tissue scale. Due to the computational cost of running simulations of nanocarriers' effect across the entire tumour (Tissue scale), we can take representative "tumour slices" and use them as a starting point in the Cell scale. To assess the effect of realistic nanocarriers parameters, as inputs, we can use the results obtained in the Molecular scale.

3. **Tissue scale—Whole tumour resolution**.

To accurately model nanocarriers' efficacy, first, we need to model a realistic tumour [57].

To generate an entire realistic tumour, we can run simulations of tumour growth from a single cell by performing agent-based modeling. A tumour can be modeled as a heterogeneous structure with the vasculature. Results of this level of simulations can be translated into input parameters for the Cell scale.

In the following sections, we will briefly describe different scale methods and models: From top to bottom (Tissue scale, Cell scale, and Molecular scale).

As evident from the title, we will focus mainly on the role of Molecular Dynamics simulations in this multiscale approach. First, we will provide a brief theoretical overview of Molecular Dynamics and some practical aspects to be considered when running the simulations. Next, we will give an overview of the basic analysis we can perform to describe our system accurately. Finally, we will discuss how all of these scales are connected and what information we can obtain.

## 2.1 Tissue Scale—Generating a Virtual Tumour

Virtual tumour growth can be simulated using PhysiCell—an open-source, agent-based 2D and 3D modeling framework [19].

It incorporates two aspects:

(1) at the scale of the individual cell, it operates as an agent-based environment. Tumour dynamics emerges as a result of a large number of interacting cells. Cells respond to the cues from their microenvironment and, in turn, influence their microenvironment. This aspect includes sub-models for cell fluid and solid volume changes, cycle progression, apoptosis, necrosis, mechanics, and motility, as well as functions for diffusion, decay, cell-based secretions/uptake, and bulk supply/uptake.

(2) Agent-based environment is coupled with the biotransport solver—BioFVM [18], a multi-substrate diffusion solver that deals with vectors of diffusing substrates. This aspect includes functions for diffusion, decay, cell-based secretions/uptake, and bulk supply/uptake for each substrate.

Although PhysiCell is suitable as-is, here we will refer to the modified version of PhyisiCell created within EVONANO platform [58] since it includes cancer stem cells and vasculature growth. Both of those elements have a substantial impact on tumour dynamics.

Cancer stem cells are a minor subpopulation of cancer cells. They are capable of unlimited self-renewal, differentiation, and tumorigenesis [14]. Additionally, they are insensitive to most cancer treatments, and their numbers within a tumour often increase after the chemotherapy [35, 67].

They also have an essential role in tumour vasculature growth.

When the tumour reaches approximately 1–2 mm, it still relies on the pre-existing vasculature for blood supply. However, it is usually not enough to support further tumour growth [27, 66]. This means that cancer cells don’t get enough nutrients or oxygen, which leads to necrotic cell death and the state of hypoxia. Under these conditions, cancer stem cells survive and produce factors that stimulate angiogenesis, promoting tumour progression [8, 67].

Thus, both of these elements are important in modeling realistic tumours and treatment.

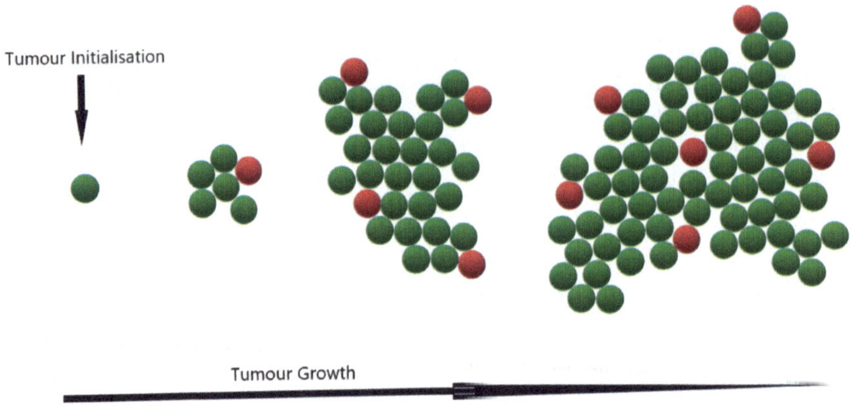


**Fig. 1** Tumour growth illustration

In simulations, tumour growth is initialized with a single cancer cell. Then, it grows until it reaches the size of approximately half a million cancer cells. At that moment, the simulation is stopped (Fig. 1).

The output of this simulation is heterogeneous, vascular tumour. This output is used to generate representative scenarios for the Cell scale simulations.

## 2.2 *Cell Scale*

At this point, we have our virtual, vascular, and heterogeneous tumour created using the higher-scale simulations. The next step is the investigation of how nanocarrier parameters influence tumour dynamics.

Another level of resolution is necessary since the modeling of nanocarriers' influence over a whole tumour would be computationally expensive. It would also limit the degree of the dynamics we aim to investigate.

To circumvent this, we can use the output from the Tissue scale simulations and systematically select representative sections of tissue—"tumour slices" to test the chosen nanocarriers.

For the investigation of nanocarrier-cell interactions and movement of nanocarriers across the tissue, we can use STochastic Engine for Pathways Simulations (STEPS)—an open-source stochastic simulator of reaction-diffusion systems in arbitrarily complex 3D geometries [10, 28, 70].

In this model, the Michaelis–Menten reaction network is used to model nanocarrier-cell interactions using the following equation [22]:

$$NP_f + R\overrightarrow{K_a},_{K_d} C\overrightarrow{K_i} NP_i + R \tag{1}$$

where $NP_f$ represents the free nanocarrier, $R$—receptors on the cell membrane, $K_a$ and $K_d$ represent the binding rate of the nanocarriers to the receptors and the dissociation rate of nanocarriers release from the receptors, respectively. $C$ represents nanocarrier-receptor complexes, $NP_i$ is internalized nanocarrier, and $K_i$ is nanocarrier internalization rate.

Nanocarriers can move with the probability:

$$D/L^2 \tag{2}$$

where $D$ is the diffusion coefficient, and $L$ is the length of a cell.

The amount of nanocarriers that can extravasate into the tumour within an idealized murine model, $NP0$, is calculated using the following equation:

$$NP0 = IDW\frac{PIDS^2L}{V_t} \times \frac{N_A}{ME} \tag{3}$$

where $ID$ is the total injected dose of nanocarriers, $W$—the weight of the mouse, $PID$ represents the percentage of the injected dose that reaches the tumour, $S$—the characteristic length scale of the cell, $L$—the total required penetration depth, $V_t$—the tumour volume, $M$ the molecular mass of the nanocarrier payload, $E$ the total number of payload molecules, and $N_A$—Avogadro's constant.

The following equation gives the threshold of nanocarriers required to induce cell death:

$$NP_{max} = \frac{PS^3N_A}{E} \tag{4}$$

where $P$ is Potency and is represented by the IC90 of the anti-cancer drug. We assume that the nanocarriers' cytotoxic effect is a function of the total carried payload, and that the nanocarrier vector does not alter the properties of the drug.

Since the overall idea is to test the influence of realistic nanocarriers, input parameters for Virtual cell scale are obtained from the Molecular Dynamics simulations.

## 3 Molecular Dynamics

Since the focus of this chapter is Molecular dynamics, first, we will provide a brief theoretical introduction. Next, we will describe the general simulation procedure and some practical aspects to be considered. Then we will go over basic analysis which can be used to characterize nanocarriers. Finally, we will show how we can incorporate obtained parameters in previously described Cell scale and Tissue scale models.

### The overall concept of Molecular Dynamics

We start by setting the system's initial conditions—a list of all atoms' positions, connections, and velocities. Then we specify the interaction potential (force field), which will allow us to derive forces among them. By using Newton's second law, we can understand how those forces affect their motions. Finally, by allowing the atoms to interact for a certain amount of time, we get the dynamic evolution of the investigated system.

The collective behavior of the atoms in the system provides the connection between the molecular scale to meso/macro phenomena. Once we know the microscopic states of the system at a particular time, we can calculate the macroscopic properties using statistical mechanics.

### What are some advantages of performing this type of simulations?

In general, Molecular Dynamics (MD) simulations can be used to study the structure and dynamics of the nanocarriers, various molecular perturbations and to observe dynamic processes over time [29].

Molecular Dynamics provides dynamic evolution of the system on a scale of a single atom. With MD, we can see the precise position of atoms at each point in time. This is impossible to achieve experimentally.

Atomistic molecular dynamics simulations accurately represent physical, chemical, and thermodynamical properties. Thus, they can provide insight into how different parameters affect the structure and surface properties of the nanocarriers.

Moreover, we can set and precisely control the conditions of the simulations. This allows us to run simulations under various conditions, meaning we can discover the effects of a wide range of perturbations.

Virtual "synthesis" of the nanocarriers is quick, easy, and requires substantially less financial resources than its experimental counterpart. Furthermore, by using this method, changes in the composition can be easily introduced, and a large number of different systems can easily be simulated and compared.

For example, we can examine nanoparticle properties such as shape and size, mechanical or elastic properties [3, 5, 41, 63]. We can focus on the nanocarriers and investigate the effects of different properties on their structure, such as the effect of the ligand/drug hydrophobicity/hydrophilicity, the impact of ligand length and distribution, surface coverage, etc. [34, 55, 63] We can compare structures of nanocarriers in different solutions, in presence/absence or ions, etc. [25, 39]. We can go deeper into the main interactions between functionalized nanoparticle and drug or interactions between the nanocarrier and cell membrane [25, 59]. We can investigate the parameters of the protein corona by simulating interactions between the nanocarrier and blood proteins [24]. All of these aspects are very challenging to investigate with any experimental technique [29].

However, there are limitations to this method. Due to the size of the systems, we can mainly focus on small systems and short time scales [48] (Thus, the need for the coupling with the higher scale simulations).

### 3.1 Atomistic Molecular Dynamics Simulations

As mentioned, in the atomistic MD simulations, we model the motion of each atom. Each atom is represented as a ball, centered at the nucleus, with the precise mass. On this scale, electronic properties are averaged. Instead of modeling them explicitly, electrons are represented as the partial charges on the atoms. Bonds between the atoms are represented as strings with a specific force constant and resting length.

Parameters for force fields (interaction potentials) used at this scale are derived from the experiments and/or quantum mechanical calculations. General description of the force fields—FF includes bonded (associated with covalent bonds, bond angles, and bond dihedrals) and non-bonded terms (associated with van der Waals forces and electrostatic charge).

$$V_{total} = V_{bonded} + V_{nonbonded} \tag{5}$$

where

$$V_{bonded} = V_{stretch} + V_{bond} + V_{dihedral} + V_{improper} \tag{6}$$

and

$$V_{nonbonded} = V_{electrostatic} + V_{vanderWaals} \tag{7}$$

Since parameters for the force-field can be derived from different sources, there are multiple force fields that can be used.

Selecting the correct one is very important. The choice depends on the type of molecule you want to simulate and the question you would like to answer. All force fields are approximations, and their parameters are optimized to reproduce experimental data under particular conditions. For example, the force field applied to calculate the solvation free-energy of amino side chains and proteins might not be suitable for the same type of calculations of inorganic nanocarriers. Special care should be taken when considering electrostatic interactions and the functional form of the local potentials, polarization effects, the representation of the solvent, etc. [54]. The Force field has a significant impact on the reliability of the results and thus should be chosen with caution.

The next step in the simulations is the integration of Newton's equations of motion. This is done in discrete time steps. The timestep is determined by energy conservation at short times. It should be chosen, so it is significantly shorter than the shortest relevant time scale in the simulation. At each timestep, an enormous number of calculations are occurring. Apart from slowing down the simulation, at long timescales and larger systems, the number of required timesteps can become so large that simulations become exceptionally computationally expensive. Thus, the size of the system and timescale become obvious limitations for this method.

The main output of the simulations is the molecular trajectory file which contains the time-evolving coordinates of a system and optionally velocities and forces.

## 3.2 Practical Aspects

The first step in MD is the preparation of the systems we want to simulate.

In general, we start by using the coordinates of known crystal structures. X-ray or NMR structures can be a good starting point since the coordinates of most atoms are known with certainty. We might need to add some missing atoms to the structures, e.g., hydrogen atoms.

However, in this chapter, we are focused on the design of new nanocarriers. We aim to test theoretical designs before synthesis, meaning that the crystal structure is not available. Although we can run simulations of synthesized nanoparticles (which is also necessary to validate our results), the advantage of the MD is that we can simulate a large number of different systems. Based on the results, we can choose the most promising ones and start the experimental work from there to save time and resources.

The lack of crystal structure is not an issue since we can build our system from scratch. The first step is the preparation of the structure. This can be done in Avogadro software [21]. After we prepare our structure, we need to assign molecular electrostatic potential-based charges, which can be done using, for example, RED server—an open web service [65]. Since we are focusing on the nanocarriers, which usually comprise more than one element (nanoparticle, ligands, drugs), we can prepare these structures separately and then assemble them using another free software—Packmol [38].

When our system is assembled, we add appropriate solvent molecules (e.g., water, salt, ions). Then, finally, we assign the force field parameters.

In general, after we prepared the system, there are several essential steps in MD simulations: Initialization, equilibration, and the production phase (Fig. 2).

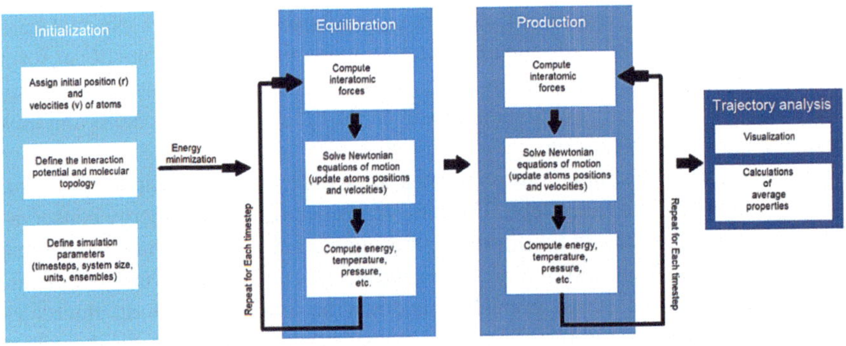


**Fig. 2** General steps in molecular dynamics simulations

1. **Initialization**. This phase requires specifying the initial coordinates and velocities of all atoms in the system (nanoparticle, water, salt, ions, etc.). Initial velocities of atoms are set so that the total kinetic energy of the system corresponds to the value of the targeted temperature. Although prepared structures are a good starting point, there could still be some overlaps of atoms in different residues. So we need to perform the Energy minimization to adjust the initial structure to the force field and relax possible steric clashes.
2. **Equilibration**. After the Initialization, the system is out of equilibrium. During the heating to the targeted temperature, kinetic energy is added to the system. So during this phase, there is an exchange between kinetic and potential energies. This means that properties we want to measure are still fluctuating to a large extent. Equilibration is achieved when the fluctuations occur about the mean value. It is difficult to know in advance how long we should run the equilibration phase. The best approach would be to observe the property we want to measure and wait until it approaches the constant value.
3. **Production**. Once the equilibrium is reached, the production phase starts. This phase is the only part of the dynamics simulation from which data is accumulated. Since the properties we are interested in represent the time average behavior of the atoms at the equilibrium, to obtain statistically significant results, simulations should be run long enough to match the kinetics of the investigated phenomena.

There is a wide range of software we can use for Molecular Dynamics simulations, such as AMBER [9, 68], GROMACS [1], CHARMM [7], LAMMPS [47], NAMD [46] etc. General simulation protocol is the same for all packages, and they all perform similar computations[29]. However, computational efficiency can vary. This depends on the hardware, whether you can use GPU, is the code parallelized, etc. When choosing the software, you should consider aspects relevant to the systems you wish to investigate. For example, it might be important whether you can model different systems and solvent environments in this software, use different force fields for the system parametrization, are there incorporated specific tools for the trajectory analyses, etc.

## 3.3 *Data Analysis*

Once we obtain molecular trajectories, many different analyses can be performed depending on the goal and the system of interest. Here, we are focusing on the most common ones since different systems require different analysis types, and going through all of the possibilities would fall out of the scope of this chapter.

### 3.3.1 Radial Distribution Function

One of the most common analyses is the calculation of the Radial distribution function, RDF.

RDF describes how the density of the surrounding particles varies as a function of distance from the reference particle. Thus, it is a valuable tool to describe the atomic structure of the simulated system. RDF describes the local structural organization around any given atom and can provide information about the dynamical change of the structure. In general, calculating RDF involves determining how many particles are within a distance of r and r+dr away from a particle.

We will illustrate this on the example from our previously published paper [34].

In this example, we have simulated gold nanoparticle 5.5 nm in diameter functionalized with two ligands types: (i) ligand which carries the hydrophobic anticancer drug Quinolinol, (ii) zwitterionic background ligand used to increase the solubility of the system. The ratio of the ligands carrying the drug to the background ligands was 1:1. Simulations were run in the aqueous solution.

In Fig. 3 we have the corresponding RDF graph.

Coordinate origin (0.0) on the graph represents the center of the gold nanoparticle core (the central gold atom is taken as a reference particle). The red line on the RDF graph describes the density of all atoms in the entire coating, and the broadness of this

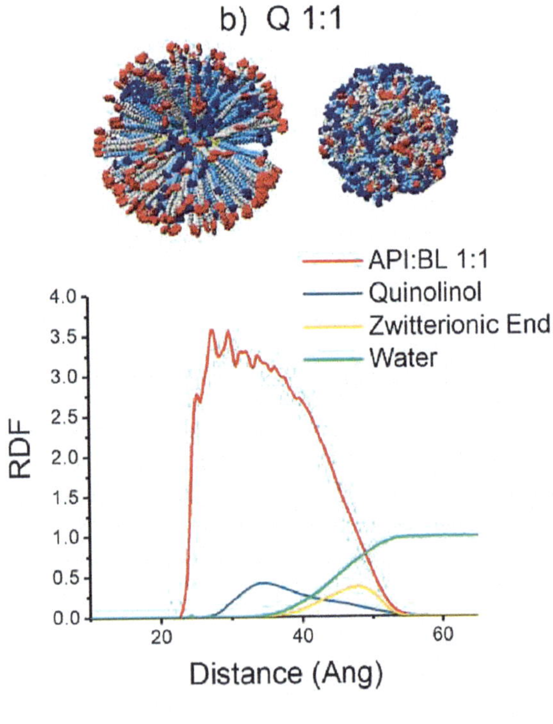


**Fig. 3** RDF plot and simulation snapshots. The first frame shows the nanocarrier structure before the simulations, while the second frame shows the equilibrated structure after the simulations. In the snapshots ligand carrying the drug is colored grey, Quinolinol is colored red, and the background ligand is colored light blue with the dark blue zwitterionic terminal end. RDF graph corresponding to the second frame is shown below the snapshots [34]

peak indicates ligand flexibility. Suppose we pay attention separately to the Quinolinol and zwitterionic terminal end. In that case, we can see that the average density of Quinolinol is closer to the gold nanoparticle surface than the average density of the zwitterionic terminal end. This tells us that a large number of the Quinolinol molecules are hidden beneath the coating, away from the surface in contact with the solvent. This is in agreement with the hydrophobic nature of the drug molecules.

The fact that most of the Quinolinol molecules are hidden below the coating is also evident from the simulation snapshots. The left image shows the theoretical structure where we can visually observe the amount of the attached drug (colored red). In the right snapshot, which represents the equilibrated structure in the aqueous solution, we can see that the blue color is dominant on the surface, which represents the background ligands.

### 3.3.2 Radius of Gyration and Asphericity

The radius of gyration, $R_g$, provides an estimate of the compactness of the investigated structure. This is an essential aspect since it reflects its stability. The more this value fluctuates, the less the structure is stable.

It is calculated as:

$$< R_g^2 >= \left( \frac{1}{M} < [\sum_{i=1}^{N} m_i [r_i] - R]^2 > \right) \tag{8}$$

where $R$ is the center of mass of the molecule, $M$ is the total mass of the molecule, $r_i$ is the position of the $i_{th}$ atom, $m_i$ is the mass of the $i_{th}$ atom, $N$ is number of atoms for the given molecule.

Apart from the radius of gyration, it is also possible to estimate the radius of gyration tensor $S$. Its eigenvalues are principal moments of the gyration tensor ($\lambda_x$, $\lambda_y$, $\lambda_z$).

$$< R_g^2 >= \lambda_x^2 + \lambda_y^2 + \lambda_z^2 \tag{9}$$

The principal moments can be combined to calculate other related shape properties such as Asphericity ($b$):

$$b = \lambda_z^2 - \frac{1}{2}(\lambda_x^2 + \lambda_y^2) \tag{10}$$

Asphericity is a measure of deviation from spherical symmetry. It is limited between values 0 and 1.

When

$$\lambda_x = \lambda_y = \lambda_z = 0 \tag{11}$$

the distribution of particles is spherically symmetric.

Asphericity can be helpful in describing the shape of the nanoparticle, e.g., under physiological conditions.

### 3.3.3 Nanoparticle Size and Coating Thickness

Besides the shape, the size of the nanocarriers is another crucial parameter since it plays a significant role in their biological behavior. Thus determining it accurately is of utmost importance.

Considering that, we chose to pay more attention to describing the role of Molecular Dynamics simulations for this purpose. To do so, we will show an example of gold nanocarriers.

For this example, we will use gold nanoparticles with the core size of 5.5 nm functionalized with one ligand type carrying the different anticancer drugs: Niclosamide (NCL), Gemcitabine (GEM), Panobinostat (PAN), and Quinolinol (OQL). Core size and the ligand carrying the drugs are the same for all systems. We wanted to illustrate how the change in just one part of the system affects the overall nanocarrier structure.

Size can be estimated even before running the simulations. The first step would be the literal "drawing" of the molecule (Fig. 4). We can do that based on our knowledge of the atoms, their sizes, types of bonds they can form with other atoms, length of those bonds, angles, etc. So any chemist drawing the same molecule will draw highly similar structures.

If we represented these structures in three dimensions, we could see how our nanocarriers look (Fig. 5). We can do this using tools such as Chimera [45], VMD [30], Avogradro etc.

Once we have prepared the desired molecular structures, we can simply measure the distance between terminal atoms on opposite sides of the nanocarriers to assess the size (Fig. 6).

Structures in Fig. 5 represent the best estimate we can get at this point. However, they are not realistic. Although they might look like this in a vacuum, the structure will probably change once we put them in a solution. This change in structures happens due to numerous interactions we cannot consider in advance, such as solvent effects and all the intra- and intermolecular interactions. This can certainly affect the overall size, shape, and coating thickness of the nanoparticle. If our systems have some degree of flexibility, the theoretical prediction of the nanocarriers' size can be over- or underestimated. Thus, this is one of the most crucial properties to determine accurately since the change in the structure can affect its function.

**The measured size of the nanocarrier** ($d_m$) can be calculated from the trajectories based on the average distance between the center of the nanocarrier and the terminal groups [26].

In this example, we also calculated the predicted size of the nanocarriers, i.e., nanocarrier size before the simulations ($d_{ext}$). This is not a necessary step in the data

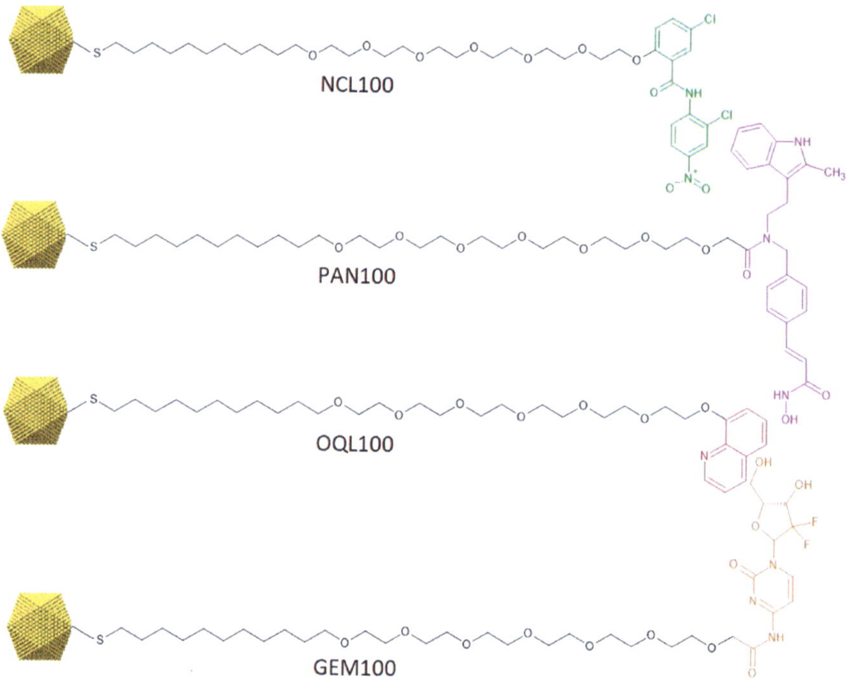


**Fig. 4** Structures of the nanocarriers used in this example. Each nanocarrier consists of a gold nanoparticle core (AuNP) 5.5 nm in diameter. Ligand carrying the drug is the same for all systems and is colored grey. NCL100: AuNP functionalized with ligand carrying anticancer drug Niclosamide. Niclosamide molecule is colored green, PAN100: AuNP functionalized with ligand carrying anticancer drug Panobinostat (colored pink), OQL100: AuNP functionalized with ligand carrying anticancer drug Quinolinol (colored red), GEM100: AuNP functionalized with ligand carrying anticancer drug Gemcitabine (colored orange)

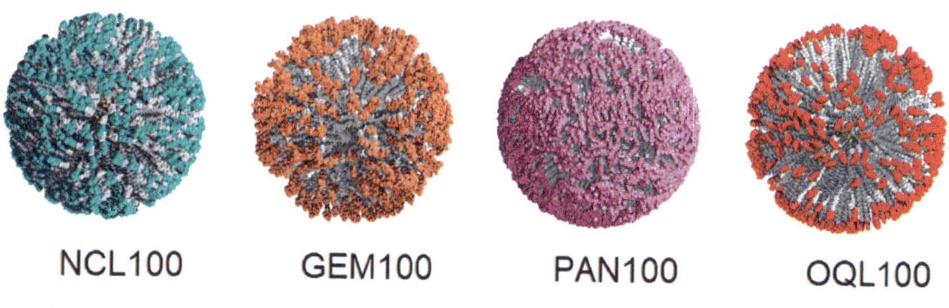


**Fig. 5** Three-dimensional representation of the theoretical design of the nanocarriers. Each nanocarrier consists of a gold nanoparticle core (AuNP) 5.5 nm in diameter. Ligand carrying the drug is same for all systems and is colored grey **a** NCL—AuNP functionalized with ligand carrying anticancer drug Niclosamide (Niclosamide molecules are colored green), **b** GEM—AuNP functionalized with ligand carrying anticancer drug Gemcitabine (colored orange), **c** PAN—AuNP functionalized with ligand carrying anticancer drug Panobinostat (colored pink), **d** AuNP functionalized with ligand carrying anticancer drug Quinolinol (colored red)

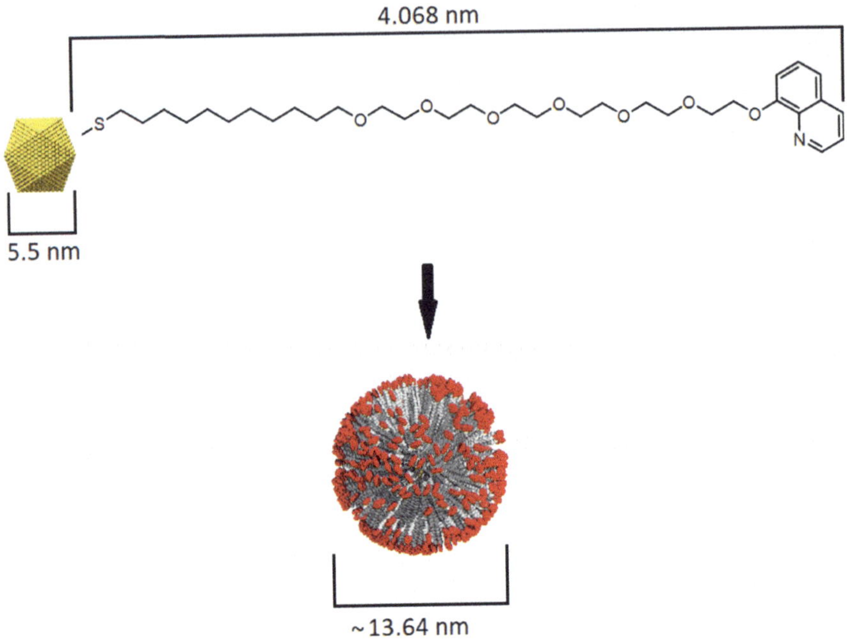


**Fig. 6** An example of theoretical prediction of the nanocarrier size

analysis. However, we wanted to demonstrate to what extent the measured size can deviate from the predicted to illustrate the importance of these calculations. **The measured size of the nanocarrier** ($d_m$) was calculated from the trajectories as the average distance between the center of the mass of the Au core and the terminal groups. **Predicted size of the nanocarriers**, i.e., nanocarrier size before the simulations ($d_{ext}$) was calculated as the average distance between the center of mass of the Au core and averaged lengths of extended ligands. Visual results are shown in Fig. 7.

As it can be observed from Fig. 7 there are significant deviations in both size and shape for all simulated nanocarriers. In this example, deviations of measured size from the predicted are in the range from 15–20%.

When it comes to nanoparticle functionalization, besides the size and shape, properties of the coating have an important role since they are linked to the dynamics and interactions between the nanocarriers [52]. **Coating thickness** can be calculated from the trajectories as the thickness which contains 97% of the coating atoms [23].

### 3.3.4 Solvent Accessible Surface Area

**The solvent accessible surface area—SASA** describes the area of the molecular surface which is exposed enough so that molecule of interest can interact with the

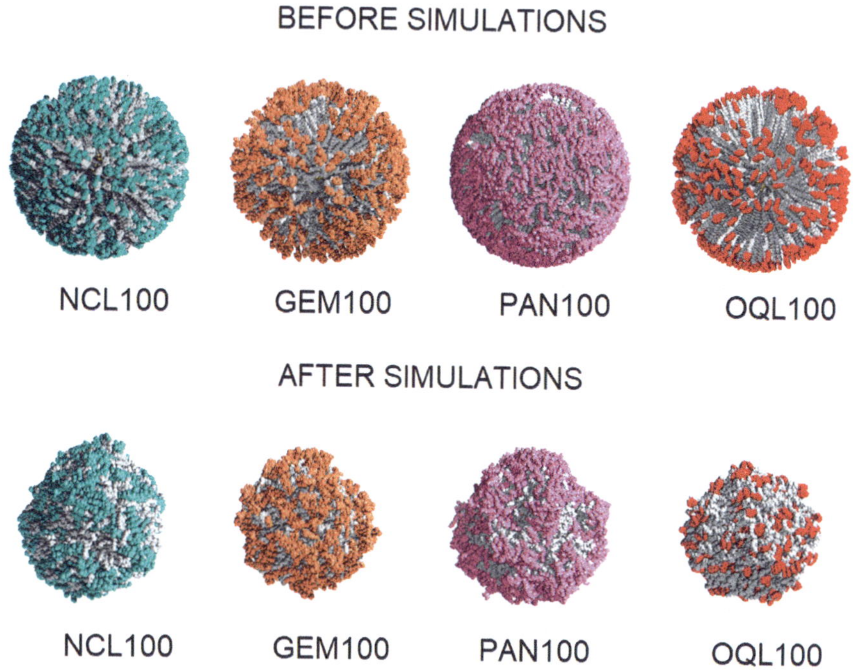


**Fig. 7** Visual comparison of the predicted versus measured nanoparticle size. First row represents the Predicted structures of nanocarriers. Second row represents the equilibrated structures in aqueous solutions after the simulations. Column **a** NCL—AuNP carrying the Niclosamide (colored green), Column **b** GEM—AuNP carrying the Gemcitabine (colored orange), Column **c** PAN—AuNP carrying the Panobinostat (colored pink), Column **d** OQL—AuNP carrying the Quinolinol (colored red)

solvent molecules. These interactions have a crucial role in the stabilization of the structure and molecule solubility.

It is most often calculated using the "rolling ball" algorithm. In this approach, a probe sphere, representing the solvent molecules, rolls over the Van der Waals surface and maps out the area. Probe estimates water molecule's radius—a typical value is 1.4Å.

SASA can also be calculated for separate fragments to see if they are adequately exposed, i.e., available for interactions with the surrounding environment.

In the case of the nanocarriers with covalently attached drugs, these SASA calculations can be used to determine the percentage of the accessible drug molecules. This value represents the "yield" of the equilibrated nanocarrier, i.e., if our starting structure before the simulation initially had X drug molecules attached, how many drug molecules are available for interactions with their environment in a solution—this value is used as Payload parameter in the Cell scale simulations.

Accurately assessing the carried payload is crucial since it can significantly affect the dosage and frequency of administration in a realistic scenario.

### 3.3.5 Diffusion Coefficient

**Diffusion coefficient** is a fundamental transport property. It plays a major role in describing mass transfer processes.

It can be calculated from the obtained trajectories using the Einstein relation:

$$2nD = \lim_{t \to \infty} \left( \frac{MSD}{t} \right) \tag{12}$$

where $n$ is the number of dimensions, $D$ is the diffusion coefficient, $MSD$ is the mean square displacement and $t$ is time.

$MSD$ is calculated by using distance traveled from the initial position.

The obtained values of the Diffusion coefficient can be directly used as input for higher scale simulations.

Since the diffusion is a dynamic property, simulation time should be carefully chosen since it is unlikely we can achieve convergence with short simulation time.

## 4 How Is It All Connected?—A Short Summary

This section will describe how to connect the above-described models and scales for efficient testing of nanocarriers against representative biological scenarios (Fig. 8).

The first step is the generation of the virtual tumour via modified PhysiCell. A tumour is set to grow from a single cell. While growing, it is developing vasculature. Simulation is stopped when a tumour grows to approximately 500.000 cells. A tumour is heterogeneous, containing both cancer cells and cancer stem cells. We use this virtual tumour to generate representative scenarios for the nanocarrier testing.

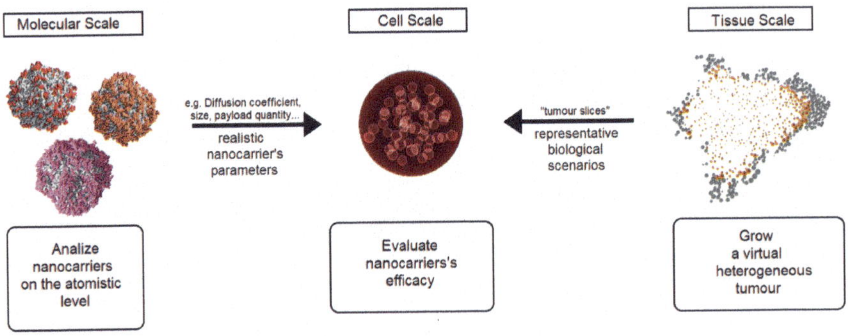


**Fig. 8** Overview of the multiscale model for evaluating nanocarrier's efficacy. At the Molecular scale, we analyze the nanocarriers structure and dynamics. We extract realistic parameters to use as inputs for the Cell Scale. At the Tissue scale we generate a virtual, heterogeneous tumour. We use the grown tumour to extract the representative biological scenarios. At the Cell scale, we test the chosen nanocarriers efficacy across the tumour slices

To obtain realistic nanocarriers' parameters, we perform Molecular Dynamics simulations. The advantage of running Molecular Dynamics simulations is that we can simulate a large number of systems and perform their detailed analyses as described above. However, testing each design in the Multiscale model can be computationally expensive. To save resources, we can use MD for the preliminary assessment of nanocarriers' structure and dynamics and choose the most promising candidates for the higher-scale simulations.

Parameters obtained at this scale such as Nanocarrier size, Diffusion Coefficient, Payload quantity etc. can be used as input parameters for the Cell scale (Specific parameters are dependent on the type of the nanocarriers we want to simulate).

In the Cell scale, we consider nanocarrier-cell interactions, nanocarriers' movement across the tissue, and nanocarrier cytotoxicity. Thus we can obtain a preliminary picture of the chosen treatments' efficacy for a heterogeneous tumour.

As input parameters for this scale, we use the combination of outputs from the Tissue scale and the Molecular scale, as described above.

As in the case of Molecular Dynamics, various analyses can be performed. For example, we can calculate the number of internalized nanocarriers, the distribution of nanocarriers across the tissue, the number of killed cancer cells or cancer stem cells. We can also assess the concentration of the nanocarriers necessary to kill the cancer cells with regard to the payload quantity. With this data, we can make a preliminary assessment of the nanocarrier efficacy. The results could also help us determine the dosage and frequency of the administration.

On the other hand, we can make crucial connections between nanocarriers' physico-chemical parameters and their effect on the tumour. We can achieve that by testing multiple nanocarriers. For example, we can test nanocarriers of different sizes, carrying different amounts of payload, nanocarriers with different diffusion coefficients, etc. and assess how the change in nanocarrier parameters influences the tumour dynamic. Analysis of these results can provide valuable insights that can guide nanocarrier design.

Since the focus of this chapter is the role of the Molecular Dynamic simulation in this Multiscale model, some details regarding the Tissue and Cell scale are omitted. For more information regarding those models and their role in the Multiscale modeling, see the work by Stillman et al. [58].

## 5 Conclusion

Nanocarriers show great potential in drug delivery. By taking advantage of their small size and unique properties, we can create novel nanocarrier designs to overcome numerous obstacles traditional cancer treatments encounter. In addition, by improving the delivery of clinically approved anticancer therapies, we can significantly reduce their side effects and increase their efficacy.

However, the rate of the translation of nanocarriers from the laboratory to the clinic has been relatively slow. Although there are standardized methods and developed pipelines for small molecules, the existing situation is different for nanocarriers. We need to solve many experimental difficulties to create a standardized pipeline and speed this process up.

Designing an effective nanocarrier implies trial and error. First, each nanocarrier needs to be synthesized and tested to assess its efficacy. Then, based on the efficacy assessment, we take a step back and introduce modifications to improve the design. This is extremely expensive and time-consuming.

During the testing phase, besides the accurate description of their properties on the nanoscale, we need to understand how those properties relate to the tumour dynamics. Obtaining this knowledge requires connecting different length and time scales. However, a common disadvantage of in vitro and in vivo methods is the lack of the resolution needed for acquiring precise and accurate data.

A computational approach can be used to complement the experiments. During the years, many models have been developed to describe various phenomena with precision inaccessible to the experimental methods spanning over all relevant length and time scales. Multiscale simulations can connect different levels of resolution, thus providing a preliminary assessment of nanocarrier efficacy. This approach is significantly quicker and requires much fewer resources than the purely experimental approach.

In this chapter, we gave a brief overview of three different levels of resolution: Molecular, Cell, and Tissue. Next, we described how they are all connected in the presented Multiscale Model.

We paid particular attention to the Molecular Dynamics simulations and their role within this model.

Molecular Dynamics by itself is gaining more attention in the field of Drug delivery. It operates on a resolution of a single atom, which can lead to important insights into the structure and dynamics of the nanocarriers.

Virtual "synthesis" of the nanocarriers is more straightforward and less time-consuming than its experimental counterpart. By using this method, changes in the system's composition can be much more easily introduced. This allows for a fast preliminary investigation of a large number of systems.

In order to be confident in the method and the obtained results, validation against the experimental values is necessary. From this necessity, we can conclude that the description obtained with Molecular Dynamics simulations is realistic.

Outputs from MD simulations can be used as realistic parameters for the higher-scale simulations. By integrating the molecular level of resolution with the Cell scale and the Tissue scale, we create a computational pipeline for quick preliminary testing of the nanocarriers' efficacy, shortening the path from the bench to the clinic.

**Acknowledgements** I would like to thank my supervisor dr. Igor Balaz for his consistent support and guidance during the writing of this chapter. This work was supported by the European union's Horizon 2020 research and innovation programme under grant agreement No 800983.

## References


1. Abraham, M.J., Murtola, T., Schulz, R., Pall, S., Smith, J.C., Hess, B., Lindahl, E.: GROMACS: high performance molecular simulations through multi-level parallelism from laptops to supercomputers. Software X (2015). https://doi.org/10.1016/j.softx.2015.06.001
2. Adiseshalah, P.P., Hall, J.B., McNeil, S.E.: Nanomaterial standards for efficacy and toxicity assessment. WIREs Nanomed. Nanobiotechnol. (2009). https://doi.org/10.1002/wnan.66
3. Adnan, A., Sun, C.T., Mahfux, H.: A molecular dynamics simulation study to investigate the effect of filler size on elastic properties of polymer nanocomposites. Compos. Sci. Technol. (2007). https://doi.org/10.1016/j.compscitech.2006.09.015
4. Ahmad, A., Khan, F., Mishra, R.K., Khan, R.: Precision cancer nanotherapy: evolving role of multifunctional nanoparticles for cancer active targeting. J. Med. Chem. (2019). https://doi.org/10.1021/acs.jmedchem.9b00511
5. Blazhynska, M.M., Kyrychenko, A., Kalugin, O.N.: Molecular dynamics simulation of the size-dependent morphological stability of cubic shape silver nanoparticles. Mol. Simul. (2018). https://doi.org/10.1080/08927022.2018.1469751
6. Bouzo, B.L., Calvelo, M., Martin-Pastor, M., Garcia-Fandino, R., de la Fuente, M.: In vitro-In silico modeling approach to rationally designed simple and versatile drug delivery systems. J. Phys. Chem. B (2020). https://doi.org/10.1021/acs.jpcb.0c02731
7. Brooks, B.R., Brooks, C.L.III., MacKerell, A.D. Jr., Nilsson, L., Petrella, R.J., Roux, B., Won, Y., Archontis, G., Bartels, C., Boresch, S., Caflisch, A., Caves, L., Cui, Q., Dinner, A.R., Feig, M., Fischer, S., Gao, J., Hodoscek, M., Im, W., Lazaridis, T., Ma, J., Ovchnnikov, V., Paci, E., PAstor, R.W., Post, C.B., Pu, J.Z., Scaefer, M., Tidor, B., Veneable, R.M., Woodcock, H.L., Wu, X., Yang, W., York, D.M., Karplus, M.: CHARMM: the biomolecular simulation program. J. Comput. Chem. (2010). https://doi.org/10.1002/jcc.21287
8. Cabarcas, S.M., Mathews, L.A., Farrar, W.L.: The cancer stem cell niche-there goes the neighborhood? Int. J. Cancer (2011). https://doi.org/10.1002/ijc.26312
9. Case, D.A., Aktulga, H.M., Belfon, K., Ben-Shalom, I.Y., Brozell, S.R., Cerutti, D.S., Cheatham, T.E., III., Cruzeiro, V.W.D., Darden, T.A., Duke, R.E., Giambasu, G., Gilson, M.K., Gohlke, H., Goetz, A.W., Harris, R., Izadi, S., Izmailov, S.A., Jin, C., Kasavajhala, K., Kaymak, M.C., King, E., Kovalenko, A., Kurtzman, T., Lee, T.S., LeGrand, S., Li, P., Lin, C., Liu, J., Luchko, T., Luo, R., Machado, M., Man, V., Manathunga, M., Merz, K.M., Miao, Y., Mikhailovskii, O., Monard, G., Nguyen, H., O'Hearn, K.A., Onufriev, A., Pan, F., Pantano, S., Qi, R., Rahnamoun, A., Roe, D.R., Roitberg, A., Sagui, C., Schott-Verdugo, S., Shen, J., Simmerling, C.L., Skrynnikov, N.R., Smith, J., Swails, J., Walker, R.C., Wang, J., Wei, H., Wolf, R.M., Wu, X., Xue, Y., York, D.M., Zhao, S., Kollman, P.A.: Amber 2021. University of California, San Francisco (2021)
10. Chen, Q., De Schutter, E.: Parallel STEPS: large scale stochastic spatial reaction-diffusion simulation with high performance computers. Front. Neuroinform. (2017). https://doi.org/10.3389/fninf.2017.00013
11. Chrastina, A., Massey, A.M., Schnitzer, J.E.: Overcoming in vivo barriers to targeted nanodelivery. Wiley Interdiscip. Rev. Nanomed. Nanobiotechnol. (2011). https://doi.org/10.1002/wnan.143
12. Crist, R.M., Grossman, J.H., Patri, A.K., Stern, S.T., Dobrovolskaia, M.A., Adiseshaiah, P.P., Clogston, J.D., McNeil, S.E.: Common pitfalls in nanotechnology: lesson learned from NCI's nanotechnology characterization laboratory. Integr. Biol. (Camb.) (2013). https://doi.org/10.1039/c2ib20117h
13. de la Fuente, J.M., Grazu, V.: Nanobiotechnology: Inorganic Nanoparticles vs Organic Nanoparticles. Elsevier, Amsterdam (2012)
14. Dean, M., Fojo, T., Bates, S.: Tumour stem cells and drug resistance. Nat. Rev. Cancer (2005). https://doi.org/10.1038/nrc1590
15. Deisboeck, T.S., Want, Z., Mackling, P., Cristini, V.: Multiscale cancer modeling. Annu. Rev. Biomed. Eng. (2011). https://doi.org/10.1146/annurev-bioeng-071910-124729

16. Fanciullino, R., Ciccolini, J., Milano, G.: Challenges, expectations and limits for nanoparticles-based therapeutics in cancer: a focus on nano-albumin-bound drugs. Crit. Rev. Oncol. Hematol. (2013). https://doi.org/10.1016/j.critrevonc.2013.06.010
17. Ford Versypt, A.N.: Multiscale modeling in disease. Curr. Opin. Syst. Biol. (2021). https://doi.org/10.1016/j.coisb.2021.05.001
18. Ghaffarizadeh, A., Friedman, S.H., Macklin, P.: BioFVM: an efficient, parallelized diffusive transport solver for 3-D biological simulations. Bioinformatics (2016). https://doi.org/10.1093/bioinformatics/btv730
19. Ghaffarizadeh, A., Heiland, R., Friedman, S.H., Mumenthaler, S.M., Macklin, P.: PhysiCell: an open source physics-based cell simulator for 3-D multicellular systems. PLoS Comput. Biol. (2018). https://doi.org/10.1371/journal.pcbi.1005991
20. Grossman, J.H., Crist, R.M., Clogston, J.D.: Early development challenges for drug products containing nanomaterials. AAPS J. (2016). https://doi.org/10.1208/s12248-016-9980-4
21. Hanwell, M.D., Curits, D.E., Lonie, D.C., Vandermeersch, T., Zurek, E., Hutchison, G.R.: Avogadro: an advanced semantic chemical editor, visualization, and analysis platform. J. Cheminformatics (2012). https://doi.org/10.1186/1758-2946-4-17
22. Hauert, S., Berman, S., Nagpal, R., Bhatia, S.N.: A computational framework for identifying design guidelines to increase the penetration of targeted nanoparticles into tumors. Nano Today (2013). https://doi.org/10.1016/j.nantod.2013.11.001
23. Haume, K., Mason, N.J., Solov'yov, A.V.: Modeling of nanoparticle coatings for medical applications. Eur. Phys. J. D (2016). https://doi.org/10.1140/epjd/e2016-70282-6
24. Hazarika, Z., Jha, A.N.: Computational analysis of the silver nanoparticle-human serum albumin complex. ACS Omega (2020). https://doi.org/10.1021/acsomega.9b02340
25. Heikkila, E., Martinez-Sara, H., Gurtovenko, A.A., Vattulainen, I., Akola, J.: Atomistic simulations of anionic Au144(SR)60 nanoparticles interacting with asymmetric model lipid membranes. Biochim. et Biophys. Acta (BBA) - Biomembr. (2014). https://doi.org/10.1016/j.bbamem.2014.07.027
26. Heikkila, E., Gurtovenko, A.A., Martinez-Seara, H., Hakkinen, H., Vattulainen, I., Akola, J.: Atomistic simulations of functional Au144(SR)60 gold nanoparticles in aqueous environment. J. Phys. Chem. C (2012). https://doi.org/10.1021/jp301094m
27. Helczynska, K., Kronblad, A., Jogi, A., Nilsson, E., Beckman, S., Landberg, G., Pahlman, S.: Hypoxia promotes a dedifferentiated phenotype in ductal breast carcinoma in situ. Cancer Res.h (2003)
28. Hepburn, I., Chen, W., Wils, S., De Schutter, E.: STEPS: efficient simulation of stochastic reaction-diffusion models in realistic morphologies. BMC Syst. Biol. (2012). https://doi.org/10.1186/1752-0509-6-36
29. Hollingsworth, S.A., Dror, R.O.: Molecular dynamics simulation for all. Neuron (2019). https://doi.org/10.1016/j.neuron.2018.08.011
30. Humphrey, W., Dalke, A. and Schulten, K.: MD - visual molecular dynamics. J. Mol. Graph. (1996). https://doi.org/10.1016/0263-7855(96)00018-5
31. Khan, I., Saeed, K., Khan, I.: Nanoparticles: properties, applications and toxicities. Arab. J. Chem. (2019). https://doi.org/10.1016/j.arabjc.2017.05.011
32. Kim, D., Shin, K., Gu Kwon, S., Hyeon, T.: Synthesis and biomedical applications of multifunctional nanoparticles. Adv. Mater. Interfaces (2018). https://doi.org/10.1002/adma.201802309
33. Kim, K.-S., Han, J.-H., Kim, H.-K., Choi, H.S., Kim, G.R., Song, H., An, H.H., Han, D.K., Park, W., Park, K.-S.: Multifunctional nanoparticles for genetic engineering and bioimaging of natural killer (NK) cell therapeutics. Biomaterials (2019). https://doi.org/10.1016/j.biomaterials.2019.119418
34. Kovacevic, M., Balaz, I., Marson, D., Laurini, E., Jovic, B.: Mixed-monolayer functionalized gold nanoparticles for cancer treatment: Atomistic molecular dynamics simulations study. Biosystems (2021). https://doi.org/10.1016/j.biosystems.2021.104354
35. Kurtova, A.V., Xiao, J., Mo, Q., Pazhanisamy, S., Krasnow, R., Lerner, S.P., Chen, F., Roh, T.T., Lay, E., Levy Ho, P., Syson Chan, K.: Blocking PGE2-induced tumour repopulation abrogates bladder cancer chemoresistance. Nature (2015). https://doi.org/10.1038/nature14034

36. Liu, Y., Qiao, L., Zhang, S., Wan, G., Chen, B., Zhou, P., Zhang, N., Wang, Y.: Dual pH-responsive multifunctional nanoparticles for targeted treatment of breast cancer by combining immunotherapy and chemotherapy. Acta Biomater. (2018). https://doi.org/10.1016/j.actbio.2017.11.010
37. Marson, D., Guida, F., Sologan, M., Boccardo, S., Pengo, P., Perissinotto, F., Iacuzzi, V., Pellizzoni, E., Polizzi, S., Casalis, L., Pasquato, L., Pacor, S., Tossi, A., Possoco, P.: Mixed Fluorinated/Hydrogenated self-assembled monolayer-protected gold nanoparticles: in silico and in vitro behaviour. Small (2019). https://doi.org/10.1002/smll.201900323
38. Martinez, L., Birgin, E.G., Martinez, J.M.: PACKMOL: a package for building initial configurations for molecular dynamics simulations. J. Comput. Chem. (2009). https://doi.org/10.1002/jcc.21224
39. Maskey, S., Lane, J.M.D., Perahia, D., Grest, G.S.: Structure of rigid polymers confined to nanoparticles: molecular dynamics simulations insight. Langmuir (2016). https://doi.org/10.1021/acs.langmuir.5b04568
40. Mellor, H.R., Callaghan, R.: Accumulation and distribution of doxorubicin in tumour spheroids: the influence of acidity and expression of P-glycoprotein. Cancer Chemother. Pharmacol. (2021). https://doi.org/10.1007/s00280-011-1598-8
41. Moayyer, H.A., Naderi, M., Aghazadeh Mohandesi, J., Ramazani, A.: Molecular dynamics simulation of mechanical properties of polystyrene nanoparticles under uniaxial compression test. Comput. Mater. Sci. (2020). https://doi.org/10.1016/j.commatsci.2020.109553
42. Murugan, C., Rayappan, K., Thangam, R., Bhanumathi, R., Shanti, K., Vivek, R., Thirumurugan, R., Bhattacharyya, A., Sivasubramanian, S., Gunesekeran, P., Kannan, S.: Combinatorial nanocarrier based drug delivery approach for amalgamation of anti-tumor agents in breast cancer cells: an improved nanomedicine strategy. Sci. Rep. (2016). https://doi.org/10.1038/srep34053
43. Nichols, J.W., Bae, Y.H.: Odyssey of a cancer nanoparticle: from injection site to site of action. Nano Today (2012). https://doi.org/10.1016/j.nantod.2012.10.010
44. Peppas, N.A.: Receptors: models for binding, trafficking, and signaling. J. Control. Release (1996)
45. Pettersen, E.F., Goddard, T.D., Huang, C.C., Couch, G.S., Greenblatt, D.M., Meng, E.C., Ferrin, E.: UCSF chimera-a visualization system for exploratory research and analysis. Comput. Chem. (2004). https://doi.org/10.1002/jcc.20084
46. Phillips, J.C., Hardy, D.J., Maia, J.D.C., Stone, J.E., Ribeiro, J.V., Bernardi, R.C., Buch, R., Fiorin, G., Henin, J., Jiang, W., McGreevy, R., Melo, M.C.R., Radak, B.K., Skeel, R.D., Singharoy, A., Wang, Y., Roux, B., Aksimentiev, A., Luthey-Schulten, Z., Kale, L.V., Schulten, K., Chipot, C., Tajkhorshid, E.: Scalable molecular dynamics on CPU and GPU architectures with NAMD. J. Chem. Phys. (2020). https://doi.org/10.1063/5.0014475
47. Plimpton, S.: Fast parallel algorithms for short-range molecular dynamics. J. Comput. Phys. (1995). https://doi.org/10.1006/jcph.1995.1039
48. Ramezanpour, M., Leung, S.S.W., Delgado-Magnero, K.H., Bashe, B.Y.M., Thewalt, J., Tieleman, D.P.: Computational and experimental approaches for investigating nanoparticle-based drug delivery systems. Biochim. et Biophys. Acta - Biomembr. (2016). https://doi.org/10.1016/j.bbamem.2016.02.028
49. Rana, S., Bajaj, A., Mout, R., Rotello, V.M.: Monolayer coated fold nanoparticles for delivery applications. Adv. Drug Deliv. Rev. (2012). https://doi.org/10.1016/j.addr.2011.08.006
50. Ray, P., Alhalhooly, L., Ghosh, A., Choi, Y., Banarjee, S., Mallik, S., Banarjee, S., Quadir, M.: Size-transformable, multifunctional nanoparticles from hyperbranched polymers for environment-specific therapeutic delivery. ACS Biomater. Sci. Eng. (2019). https://doi.org/10.1021/acsbiomaterials.8b01608
51. Rennick, J.J., Johnston, A.P.R., Parton, R.G.: Key principles and methods for studying the endocytosis of biological and nanoparticle therapeutics. Nat. Nanotechnol. (2021). https://doi.org/10.1038/s41565-021-00858-8
52. Salerno, K.M., Ismail, A.E., Lane, J.M.D., Grest, G.S.: Coating thickness and coverage effects on the forces between silica nanoparticles in water. J. Chem. Phys. (2014). https://doi.org/10.1063/1.4874638

53. Sau, aA groundbreaking approach for reprogramming malfunctioned tumor environment. J. Control. Release (2018). https://doi.org/10.1016/j.jconrel.2018.01.028
54. Schlick, T.: Molecular Modeling and Simulation: An Interdisciplinary Guide. Springer, Berlin (2002)
55. Sridhar, D.B., Gupta, R., Rai.B.: Effect of surface coverage and chemistry on self-assembly of monolayer protected gold nanoparticles: a molecular dynamics simulation study. Phys. Chem. Chem. Phys. (2018). https://doi.org/10.1039/C8CP04044C
56. Stallard, Morrison, J.G., George, W.D., Kaye, S.B.: Distribution of doxorubicin to normal breast and tumour tissue in patients undergoing mastectomy. Cancer Chemother. Pharmacol. (1990). https://doi.org/10.1007/bf00684887
57. Stillman, N.R., Kovacevic, M., Balaz, I., Hauert, S.: In silico modelling of cancer nanomedicine, across scales and transport barriers. NPJ Comput. Mater. (2020). https://doi.org/10.1038/s41524-020-00366-8
58. Stillman, N., Balaz, I., Tsompanas, M.-A., Kovacevic, M., Azimi, S., Lafond, S., Adamtzky, A., Hauert, S.: Evolutionary computational platform for the automatic discovery of nanocarriers for cancer treatment. NPJ Comput. Mater. (2021). https://doi.org/10.1038/s41524-021-00614-5
59. Stipa, P., Marano, S., Galeazzi, R., Minnelli, C., Mobbil, G., Ludadio, E.: Prediction of drug-carrier interactions of PLA and PLGA drug-loaded nanoparticles by molecular dynamics simulations. Eur. Polymer J. (2021). https://doi.org/10.1016/j.eurpolymj.2021.110292
60. Swierczewska, M., Crist, R.M., McNeil, S.E.: Evaluating nanomedicines: obstacles and advancements. Methods in Molecular Biology (Clifton, N.J.) (2018). https://doi.org/10.1007/978-1-4939-7352-1_1
61. Torrice, M.: Does nanomedicine have a delivery problem? ACS Cent. Sci. (2016). https://doi.org/10.1021/acscentsci.6b00190
62. van der Meel, R., Sulheim, E., Shi, Y., Kiessling, F., Mulder, W.J.M., Lammers, T.: Smart cancer nanomedicine. Nat. Nanotechnol. (2019). https://doi.org/10.1038/s41565-019-0567-y
63. Van Lehn, R.C., Alexander-Katz, A.: Structure of mixed-monolayer-protected nanoparticles in aqueous salt solution from atomistic molecular dynamics simulations. J. Phys. Chem. (2013). https://doi.org/10.1021/jp406035e
64. Van Lehn, R.C., Alexander-Katz, A.: Structure of mixed-monolayer-protected nanoparticles in aqueous salt solution from atomistic molecular dynamics simulations. J. Phys. Chem. C (2013). https://doi.org/10.1021/jp406035e
65. Vanquelef, E., Simon, S., Marquant, G., Garcia, E., Klimerak, G., Delepine, J.C., Cieplak, P., Dupradeau, F.-Y.: R.E.D. Server: a web service for deriving RESP and ESP charges and building force field libraries for new molecules and molecular fragments. Nucl. Acids Res. (2011). https://doi.org/10.1093/nar/gkr288
66. Vaupel, P., Harrison, L.: Tumor hypoxia: causative factors, compensatory mechanisms, and cellular response. Oncologist (2004). https://doi.org/10.1634/theoncologist.9-90005-4
67. Vinogradov, S., Wei, X.: Cancer stem cells and drug resistance: the potential of nanomedicine. Nanomedicine (2012). https://doi.org/10.2217/nnm.12.22
68. Wang, J., Wolf, R.M., Caldwell, J.W., Kollman, P.A., Case, D.A.: Development and testing of a general amber force field. Comput. Chem. (2004). https://doi.org/10.1002/jcc.20035
69. Wei, A., Mehtala, J.G., Patri, A.K.: Challenges and opportunities in the advancement of nanomedicines. J. Control. Release (2012). https://doi.org/10.1016/j.jconrel.2012.10.007
70. Wils, S., De Schutter, E.: Reaction-diffusion in complex 3D geometries: mesh construction and stochastic simulation with STEPS. BMC Neurosci. (2008). https://doi.org/10.1186/1471-2202-9-S1-P85
71. Xu, S., Nie, Z., Seo, M., Lewis, P., Kumacheva, E., Stone, H.A., Garstecki, P., Weibel, D.B., Gitlin, I., Whitesides, G.M.: Generation of monodisperse particles by using microfluidics: control over size, shape, and composition. Angew. Chem. Int. Ed. Engl. (2005). https://doi.org/10.1002/anie.200462226
72. Yamashita, F., Hashida, M.: Pharmacokinetic considerations for targeted drug delivery. Adv. Drug Deliv. Rev. (2013). https://doi.org/10.1016/j.addr.2012.11.006

73. Yang, S., Chen, L., Zhou, X., Sun, P., Fu, L., You, Y., Xu, M., You, Z.: Tumor-targeted biodegradable multifunctional nanoparticles for cancer theranostics. Chem. Eng. J. (2019). https://doi.org/10.1016/j.cej.2019.122171
74. Yetisgin, A.A., Cetinel, S., Zuvin, M., Kosar, A., Kutlu, O.: Therapeutic nanoparticles and their targeted delivery applicatioins. Molecules (2020). https://doi.org/10.3390/molecules25092193
75. Zamboni, W.C., Torchilin, V., Patri, A.K., Hrkach, J., Stern, S., Lee, R., Nel, A., Panaro, N.J., Grodzinski, P.: Best practices in cancer nanotechnology: perspective from NCI nanotechnology alliance. Clin. Cancer Res. (2012). https://doi.org/10.1158/1078-0432.CCR-11-2938
76. Zhao, M., van Straten, D., Broekman, M.L.D., Preatm, V., Schiffelers, R.M.: Nanocarrier-based drug combination therapy for glioblastoma. Theranostics (2020). https://doi.org/10.7150/thno.38147

# A Haploid-Diploid Evolutionary Algorithm Optimizing Nanoparticle Based Cancer Treatments

**Michail-Antisthenis Tsompanas, Larry Bull, Andrew Adamatzky, and Igor Balaz**

**Abstract** This paper uses a recent explanation for the fundamental haploid-diploid lifecycle of eukaryotic organisms to present a new evolutionary algorithm that differs from all previous known work using diploid representations. A form of the Baldwin effect has been identified as inherent to the evolutionary mechanisms of eukaryotes and a simplified version is presented here which maintains such behaviour. Using a well-known abstract tuneable model, it is shown that varying fitness landscape ruggedness varies the benefit of haploid-diploid algorithms. Moreover, the methodology is applied to optimise the targeted delivery of a therapeutic compound utilizing nano-particles to cancerous tumour cells with the multicellular simulator PhysiCell.

## 1 Introduction

The vast majority of work within evolutionary computation has used an underlying haploid representation scheme; individuals are each one solution to the given problem [8]. Typically, bacteria contain one set of genes, whereas the more complex eukaryotic organisms—such as plants and animals—are predominantly diploid, containing two sets of genes [22]. A small body of work exists using a diploid representation scheme, i.e., individuals carry two solutions to the given problem. In such cases recombination typically occurs between corresponding haploids/genes in each parent, essentially doubling the standard process, and a dominance scheme is utilized

M.-A. Tsompanas (✉) · A. Adamatzky
Unconventional Computing Laboratory, University of the West of England,
Bristol BS16 1QY, UK
e-mail: antisthenis.tsompanas@uwe.ac.uk

L. Bull
Department of Computer Science and Creative Technologies, University of the West of England,
Bristol BS16 1QY, UK

I. Balaz
Laboratory for Meteorology, Physics and Biophysics, Faculty of Agriculture,
Trg Dositeja Obradovica 8, University of Novi Sad, 21000 Novi Sad, Serbia



I. Balaz and A. Adamatzky (eds.), *Cancer, Complexity, Computation*,
Emergence, Complexity and Computation 46,
https://doi.org/10.1007/978-3-031-04379-6_10

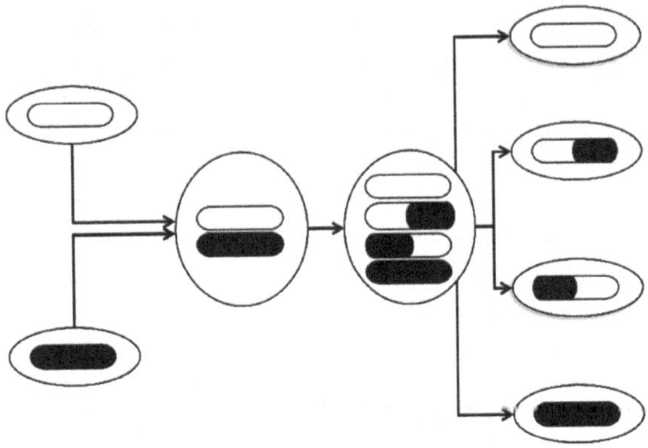

**Fig. 1** Two-step meiosis with recombination under haploid-diploid reproduction as seen in most eukaryotic organisms (after [19])

to reduce the diploid down to a traditional haploid solution for evaluation. That is, as individuals carry two sets of genes/variables, a heuristic is included to choose which of the genes to use (see [4] for a review).

Eukaryotes exploit a so-called haploid-diploid cycle where haploid cells are brought together to form the diploid cell/organism. At the point of reproduction by the cell/organism, the haploid genomes within the diploid each form haploid gamete cells that (may) join with a haploid gamete from another cell/organism to form a diploid (Fig. 1). Specifically, each of the two genomes in an organism is replicated, with one copy of each genome being crossed over. In this way copies of the original pair of genomes may be passed on, mutations aside, along with two versions containing a mixture of genes from each. Previous explanations for the emergence of the alternation between the haploid and diploid states are typically based upon its being driven by changes in the environment (after [16]). Recently, an explanation for the haploid-diploid cycle in eukaryotes has been presented [6] which also explained other aspects of their sexual reproduction, including the use of recombination, based upon the Baldwin effect [2]. The Baldwin effect is here defined as the existence of phenotypic plasticity that enables an organism to exhibit a different (better) fitness than its genome directly represents. Over time, as evolution is guided towards such regions under selection, higher fitness alleles/genomes which rely less upon the phenotypic plasticity can be discovered and become assimilated into the population.

The rest of the paper is arranged as follows: the next section presents the new understanding of how eukaryotic organisms evolve. A new simplified haploid-diploid algorithm is then presented, which maintains the basic mechanisms of the natural case. Finally, the new approach is applied to a high-throughput multicellular simulator to find potentially new therapeutic designs that maximise cancer tumour regression.

## 2 Eukaryotic Evolution and the Baldwin Effect

Key to the new explanation for the evolution of eukaryotes is to view the process from the perspective of the constituent haploids: a diploid organism may be seen to simultaneously represent two points in the underlying haploid fitness landscape. The fitness associated with those two haploids is therefore that achieved in their combined form as a diploid; each haploid genome will have the same fitness value and that will almost certainly differ from that of their corresponding haploid organism due to the interactions between the two genomes. That is, the effects of haploid genome combination into a diploid can be seen as a simple form of phenotypic plasticity for the individual haploids before they revert to a solitary state during reproduction. In this way evolution can be seen to be both assigning a single fitness value to the region of the landscape between the two points represented by a diploid's constituent haploid genomes and altering the shape of the haploid fitness landscape. In particular, the latter enables the landscape to be smoothed under a rudimentary Baldwin effect process [11], whilst the former can be seen to represent a simple form of generalization over the landscape. Note this is in direct contrast to typical cases of bacteria—and evolutionary algorithms—where individuals represent a single point in the (haploid) fitness landscape only and the fitness assigned is that due solely to their given gene combination. The same is also true in all known previous diploid representation schemes.

Numerous explanations exist for the benefits of recombination in both natural (e.g., [3]) and artificial systems (e.g., [20]). The latter focusing solely upon haploid genomes and neither considering the potential Baldwin effect under the haploid-diploid cycle. The role becomes clear under the new view: recombination facilitates genetic assimilation within the simple form of the Baldwin effect. If the haploid pairing is beneficial and the diploid is chosen under selection to reproduce, the recombination process can bring an assortment of those partnered genes together into new haploid genomes. In this way the fitter allele values from the pair of partnered haploids may come to exist within individual haploids more quickly than the under mutation alone (see [6] for full details). Hence, in the emergence of more complex organisms, natural evolution appears to have discovered a more sophisticated approach to navigating their fitness landscapes.

The Baldwin effect has long been used within evolutionary computation (after [11]). This paper aims to show how the benefits of a haploid-diploid cycle can be exploited as a form of evolutionary computation. However, rather than just adopt nature's scheme under which a single individual requires both haploid genomes to be evaluated (as in [6]), a simpler scheme is proposed as a form of post-processing for a traditional evolutionary algorithm. This is first explored using the NK model of fitness landscapes.

## 3 The NK Model

The NK model [13] was introduced to allow the systematic study of various aspects of fitness landscapes (see [14] for an overview). In the standard model, the features of the fitness landscapes are specified by two parameters: $N$, the length of the genome; and $K$, the number of genes that has an effect on the fitness contribution of each (binary) gene. Thus, increasing $K$ with respect to $N$ increases the epistatic linkage, increasing the ruggedness of the fitness landscape. The increase in epistasis increases the number of optima, increases the steepness of their sides, and decreases their correlation.

The model assumes all intragenome interactions are so complex that it is only appropriate to assign random values to their effects on fitness. Therefore for each of the possible $K$ interactions a table of $2^{(K+1)}$ fitnesses is created for each gene with all entries in the range 0.0 to 1.0, such that there is one fitness for each combination of traits (Fig. 2). The fitness contribution of each gene is found from its table. These fitnesses are then summed and normalized by $N$ to give the selective fitness of the total genome. The results reported in the next section are the average of 10 runs (random starting populations) on each of 10 NK functions, i.e. 100 runs, for 20,000 generations. Here $0 \leq K \leq 15$, for $N = 50$ and $N = 100$.

## 4 A Simple Haploid-Diploid Algorithm

Figure 3a shows a schematic of a traditional evolutionary algorithm (EA) which exploits binary tournament selection, one-point recombination, single-point mutation (randomly chosen gene), and creates one offspring per cycle (steady state) which is evaluated and replaces the worst individual in the population here. Figure 3b shows how the learning mechanism described above is implemented on top of that process. As can be seen: a traditional population of evaluated haploid individuals is maintained (A); a temporary population of diploid solutions is created from them by copying

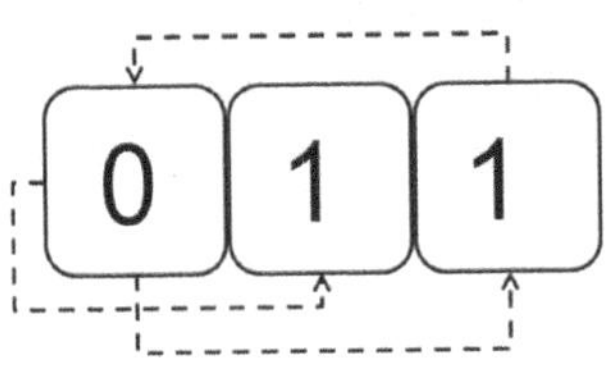


| $K_0$ $K_1$ | Fitness |
|---|---|
| 0 0 | 0.76 |
| 0 1 | 0.23 |
| **1 0** | **0.56** |
| 1 1 | 0.98 |

**Fig. 2** An example NK model ($N = 3$, $K = 1$) showing how the fitness contribution of each gene depends on $K$ random genes (left). Therefore there are $2^{(K+1)}$ possible allele combinations per gene, each of which is assigned a random fitness. Each gene of the genome has such a table created for it (right, centre gene shown). Total fitness is the normalized sum of these values

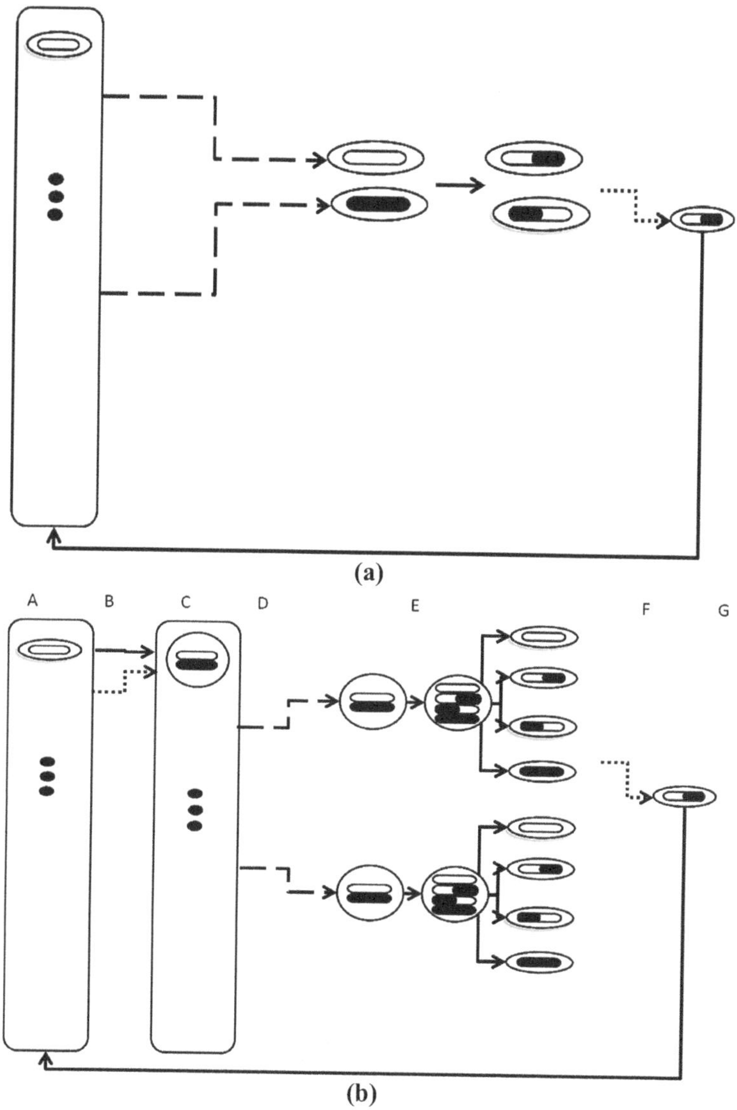


**Fig. 3** A schematic of the traditional evolutionary algorithm (**a**) and of the simple haploid-diploid algorithm (**b**)

each haploid individual and then another haploid is chosen at random (B), with the fitness of the two haploids averaged (C); binary tournament selection then uses those fitnesses to pick two diploid parents (D); the haploid-diploid reproduction cycle with two-step meiosis as shown in Fig. 1 is then used for the two chosen parents (E); one of the resulting haploids is chosen at random, mutated (single-point), and evaluated (F); the offspring haploid is inserted into the original population replacing the worst individual (G).

Figure 4 shows example results from running both the standard EA and the haploid-diploid EA (HDEA) on various NK fitness landscapes. Here population size $P = 30$. As can be seen, when $K > 4$, the HDEA performs best for $N = 50$ and $K > 2$ for $N = 100$ (T-test, $p < 0.05$). Thus, as anticipated, the simple Baldwin effect process proves beneficial with increased fitness landscape ruggedness due to its ability to smooth the underlying shape. Figure 5 shows examples of how this is

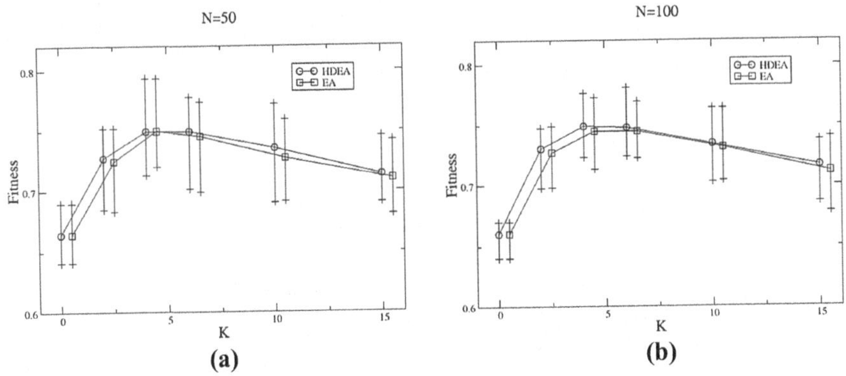


**Fig. 4** Showing examples of the fitness reached after 20,000 generations on landscapes of various size ($N$) and ruggedness ($K$). Error bars show min and max values

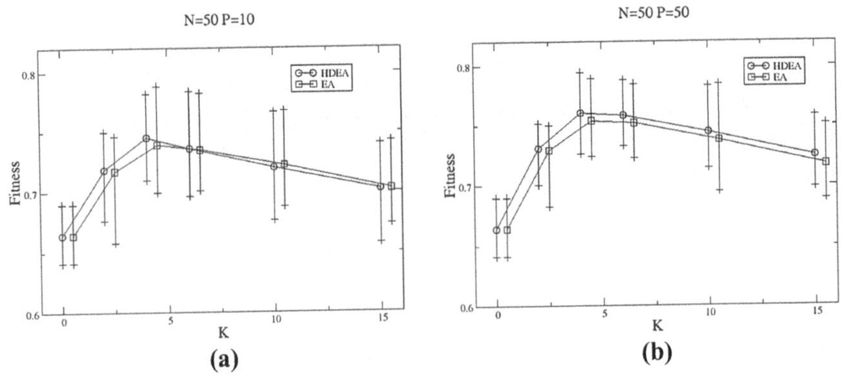


**Fig. 5** Showing examples of the fitness reached after 20,000 generations with differing population sizes ($P$)

also true for different $P$, although the benefit is lost for higher $K$ when $P = 10$. Related to this, since the HDEA makes a temporary population of diploids containing extra copies of randomly chosen haploid solutions, it might be argued that a larger population is available to selection than in the standard EA. Moreover, as the underlying traditional haploid population converges upon higher fitness solutions, the random sampling could be increasing their number and, thereby, altering the comparative selection pressure over time. However, results from simply creating a temporary haploid population of size $2P$, in the same way as the temporary diploid population, does not alter performance significantly (not shown here, e.g., see [12] for discussions of dynamic population sizing in general).

Whilst it is beyond the scope of this paper to exhaustively review previous diploid EAs (DEAs)(see [4]), a simple diploid version of the traditional EA has been created and explored. Here the fitness of the diploid is the average of its two constituent haploids and the corresponding haploids undergo one-point recombination. For comparison, without a dominance mechanism, populations contain either half as many individuals or are run for half as many generations as the EA and HDEA to maintain the same number of function evaluations. The results find no significant difference in performance over the traditional EA for any of the parameters explored here (not shown).

The findings with this abstract model are now explored in the context of simulating nano-particle therapy delivery for cancer tumour regression within PhysiCell v.1.5.1 (see [17] for an overview of computational modelling in cancer biology).

## 5 PhysiCell: A Physics-based Multicellular Simulator

Among an increasing amount of computational models [17, 21] studying different aspects of cancer physiology, PhysiCell [10] is one of the leading ones. The open source simulator is based on a biotransport solver (BioFVM [9]) and simulates a multicellular environment. While PhysiCell simulates cell cycling, death states, volume changes, mechanics, orientation and motility, it relies on BioFVM to simulate substrate secretion, diffusion, uptake, and decay. A significant advantage of PhysiCell is its open-source code that enables addition of new environmental substrates, cell types, and systems of cells, resulting in a general-purpose tool for investigating systems with multiple kinds of cells. This includes the ability to design cell-cell interaction rules to create a multicellular cargo delivery system that actively delivers a cancer therapeutic compound beyond regular drug transport limits to hypoxic cancer regions. We are currently exploring the use of evolutionary computing and other related techniques to optimise the design of such nano-particle (NP) delivery systems [1, 18, 23–27].

To evaluate the efficiency of the design of these NP delivery systems, the 2-D anti-cancer biorobots scenario of PhysiCell v.1.5.1 [10] was studied. This scenario utilizes three types of agents to simulate a high-throughput testing of a simple targeted drug delivery therapy. Namely, these types of agents are cancer cells, worker cells and cargo cells. Cancer cells consume oxygen and secrete a chemoattractant. The resulted gradient in oxygen concentration is employed to steer NPs, simulated as worker cells. These worker cells can be bonded with cargo cells, simulating the therapeutic compound. When a worker cell carries a cargo cell, it executes a random walk (migration) towards the gradient of the oxygen and, thus, towards accumulation of cancer cells. Whereas, when a worker cell does not carry a cargo cell it executes a random walk towards the area of the cargo cells. These random walks or migrations are controlled by input parameters of the simulator, in the range [0, 1], with 0 representing Brownian motion and 1 deterministic motion.

Finally, cargo cells simulating the therapeutic compound, can attract worker cells by exuding another simulated chemoattractant (which diffuses under BioFVM rules). As described before, worker cells can carry the cargo cells and deposit them in the affinity of cancer cells, resulting in apoptosis of these cells. The specific proximity is given by the parameter defined as cargo release $O_2$ threshold.

As per the initial example [10] and other relevant studies [18], in the 2-D anti-cancer biorobots scenario an initial 200 μm radius tumour is simulated to grow for 7 days. Then, 450 cargo cells and 50 worker cells are added in a simulated vein close to the tumour. Note here that while in previous studies a random number of each type of cells with its mean as in the aforementioned was added, here we add exactly 450 and 50 cells for every simulation to alleviate one factor of stochasticity. The simulated drug delivery system is simulated for 3 more days and then the results are analyzed.

One paradigm of this simulation (whole 10 days) takes approximately 5 min of wall-clock time on an Intel® Xeon® CPU E5-2650 at 2.20GHz with 64GB RAM using 8 of the 48 cores. To accelerate the computations and further alleviate the effect of the stochastic nature of the simulator on the results, a single tumour was used for testing every possible individual in the search space. For each test, one pre-grown tumour (for 7 days) was loaded to the simulator and the treatment was applied immediately. The test was finalized after 3 days from the introduction of the treatment, resulting in a minimization of wall-clock time to approximately 1,5 min. A static sampling approach is used, where the average of the outputs after 5 runs of the simulator with the same set of parameters was examined. The objective was determined as the remaining amount of cancer cells in the simulated area after the 3 days of treatment.

The search space was defined as a 6-dimensional space, with the 6 most prolific parameters for the behaviour of worker cells (or simulated NPs). Namely, the parameters under investigation were: the attached worker migration bias [0,1]; the unattached worker migration bias [0,1]; worker relative adhesion [0,10]; worker relative repulsion [0,10]; worker motility persistence time (minutes) [0,10]; and the cargo release $O_2$ threshold ($mmHg$) [0,20]. The rest of the parameters on the simulator are not altered from the initial distribution of the simulator [10] and depicted in Table 1.

**Table 1** Unaltered parameters of PhysiCell simulator

| Parameter | Value |
|---|---|
| Maximum attachment distance | 18 μm |
| Minimum attachment distance | 14 μm |
| Worker apoptosis rate | 0 $min^{-1}$ |
| Worker migration speed | 2 μm/min |
| Worker $O_2$ relative uptake | 0.1 $min^{-1}$ |
| Cargo $O_2$ relative uptake | 0.1 $min^{-1}$ |
| Cargo apoptosis rate | 4.065e-5 $min^{-1}$ |
| Maximum relative cell adhesion distance | 1.25 |
| Maximum elastic displacement | 50 μm |
| Damage rate | 0.03333 $min^{-1}$ |
| Repair rate | 0.004167 $min^{-1}$ |
| Drug death rate | 0.004167 $min^{-1}$ |
| Cargo relative adhesion | 0 |
| Cargo relative repulsion | 5 |
| Elastic coefficient | 0.05 $min^{-1}$ |
| Motility shutdown detection threshold | 0.001 |
| Attachment receptor threshold | 0.1 |

## 6 Results of HDEA Optimization on PhysiCell

Initially the option to load a tumour rather than simulate its growth for 7 days was investigated. In Fig. 6 the boxplot of 100 simulations for each initialization option with the same input parameters is illustrated. When comparing the initial growth (7 days tumour growth and 3 days treatment simulation, mean = 475.06, SD = 32.9, median = 480, kurtosis = 3.3515) with the loading tumour alternative (loading a tumour and 3 days treatment simulation, mean = 494.12, SD = 29.11, median = 491, kurtosis = 2.7698), the latter produces more consistent results (based on smaller standard deviation and kurtosis). Additional to the aforementioned acceleration of computations (from 5 min to 1,5 min) the loading tumour was selected for the tests presented in the following.

To study the performance of HDEA, another control algorithm was utilized to optimize the behaviour/design of worker cells, namely a steady-state genetic EA. The population size was set to $P = 50$, the selection and replacement tournament size to $T = 3$, a uniform crossover probability to $X = 80\%$ and a per allele mutation rate to $\mu = 20\%$ with a uniform random step size of range $s = [-5, 5]\%$. The HDEA was set up with the same parameters as the EA in order for the comparison to be meaningful. All comparison runs started by evaluating a randomly produced, same for each run, initial population ($P = 50$) under PhysiCell simulator, and then using the corresponding EA to evolve the design of worker cells, with a computational

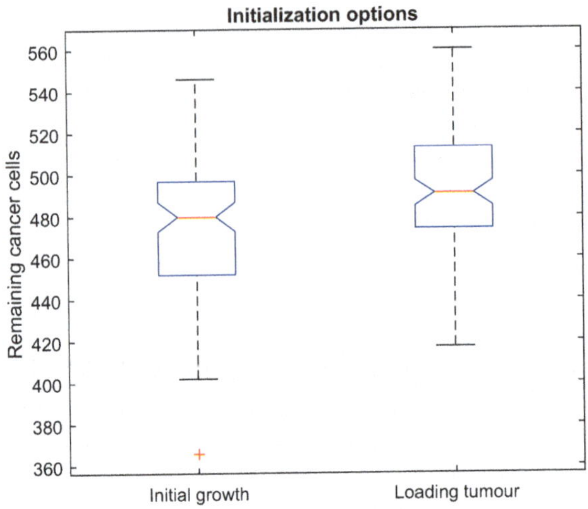


**Fig. 6** Boxplot of 100 samples for each of the initialization options

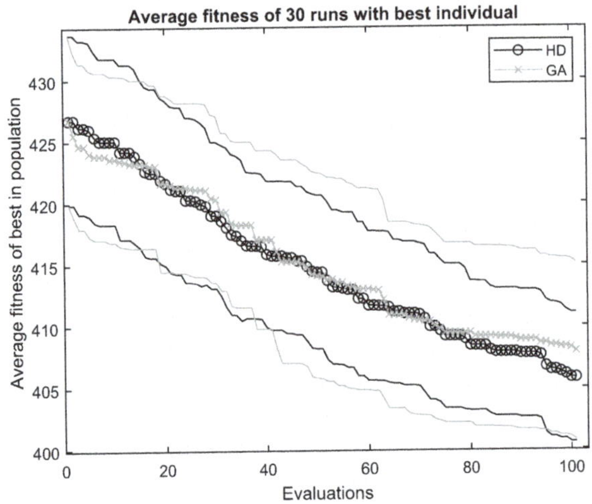


**Fig. 7** Average and confidence levels (95%) of the best individuals per evolution step for both algorithms for all 30 runs

budget of 100 individual evaluations (100 individuals $\times$ 5 samples = 500 PhysiCell simulations). In total, 30 comparison runs were executed.

In Fig. 7 the evolution of the best individuals found by the two algorithms is illustrated. Specifically, the average and confidence level at 95% for the best individuals in all 30 runs are considered. Throughout the evolution steps it is apparent that the HDEA algorithm is generally finding better solutions faster (it learns faster). Moreover, the final average of best solutions found by HDEA is better than the one by the genetic algorithm. The smaller range of the 95% confidence levels of HDEA reveal a better consistency in the solutions found by this algorithm.

Figure 8 shows the relative performance of the average solutions over time for both approaches. As can be seen, the HDEA finds fitter solutions. Note the zoomed

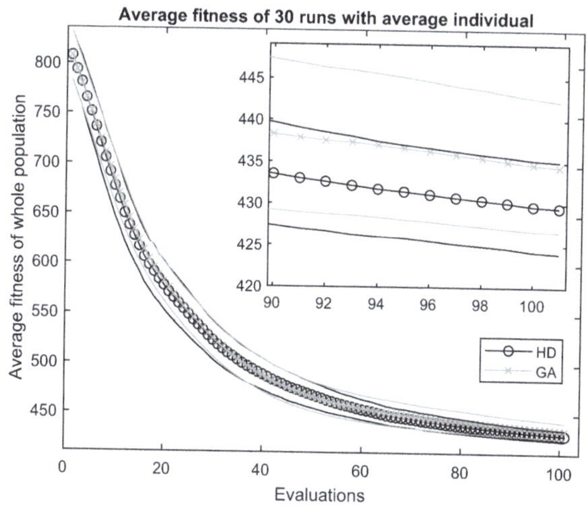


**Fig. 8** Average and confidence levels (95%) of all the individuals per evolution step for both algorithms for all 30 runs

in region of evaluations 90–100 for a clearer comparison. Although, after 100 evaluations, the best solutions (Fig. 7) are not statistically significantly better (Wilcoxon signed-rank test, $p = 0.3763$), the average solutions (Fig. 8) are better (Wilcoxon signed-rank test, $p = 0.0256$). It can be noted that the best solution found by the HDEA was significantly better (Wilcoxon signed-rank test, $p = 0.0215$) for the first ten runs of the thirty shown here.

In Figs. 9 and 10 the boxplots of the parameters of the best individual discovered during the 30 runs by GA and HDEA, respectively, are presented. In Figs. 11 and 12 the scatter plots of the parameters of the best individual discovered during the 30 runs by GA and HDEA, respectively, are depicted. It is clear that the most prolific parameter value for optimizing the design of NPs is the cargo release $O_2$ threshold

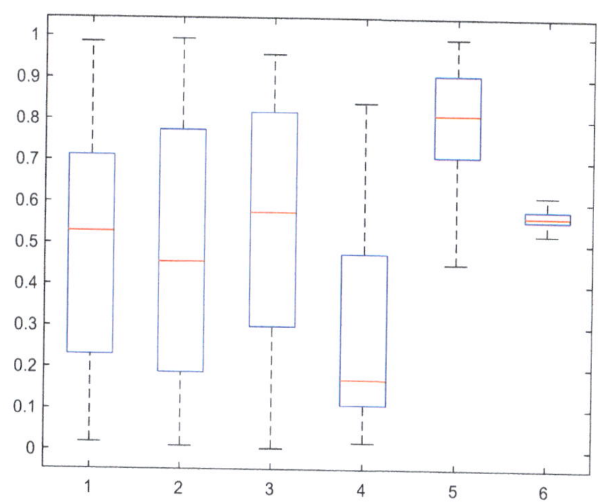

**Fig. 9** Boxplot of parameters of best individuals found by GA (in normalized ranges). Parameters: (1) attached worker migration bias (2) the unattached worker migration bias (3) worker relative adhesion (4) worker relative repulsion (5) worker motility persistence time (6) and the cargo release $O_2$ threshold

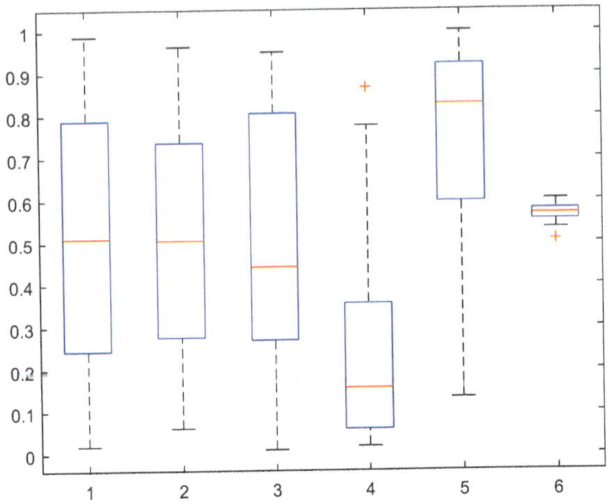


**Fig. 10** Boxplot of parameters of best individuals found by HDEA (in normalized ranges). Parameters: (1) attached worker migration bias (2) the unattached worker migration bias (3) worker relative adhesion (4) worker relative repulsion (5) worker motility persistence time (6) and the cargo release $O_2$ threshold

parameter. The majority of solutions are quite close to 11 $mmHg$, similar to findings from previous works [10, 18]. Although, for three of the parameters the results can not be conclusive (namely, attached and unattached worker migration bias and worker relative adhesion having almost uniform distribution like boxplots), the graphs for the other two parameters can convey the fact of the solutions being skewed towards smaller values for worker relative repulsion and higher values for worker motility persistence time.

## 7 Conclusion

In the standard evolutionary computing approach each individual solution can be seen to represent a single point in the fitness landscape. Typically, the same is true of bacteria in natural evolution. It has recently been suggested that natural evolution is using a more sophisticated approach with eukaryotes, exploiting a generalization process, whereby each individual represents a region in the fitness landscape [7]. Of course, landscape smoothing can be achieved by numerous mechanisms (after [11]) but they all require extra fitness evaluations. The scheme presented in this paper is intended to exploit the Baldwin effect through what is essentially simple population manipulation rather than through altering the underlying representation and evaluations of the standard evolutionary computing approach.

It can also be noted that the shape of the fitness landscape varies based upon the haploid genomes, which exist within a given population at any time and how they are paired. This is significant since, as has been pointed out for coevolutionary fitness landscapes [5], such movement potentially enables the temporary creation of neutral paths, where the benefits of (static) landscape neutrality are well-established (after [15]).

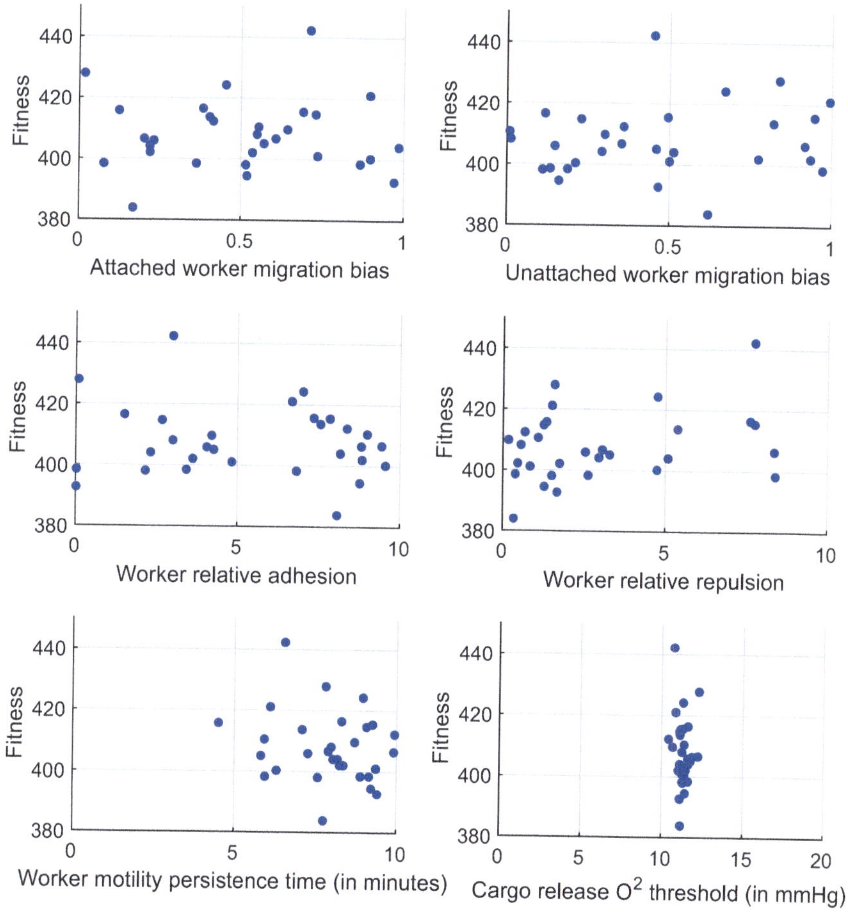


**Fig. 11** Scatter plot of parameters of best individuals found by GA for all 30 runs

The proposed HDEA method was also compared with a simple haploid EA in a more complicated simulator (PhysiCell). Again, the HDEA seems to perform better than the traditional and well-established haploid method. After analyzing the results of the methodology on the cancer treatment simulator, it can be concluded that it reaches fitter solutions faster, despite the high stochasticity injected into the fitness landscape to capture some of the dynamics of the biology. Current work is exploring the inclusion of a sexual selection-like process into the HDEA to further improve performance.

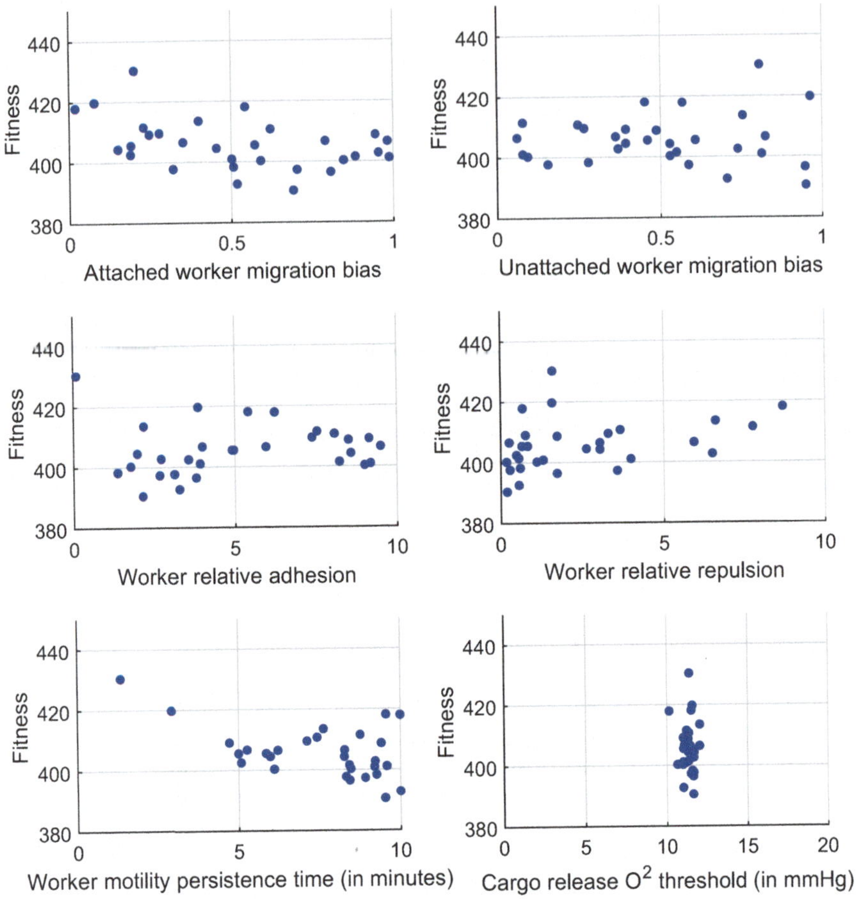


**Fig. 12** Scatter plot of parameters of best individuals found by HDEA for all 30 runs

**Acknowledgements** This work was supported by the European Research Council under the European Union's Horizon 2020 research and innovation programme under grant agreement No. 800983.

## References

1. Balaz, I., Petrić, T., Kovacevic, M., Tsompanas, M.A., Stillman, N.: Harnessing adaptive novelty for automated generation of cancer treatments. Biosystems **199**, 104290 (2021)
2. Baldwin, J.M.: A new factor in evolution. Am. Nat. **30**(354), 441–451 (1896)
3. Bernstein, H., Bernstein, C.: Evolutionary origin of recombination during meiosis. Bioscience **60**(7), 498–505 (2010)
4. Bhasin, H., Mehta, S.: On the applicability of diploid genetic algorithms. AI & Soc. **31**(2), 265–274 (2016)
5. Bull, L.: On coevolutionary genetic algorithms. Soft Comput. **5**(3), 201–207 (2001)

6. Bull, L.: The evolution of sex through the Baldwin effect. Artif. Life **23**(4), 481–492 (2017)
7. Bull, L.: The Evolution of Complexity: Simple Simulations of Major Innovations. Emergence, Complexity and Computation. Springer International Publishing (2020). https://books.google.gr/books?id=-iPNDwAAQBAJ
8. Eiben, A.E., Smith, J.E., et al.: Introduction to Evolutionary Computing, vol. 53. Springer (2003)
9. Ghaffarizadeh, A., Friedman, S.H., Macklin, P.: Biofvm: an efficient, parallelized diffusive transport solver for 3-d biological simulations. Bioinformatics **32**(8), 1256–1258 (2015)
10. Ghaffarizadeh, A., Heiland, R., Friedman, S.H., Mumenthaler, S.M., Macklin, P.: Physicell: an open source physics-based cell simulator for 3-d multicellular systems. PLoS Comput. Biol. **14**(2), e1005991 (2018)
11. Hinton, G.E., Nowlan, S.J.: How learning can guide evolution. Complex Syst. **1**(3), 495–502 (1987)
12. Karafotias, G., Hoogendoorn, M., Eiben, Á.E.: Parameter control in evolutionary algorithms: trends and challenges. IEEE Trans. Evol. Comput. **19**(2), 167–187 (2014)
13. Kauffman, S., Levin, S.: Towards a general theory of adaptive walks on rugged landscapes. J. Theor. Biol. **128**(1), 11–45 (1987)
14. Kauffman, S.A.: The Origins of Order: Self-Organization and Selection in Evolution. OUP USA (1993)
15. Kimura, M.: The Neutral Theory of Molecular Evolution. Cambridge University Press, Cambridge (1983)
16. Margulis, L., Sagan, D.: Origins of Sex: Three Billion Years of Genetic Recombination (1986)
17. Metzcar, J., Wang, Y., Heiland, R., Macklin, P.: A review of cell-based computational modeling in cancer biology. JCO Clin. Cancer Informatics **2**, 1–13 (2019)
18. Preen, R.J., Bull, L., Adamatzky, A.: Towards an evolvable cancer treatment simulator. Biosystems **182**, 1–7 (2019)
19. Smith, J.M., Szathmary, E.: The Major Transitions in Evolution. Oxford University Press, Oxford (1997)
20. Spears, W.M.: Evolutionary Algorithms: The Role of Mutation and Recombination. Springer Science & Business Media (2013)
21. Stillman, N.R., Balaz, I., Tsompanas, M.A., Kovacevic, M., Azimi, S., Lafond, S., Adamatzky, A., Hauert, S.: Evolutionary computational platform for the automatic discovery of nanocarriers for cancer treatment. NPJ Comput. Mater. **7**(1), 1–12 (2021)
22. Trun, N., Trempy, J.: Fundamental Bacterial Genetics. Wiley (2009)
23. Tsompanas, M.A., Bull, L., Adamatzky, A., Balaz, I.: Evolutionary algorithms designing nanoparticle cancer treatments with multiple particle types [application notes]. IEEE Comput. Intell. Mag. **16**(4), 85–99 (2021)
24. Tsompanas, M.A., Bull, L., Adamatzky, A., Balaz, I.: Novelty search employed into the development of cancer treatment simulations. Informatics Med. Unlocked **19**, 100347 (2020)
25. Tsompanas, M.A., Bull, L., Adamatzky, A., Balaz, I.: In silico optimization of cancer therapies with multiple types of nanoparticles applied at different times. Comput. Methods Programs Biomed. **200**, 105886 (2021)
26. Tsompanas, M.A., Bull, L., Adamatzky, A., Balaz, I.: Metameric representations on optimization of nano particle cancer treatment. Biocybern. Biomed. Eng. **41**(2), 352–361 (2021)
27. Tsompanas, M.A., Bull, L., Adamatzky, A., Balaz, I.: Utilizing differential evolution into optimizing targeted cancer treatments. In: Modern Trends in Controlled Stochastic Processes, pp. 328–340. Springer (2021)

# Drawbacks of Bench to Bed Translation of Nanomedicines for Cancer Treatment

**Petra Gener, Anna Ulldemolins, and Simo Schwartz Jr.**

## 1 State of the Art

Resistant metastatic cancer is at the moment an incurable disease and an unmet clinical need. It is one of the most lethal diseases and is causing thousands of deaths annually throughout the world. Treatment resistance and metastasis in vital organs account for 90% of cancer related deaths [1, 2]. The current treatments involve resection and the use of chemotherapeutics and radiations. However, these modalities are commonly associated with limited therapeutic effectiveness, recurrence, and side effects.

In this sense, nanomedicine-based drug delivery systems have brought a new hope for cancer patients. The word "nano" has always suggested an innovation. It is opening the cancer field to a whole new perspective, approaches and possibilities. Nanotechnology-based drug delivery systems (nano-DDS) are able to incorporate drugs or gene products with active anti-cancer activity but poor solubility, low bioavailability or inadequate toxicological profile. This leads to an improved efficacy and a superior bioavailability/biodistribution of the carried compound. This widens the therapeutic window of many cytotoxic drugs allowing the use of higher doses, while at the same time, lowering adverse effects of the treatment. Besides, the nanoparticles shell may protect the DNA, RNA content from degradation, thus securing its correct delivery to target cells. Nanocarriers also permit the combination of different treatment compounds or strategies within the same platform offering potential blocks of various parallel pathways activated in cancer resistant cells [3].

P. Gener (✉) · A. Ulldemolins · S. Schwartz Jr.
Drug Delivery and Targeting, Vall d'Hebron Institut de Recerca (VHIR), CIBBIM-Nanomedicine, Universitat Autònoma de Barcelona (UAB), 08035 Barcelona, Spain
e-mail: petra.gener@medsir.org

Networking Research Center on Bioengineering, Biomaterials and Nanomedicine (CIBER-BBN), 08035 Barcelona, Spain

 
I. Balaz and A. Adamatzky (eds.), *Cancer, Complexity, Computation*, Emergence, Complexity and Computation 46,
https://doi.org/10.1007/978-3-031-04379-6_11

An additional benefit of nanomedicines is their ability to bypass drug efflux pumps as they are internalized by endocytosis. This internalization process increases the intracellular accumulation of the drugs and ensures its release in the perinuclear region, away from membrane transporters [4].

Even though the potential of nanomedicines to ameliorate solubility, pharmacological profile and to reduce adverse effects of chemotherapeutic drugs is highly regarded and its use in the clinics augments over the last decade, its clinical translation is still suboptimal [5, 6]. Thousands scientific papers have been published, and billions of dollars have been invested in cancer nanomedicine in the last decades, however just few nanomedicines have reached clinical trials (only 2%) and even less has been commercialized (#15) for cancer treatment (Table 1) [3]. The main problems of these nanocarriers have been frequent adverse immune responses, unspecific targeting capacity, a poor cost-effectiveness ratio, and a variety of manufacturing and scale-up difficulties. Besides, the lack of clear regulatory standard and use of inadequate preclinical models complicate the translational process [7]. A new focus and new paradigm of this prospective scientific field is thus urgently needed. Nanomedicine for cancer treatment has a great potential but we should learn the lesson and to circumvent the problems during its preclinical development, in order to circumvent the vain clinical testing.

## 2 Cancer Biology

There are several reasons for the limited clinical translation of targeted cancer nanomedicines. Some of them include the poor understanding of the biology of cancer. There are many gaps in understanding of cellular and molecular processes that modulate NPs behaviour and fate in vivo efficacy.

Until the nineties, the initiation and progression of a tumour was explained by the clonal cancer model, in which cancer was thought to be driven by accumulated somatic mutations that confer uncontrolled growth, a more aggressive behavior and higher fitness to a malignant transformed cell [8, 9]. All cancer cells were considered equally malignant and equally responsive to treatment that was mostly attacking cell division machinery.

Later it was shown that not all the cells within the tumour presented the same tumourigenic potential. With this knowledge, a hierarchical model (also referred to as the cancer stem cell (CSC model)) has been described. Accordingly, only a small and distinct subpopulation of Cancer stem cells (CSCs) is alleged to have the capacity to generate and maintain tumour growth [10, 11]. In this model, cancer cells are created from a precursor cell (being a CSC the cell of origin), which went through either symmetric (generating two CSCs or two differentiated cancer cells (DCCs)) or asymmetric (generating a CSCs and a DCCs) divisions [11, 12]. Differentiated cancer cells (DCCs) do not present the ability to self-renew indefinitely and can only generate cells of their same type. On the other hand, CSCs can generate multiple and heterogeneous tumour subpopulations that differentiate into different lineages [13,

**Table 1** Ulldemolins et al. [3] Clinically approved nanoformulations for oncology in Europe and United States ordered by year of approval

| Name | Formulation | Type | Indications | Year |
|---|---|---|---|---|
| Doxil/Caelyx | PEG-ylated liposomal doxorubicin | Liposome | Kaposi sarcoma, Ovarian cancer, Multiple myeloma | 1995 (FDA) |
| DaunoXome | Liposomal daunorubicin | Liposome | Karposi sarcoma | 1996 (FDA) |
| DepoCyt | Liposomal cytarabine | Liposome | Lymphoma, leukemia | 1999 (FDA) |
| Myocet | Liposomal doxorubicin | Liposome | Breast cancer | 2000 (EMA) |
| Eligard | Leuprolide acetate and [PLGH (poly (dl-lactide-coglycolide)] polymer | Polymeric nanoparticle | Prostate cancer | 2004 (FDA) |
| Abraxane | Albumin-bound paclitaxel nanoparticle | Albumin-bound nanoparticle | Breast cancer, Non-small cell lung cancer, pancreatic cancer | 2005 (FDA) |
| Oncaspar | PEG-ylated L-asparaginase conjugate | Protein nanoparticle | Acute lymphoblastic leukemia (ALL) | 2006 (FDA) |
| Ontak | Interleukin (IL)-2 receptor antagonist with diphtheria toxin | Protein nanoparticle | Cutaneous T-cell lymphoma | 2008 (FDA) |
| Mepact | Liposomal mifamurtide | Liposome | Osteogenic sarcoma | 2009 (EMA) |
| NanoTherm | Iron Oxide Nanoparticles | Metallic nanoparticle | Brain tumours | 2011 (EMA) |
| Sylatron | PEGlyated interferon alfa-2b | Protein nanoparticle | Melanoma | 2011 (FDA) |
| Adcetris | CD30- targeted antibody (Brentuximab) and MMAE conjugate | Antibody–drug conjugate | Non-Hodgkin lymphoma | 2011 (FDA) |
| Marqibo | Liposomal vincirstine sulfate | Liposome | Acute lymphoblastic leukemia | 2012 (FDA) |
| Kadcyla | HER2-targeted antibody (Trastuzumab emtansine) and microtubule inhibitor conjugate | Antibody–drug conjugate | HER2-positive, metastatic breast cancer | 2013 (FDA) |
| Oniyde | Liposomal irinotecan | Liposome | Pancreatic cancer | 2015 (FDA) |

(continued)

**Table 1** (continued)

| Name | Formulation | Type | Indications | Year |
|---|---|---|---|---|
| Vyxeos | Liposomal daunorubicin and cytrabine | Liposome | Acute myeloid leukemia (AML) | 2017 (FDA) |
| Apalea | Paclitaxel micellar | Micelle nanoparticle | Ovarian cancer | 2018 (EMA) |
| Hensify | Hafnium oxide nanoparticles | Metallic nanoparticle | Soft tissue sarcoma | 2019 (EMA) |

14]. In terms of cancer treatment, according to the hierarchical model, the complete eradication of the CSCs population should be enough to eradicate the tumour and prevent disease relapse [15, 16]. Therefore, strong efforts have been invested, over the past decade, in the identification of CSCs within the tumour for targeted treatments. It was noted that CSCs are more permissive to DNA damage and express numerous multidrug resistant channels on their cell membrane and many currently used chemotherapeutics and radiation techniques are ineffective against them [3]. Besides, they can enter an undetectable quiescence state when the conditions of the TME are not favourable and proliferate afterwards [17].

Hence, CSCs sustain very effectively tumour growth and are responsible for cancer recurrence, metastatic spread, and current treatment failure [3, 17–21]. After treatment, the amount of CSCs frequently rises when compared to other tumour cells types. Of note only a few CSCs are needed for tumour regeneration in vivo. In this context, even though clinicians often observe tumour shrinkage at the first stages of multimodal treatment, often remaining resistant clones of CSCs survive and eventually cause tumour relapse and very aggressive tumour types with limited treatment alternatives [3, 17–21].

Some inconsistencies of the hierarchic model came after the observation that even after successful CSCs eradication, tumour growth and spread may be evidenced [22]. Thus a new dynamic stochastic model has been suggested and CSC eradication strategies become even more complex [11, 15, 17, 18, 22, 23]. This model postulates the possibility that every tumour cell can become a tumour-initiating cell, considering the possibility of inter-conversion among cell phenotypes. According to this model, tumour is composed by different cell populations that maintain a stable communication among them [15, 18]. Through this communication, they can “sense” if one specific subpopulation of the tumour cells is redundant, absent or has been depleted. When this happens, cells of this type are generated, allowing each cell subpopulation to return to their fixed phenotypic equilibrium. In this model, the amount of CSCs seems to remain constant through time, to maintain the mentioned equilibrium within the tumour micro environment (TME). And all this happens through interconversion between DCCs and CSCs [15, 17, 18]. CSCs may differentiate into DCCs (giving heterogeneity to the tumour) and DCCs can be reprogrammed and gain CSC properties, such as self-renewal capacity and treatment resistance, as well as increased expression of stemness genes [15, 22]. According to a stochastic model, any cell of the tumour can initiate the progression of the disease due to the existing phenotypic plasticity, and further, any cancer cell can recover stem cell-like phenotype by dedifferentiation [15, 17, 24]. The dynamism of phenotypic reversion seems to be comparable to the dynamism observed during epithelial-mesenchymal transition (EMT) processes, in which cancer cells can switch between epithelial and mesenchymal phenotypes. Consequently, CSCs present generally a mesenchymal phenotype, while other cancer cells show mostly epithelial characters [18].

Although the hierarchical model and the stochastic model have different considerations regarding the importance of CSCs in tumour initiation and progression are not mutually exclusive thanks to cellular plasticity (Fig. 1) [17]. Basically, the tumour is formed in a hierarchical manner, which is unstable since constant stochastic actions allow the introduction of newly hierarchically organized cell populations [15–17].

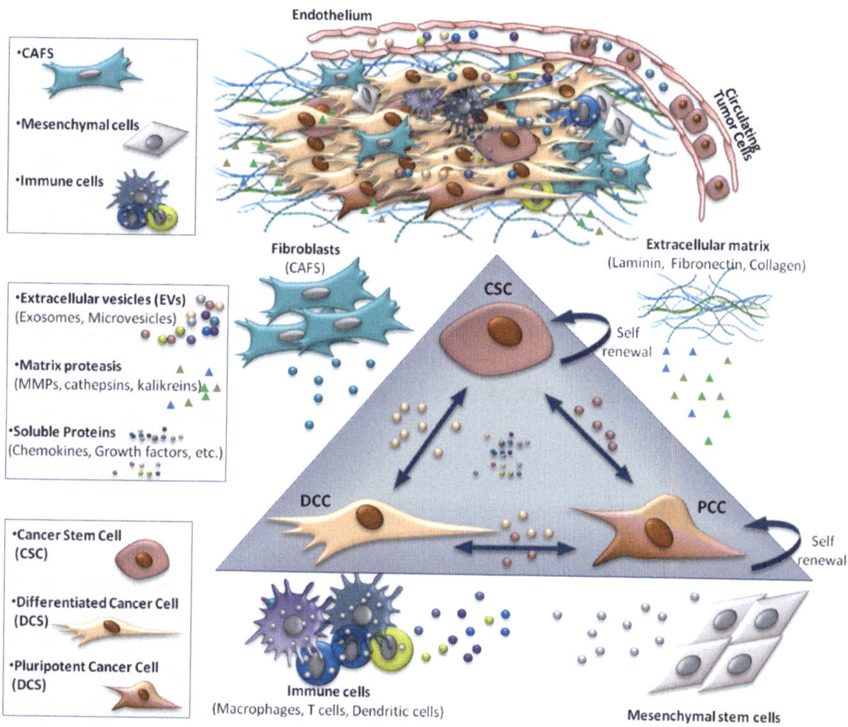


**Fig. 1** Dynamic CSC model [17]. A tumour is highly complex heterogenic dynamic entity that involves cellular components (CSCs, DCSs, fibroblasts, immune cells, mesenchymal cells etc.), cell-secreted functional molecules (growth factors, hormones, metabolites, cytokines, chemokines etc.), extracellular vesicles (exosomes, vesicles etc.) and extracellular matrix (ECM). Cell sub-populations within tumours are characterized by their respective densities and rates, and the alteration of these densities and rates alters the equilibrium state of the tumour as a whole biological system. Self-renewal and differentiation rates of cancer stem cells are finely tuned in order to ensure tumour progression

In summary, tumour should be considered as a highly complex and heterogenic dynamic entity that evolves in time, adapting and therefore surviving to adverse conditions [3, 17]. To design new effective treatment it is thus important to think in a wide perspective and not to focus on one single target. A whole tumour mass and its ability to adapt and to change in adverse conditions should be considered.

A tumour is composed of a heterogeneous mix of different environments and cell types, such as already mentioned CSCs, DCCs, but also CAFs (cancer associated fibroblasts), mesenchymal cells, tumour-infiltrating immune cells, endothelial cells and stromal cells. All of them are located within the extracellular matrix and together contribute to disease progression [16, 17, 20, 25]. They all are part of the local TME (Tumour Micro Environment) and present different mesenchymal and stem-like phenotypes. Likewise, the constant exchange of information (cytokines, growth factors etc.) among these cells is essential to guarantee survival and progression of the tumour and to orchestrate the coordination and collaboration of different cells [17, 18, 26, 27]. Cancer cells interact with other cells from the TME through direct cell-cell contact and/or by paracrine signaling. Both cellular and non-cellular components of the niche maintain stable the stemness potential of the tumour, and regulate CSC plasticity [17, 18, 26, 27].

Altering the equilibrium of the different cell populations is often translated into tumour aggressiveness and treatment resistance. In addition, epigenetic alterations contribute to the therapeutic resistance of the tumour by granting protection from therapy-induced damage to tumour cells. Elucidating the molecular mechanisms that govern cellular plasticity may be essential to overcome the challenge that current therapies face when fighting against cancer. Effective targeting therapies need to be developed to eliminate the roots of continuously evolving tumour cell populations and to avoid the regeneration of CSCs [3, 17, 27].

## 3 Biological Barriers

Nanomedicines designed based on current knowledge to treat cancer have to be able to reach targeted cells. This might be difficult however since after administration (intravenous or oral) thaymay find several barriers (gastro-intestinal tracts, endothelial cells, cell membrate etc.). Besides, nanomedicines must remain stable in the bloodstream and should be able to reduce body clearance by reticuloendothelial system (RES) and the mononuclear phagocyte system (MPS) escape. The size of the resultant nanoparticle is crucial in this sense. Of note, nanoparticles greater than 10 nm prevent premature excretion by kidneys, while nanoparticles below 200 nm are able to pass through the microcapillaries without producing embolism [28]. Cellular uptake and biodistribution is determined also by the surface charge. In general, positive charges facilitate cell internalization compared to neutral and negatively charged nanoparticles but are often more toxic [28]. Importantly, the surface charge also affects the interaction of nanoparticles with the biological fluids since in biological fluids nanoparticles' surface is surrounded by non-covalently bound protein corona [29]. The composition of protein corona is dynamic and depends on chemical composition, surface functionalization, shape, porosity, surface crystallinity, and hydrophobicity/hydrophilicity of used nanomaterial. Attachment of different surface proteins causes different interactions and/or behaviour of nanomedicines in term of hemolysis, activation of thrombocytosis, and cellular uptake. Often nano-surfaces are coated with opsonins that destine the nanoparticles to phagocytosis. Conversely, CD47 glycoprotein coating evades phagocytosis [30]. In order to bypass phagocyte-mediated cellular barriers, Poly (ethylene glycol) (PEG) is often used as surface coating of the drug-delivery system [31]. PEGylation has the ability to enhance retention time. Basically, it prolongs the circulation time by increasing its hydrophilicity and reducing the rate of glomerular filtration. Furthermore, it forms a hydrophilic shield that is able to mask the antigenic sites of the proteins and provides protection from reticuloendothelial cells, proteolytic enzymes, and phagocytes. Therefore, it delays recognition by the immune system and increases the chance of nanomedicines to target the desired tissues or cells [31]. Our current knowledge regarding protein corona is still insufficient. Theoretical modelling, such as molecular dynamics simulation, could be a powerful tool for predicting interactions and aiding mechanistic understanding of its formation [29].

In general terms, it is believed that nanomedicines tend to accumulate within the tumour tissue thanks to enhanced permeability and retention (EPR) effect. However, the presence of the EPR effect is not a general occurrence (i.e. absent in micro-metastasis) and the specific entry to cancer cells may be challenging. The EPR effect in clinical tumour therapy is likely overestimated. Besides, there is the existence of elevated tumour interstitial fluid pressure that reduces the efficacy of drug delivery by nanoparticles. The increased interstitial fluid pressure has been reported in many solid tumours, such as breast, colorectal cancers, and melanoma [32].

Another way to increase cancer cell penetration is the specific targeting of nanomedicines. Even this may look like a perfect solution; often targeted nanomedicines extravasate into the tumour vicinity and bind cells with high affinity just in the outer layer of the TME. Thus, they cannot easily penetrate to the inner parts of the tumour. This phenomenon is known as the binding site barrier (BSB) which prevents deeper penetration of nanomedicines into the tumours. Specifically, the BSB limits diffusion through the TME and results in unintended internalization of nanomedicines by stromal cells located near blood vessels. The major components of the BSB are CAFs. Besides, the proximity to the blood vessels and expression of protein receptors further complicate the penetration of nanomedicines into tumour cells [33].

Yet, the BSB and the interstitial fluid pressure have not been completely taken into account in preclinical studies due to the existence of discrepancies between animal models and human tumours [34]. The biggest issue is the disproportion between small marine tumours compared to size of human tumour that led so far into a poor translation of preclinical studies of penetration [3].

# 4 Immune Reaction

Another example of how the current preclinical standards have poor translation capacity is the existence of the potential immunogenicity of nanoparticles. It is well known that immune-mediated adverse effects might appear after administration of nanoDDS. Even though a number of clinically relevant side effects have been reported in this regard, immuno-deficient mice are commonly used for validation [35]. In this regard, the more common toxicities linked to NP failure include: erythrocyte damage, thrombogenicity (platelet aggregation, plasma coagulation), cytokine-mediated inflammation and cytokine storming, pyrogenicity, and anaphylaxis and other complement activation mediated reactions, as well as recognition and uptake by the eticuloendothelial system [36].

The immune response of any nano-matherial and/or DDS remains unpredictable. However, it is known that any physicochemical properties (size, biocompatibility, and surface chemistry) of used nanomedicine may play an important role in the activation of the immune system. The introduction of a targeting ligand on the surface may further change the immune system. Pre-clinical and clinical screening for immunotoxicities of used nanomedicines remains to be a big challenge in the frame of correct characterization of DDSs. Similar studies are based just on the estimation of immunoreactive contaminants, such as excipients and linkers [36]. Although there are current standard methods, they are insufficient to address the broad spectrum of biomarkers that indicate possible immunotoxicity of nanomedicines. In addition, there is an absence of consensus on well characterized reference materials. Therefore, preclinical studies often depend on nanomedicines with known clinical immunotoxicities (e.g. Doxil for complement activation and anaphylaxis) [36]. Besides, as mentioned before, the use of immunosuppressed animals unables to determine immunotoxicities related to nanomedicines. For this reason, in clinical studies, patients are constantly pre-medicated with immunosuppressors to prevent adverse reactions. Screening for these toxicities in preclinical development would help to prevent potentially toxic formulations [3].

Recently, immunodeficient mice engrafted with human immune systems have been established and are a powerful tool for the next generation of patient derived xenograft (PDX) models. PDX models allow small parts of the human tumour in the murine environment. In humanised PDX models, human hematopoietic stem cells (HSCs) are introduced in the immune deficient mice after gamma irradiation. As a result the pre-clinical studies may progress in complex TME within a human immune system. For example, it has been already reported a triple negative breast cancer PDX model with humanized mice that provides evidence that supports its use for the pre-clinical investigation of immune-based therapies [35, 37]. Unfortunately, the nowadays considerably expensive production of these models limits their wider use [3].

## 5 Targeting

Many hopes and efforts were deposited in the synthesis of targeted nanomedicines. Targeting is supposed to specifically identify malignant cells, and subsequently facilitate ligand-mediated cellular uptake as a result of receptor-mediated endocytosis [10]. Many nanoformulations with active targeting were studied in order to reduce drug resistance while increasing the effective amount of drug delivered to the tumour cell, however non is in the marked yet (Table 2) [3]. Engineering of DDSs with active targeting increases considerably its complexity since it makes it more difficult, time consuming. Besides, surface targeting may considerably increase potential immunogenicity of the whole system [36]. In addition, numerous limitations exist for clinical applications of some active targeting drugs because of their rapid elimination by the reticuloendothelial system and high tumour interstitial fluid pressure [38]. A lot of efforts are devoted to develop the "perfect NPs" with a wide range of specific ligands in their surface during preclinical development. It is fascinating what bioengineering is capable to achieve and all the complexity that can be designed and synthesized in a single platform. However, over the last years it has been demonstrated that translationality gets poorer when complexity gets richer. Up to date, the most clinically successful NP is Abraxane. This albumin-bound paclitaxel NP is a simply engineered NP, however very elegant, since albumin is recognized by gp60 receptors of the endothelial cells that guide the extravasation of Abraxane [39]. Many methods of functionalization are published, but most of them lack reproducibility. The functionalization process is very complicated and requires different conditions for each efficient surface modification. It involves a multi-step processing to formulate complex targeted NPs which in turn compromise final production yields [3] (Fig. 2).

Further, there is an actual unmet need to synergize passive and active targeting to improve the accumulation of nanoparticles at the desired site while at the same time enhancing their intracellular penetration [38, 40].

## 6 Cost, Regulation, Scale Up

Investment in nanomedicine in the early 2000s accelerated the development of nanoformulations that are currently available in the market. However, even with the sales success of some of them (e.g. Abraxane), there are financial challenges that hamper the development of new nanoformulations. As it has been previously explained, it is not easy to demonstrate improved efficacy and safety compared to other validated and marketed products for the same indication. Indeed, the majority of approved nanodrugs are based on currently approved drugs which face reduced financial risk because the efficacy and safety of the active ingredient had already been established [41]. On the contrary, more complex economic considerations are involved when developing a nanodrug that contains a new chemical entity. Moreover,

**Table 2** Ulldemolins et al. [3] Nanoformulations with active targeting being studied in the clinics

| Name | Formulation | Type | Targeting | Indications | Phase |
|---|---|---|---|---|---|
| BIND-014 | PL(G)A-PEGylated Docataxel | Polymeric NP | PSMA specific receptor | Metastatic Castration-Resistant Prostate Cancer | Phase II (NCT01812746) |
| SP1049-C | Pluronic-b-copolymer doxorubicin | | Pgp protein | Advanced adenocarcinoma of the esophagus | Phase II ([71]) |
| MM-302 | HER2-targeted PEGylated antibody–liposomal doxorubicin | Liposomal NP | | HER2-positive metastatic cancer | Phase II (NCT02213744) |
| MCC-465 | Liposomal Doxorubicin with F(ab')2 fragment of GAH human Mab | | GAH | Stomach cancer | Phase I ([72]) |
| 2B3-101 | Doxorubicin with glutathione | | Glutathione transporter | Brain Metastasis Menongeal carcinomatosis | Phase I/IIa (NCT01386580) Phase II (NCT01818713 |
| MBP-426 | Oxaliplatin with transferrin | | Transferrin receptor | Metastatic solid tumours | Phase I (NCT03002103) |
| anti-EGFR ILs-DOX | Doxorubicin-loaded Anti-EGFR immunoliposomes (C225-ILs-dox) | | EGFR | High-grade Gliomas | Phase I (NCT03603379) |
| IMMU-132 | Trop-2 MAb and SN-38 conjugate | Drug–antibody conjugated | trophoblastic cell-surface antigen-2 (Trop-2) | Epithelial cancers | Phase I/II (NCT01631552) |
| SGN-35 | MMAE coupled to CD30-targeted antibody | | CD30 receptor | Relapsed or Refractory Hodgkin Lymphoma (HL) | Phase II (NCT00848926) |

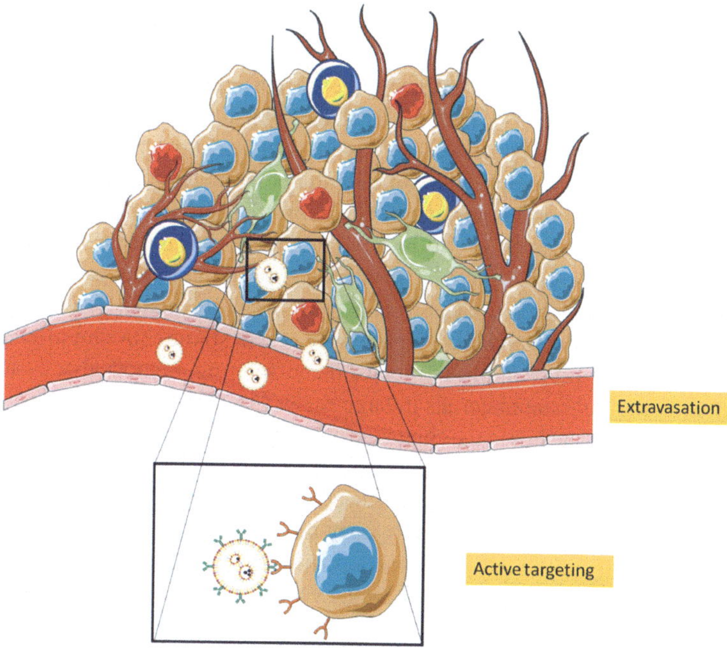


**Fig. 2** Extravasation and cell targeting [3]. The abnormally wide fenestrations in the blood vessels and the lack of lymphatic drainage facilitate extravasation of nanomedicines. Once in the TME, the targeting moiety enables its interaction with the desired cells providing and active targeting

the complexity to design a specific ligand and the conjugation techniques make the whole process more expensive with a difficult scale-up. The cost of using complex chemistry, controlled quality manufacturing and scaled production is elevated [41].

Apart from all the high costs involved, regulatory requirements also make difficult market entry. Various regulatory agencies like EMA and FDA started their discussion on the classification of nanomaterial and how to regulate them to ensure proper efficacy and safety of these materials. For example, the regulatory system in Europe allows the marketing authorization applicants to receive scientific counselling during early stages of R&D. Also, The Nanotechnology Characterization Laboratory at the National Cancer Institute in the US collects all the data on nanomedicines in oncology. Indeed, integration between materials and translational issues, such more appropriate disease models, are essential for developing accurate regulation of nanomedicines [42]. Moreover, standardized protocols for the purification and isolation processes are also needed. These protocols should be scalable to translate the techniques into large-scale EVs production under the GMP [41, 42].

The success of any product, including nanomedicines, relies on its large-scale industrial production and commercialization. Therefore, the scientific community should pay more attention to the large-scale industrial production after their successful development of the laboratory scale. A critical issue lies in the challenge of scaling-up nanomaterial synthesis and formulation from the lab to industrial scale while maintaining control over their diverse properties. Studying these phenomena early on in the development of a therapeutic agent often requires partnerships between the public and private sectors which are hard to establish [43, 44].

Part of the problem lies in the fact that nanostructured drug delivery systems have grown into increasingly complex structures with poorly understood biological fates. Size, surface charge, morphology, drug release profiles and interaction with blood components have all been shown to be key parameters that influence the biomedical outcome of nanocarriers in vivo. Controlling these numerous parameters in a reproducible manner over intricate structures can be problematic, as well as difficult to bring to industrial standards. Accordingly, recent studies have highlighted the importance of developing simple and robust manufacturing methods that ensure the proper formulation of nanomaterials in large scales [45, 46].

Another critical issue lies in the challenge of scaling-up the synthesis of a nanocarrier's constitutive materials. Guidelines from the pharmaceutical industry in terms of good manufacturing practice and pharmaceutical quality can be applied, but the nature and complexity of nanomaterials carry their extra set of challenges. Indeed, while any scale-up of laboratory processes is difficult, nanomaterials production is made more challenging by the fact that subtle variations in the synthesis and manufacturing processes can result in significantly different products. It is necessary to implement a robust quality control system in place, validating batch-to-batch consistency and biological equivalence. Identifying critical parameters that might affect the reproducibility of nanomaterial properties early on and monitoring these parameters on multiple batches is a good practice which can minimize the appearance of reproducibility issues later on [43, 45, 46].

## 7 Future Perspectives of Nanomedicine for Cancer Treatment

Cancer still remains as one of the deadliest diseases worldwide. New treatments and therapeutic approaches are thus considered necessary. In this context, drug delivery into resistant cancer cells under the shield of nanonomedicine may represent a solution to avoid the undesirably common side-effects and to promote the efficiency of the therapy. Even though the use of synthetic nanoparticles has been in the limelight for the past years, its clinical translation and consequent improved patient outcome is few and far between. There is a clear need to focus on existing nano-carriers, combination therapies, patient selection and ways to enable rapid and more efficient clinical translation in order to ameliorate the present poor clinical outcomes. Since there are a lot of unknown and unexplored outcomes, regarding the behaviour of DDS within the biological system, mathematical models and software with the help of artificial intelligence (AI) may help the clinician to choose the best possible option for each patient. It is necessary to promote the entrance of new nano-products in clinical phases including new materials (i.e. hydrogels, extracellular vesicles). Hopefully, the current and recent generalized use of nanoparticles in clinics, like the new Covid-19 vaccines, will prompt further clinical use of prospective nanomedicines also for treatment of advanced cancer.

## References


1. Arnold, M., Rutherford, M.J., Bardot, A., Ferlay, J., Andersson, T.M.L., Myklebust, T.Ã., Tervonen, H., Thursfield, V., Ransom, D., Shack, L., Woods, R.R., Turner, D., Leonfellner, S., Ryan, S., Saint-Jacques, N., De, P., McClure, C., Ramanakumar, A.V., Stuart-Panko, H., Engholm, G., Walsh, P.M., Jackson, C., Vernon, S., Morgan, E., Gavin, A., Morrison, D.S., Huws, D.W., Porter, G., Butler, J., Bryant, H., Currow, D.C., Hiom, S., Parkin, D.M., Sasieni, P., Lambert, P.C., MÃ¸ller, B., Soerjomataram, I., Bray, F.: Progress in cancer survival, mortality, and incidence in seven high-income countries 1995â€“2014 (ICBP SURVMARK-2): a population-based study. Lancet Oncol. **20**(11), 1493–1505 (2011) 2019. https://doi.org/10.1016/S1470-2045(19)30456-5
2. Pilleron, S., AUID-ORCID: https:/., Sotoâ-Pereză-deâ-Celis, E., Vignat, J., Ferlay, J., Soerjomataram, I., Bray, F.A., Sarfati, D.A.: Estimated global cancer incidence in the oldest adults in 2018 and projections to 2050. Int. J. Cancer **148**(3), 601–608 (2003). 01 Feb 2021. Epub 17 Aug 2020. https://doi.org/10.1002/ijc.33232-8
3. Ulldemolins, A., Seras-Franzoso, J., Andrade, F., Rafael, D., Abasolo, I., Gener, P., Schwartz Jr, S.: Perspectives of nano-carrier drug delivery systems to overcome cancer drug resistance in the clinics. Cancer Drug Resist **3** (2020) (Online)
4. Xue, X.F., Liang, X.J.: Overcoming drug efflux-based multidrug resistance in cancer with nanotechnology. Chin. J. Cancer **31**(2), 100–109 (2002). Feb 2012. https://doi.org/10.5732/cjc.0111.0326
5. Shi, J., Kantoff, P.W., Wooster, R., Farokhzad, O.C.: Cancer nanomedicine: progress, challenges and opportunities. Nat. Rev. Cancer **17**(1), 20–37 (2001). Jan 2017. Epub 11 Nov 2016. https://doi.org/10.1038/nrc.2016.108
6. Salvioni, L., Rizzuto, M.A., Bertolini, J.A., Pandolfi, L., Colombo, M., Prosperi, D.A.: (Basel)—Thirty years of cancer nanomedicine: success, frustration, and hope. Cancers (Basel) **11**(12) (2019). https://doi.org/10.3390/cancers11121855
7. van der Meel, R., Sulheim, E., Shi, Y., Kiessling, F., Mulder, W.J., Lammers, T.: Smart cancer nanomedicine: strategic directions to improve translation and exploitation. Nat. Nanotechnol. **14**(11), 1007–1017 (2011). Nov 2019. Epub 06 Nov 2019. https://doi.org/10.1038/s41565-019-0567-y-17
8. Lee, G., Hall, R.R., III, Ahmed, A.U.: Cancer stem cells: cellular plasticity, niche, and its clinical relevance. J. Stem Cell Res. Ther. **6**(10) (2010). Epub 2016 Oct 26–7633. https://doi.org/10.4172/2157-76331000363
9. Lengauer, C., Kinzler, K.W., Vogelstein, B.: Genetic instabilities in human cancers. Nature **396**, 643–649 (1998)
10. Gener, P., Rafael, D.F., Fernandez, Y., Ortega, J.S., Arango, D., Abasolo, I., Videira, M., Schwartz, S., Jr.: Cancer stem cells and personalized cancer nanomedicine. Nanomedicine (Lond) **11**, 307–320 (2016)
11. Shackleton, M., Quintana, E., Fearon, E.R., Morrison, S.J.: Heterogeneity in cancer: cancer stem cells versus clonal evolution. Cell **138**, 822–829 (2009)
12. Hermann, P.C., Bhaskar, S., FAU., Cioffi, M., Cioffi, M.F., Heeschen, C. Cancer stem cells in solid tumors. Semin. Cancer Biol. **20**(2), 77–84 (2002). Epub 3 Apr 2010. https://doi.org/10.1016/j.semcancer.2010.03.004
13. Sell, S.: Stem cell origin of cancer and differentiation therapy. Crit. Rev. Oncol. Hematol. **51**(1), 1–28 (2001). Jul 2004. https://doi.org/10.1016/j.critrevonc.2004.04.007-28

14. Shackleton, M.: Normal stem cells and cancer stem cells: similar and different. Semin. Cancer Biol. **20**, 85–92 (2010)
15. Olmeda, F., Ben, A.M.: Clonal pattern dynamics in tumor: the concept of cancer stem cells. Sci. Rep. **9**(1), 15607 (2001). 30 Oct 2019. https://doi.org/10.1038/s41598-019-51575-1
16. Plaks, V., Kong, N., Werb, Z.: The cancer stem cell niche: how essential is the niche in regulating stemness of tumor cells? Cell Stem Cell **16**(3), 225–38 (2003). 5 Mar 2015. https://doi.org/10.1016/j.stem.2015 02.015-38
17. Gener P, Gonzalez Callejo P, Seras-Franzoso J, Andrade F, Rafael D, Abasolo I, Schwartz S Jr. The potential of nanomedicine to alter cancer stem cell dynamics: the impact of extracellular vesicles. Nanomedicine (Lond). 2020 Dec;15(28):2785–2800. https://doi.org/10.2217/nnm-2020-0099. Epub 2020 Nov 16. PMID: 33191837
18. Gener, P., Seras-Franzoso, J., Gonzales Callego, P., Andrade, F., Rafael, D., et al.: Dynamism, Sensitivity, and Consequences of Mesenchymal and Stem-Like Phenotype of Cancer Cells. Stem Cells Int. (2018). (in press)
19. Shibue, T., Weinberg, R.A.: EMT, CSCs, and drug resistance: the mechanistic link and clinical implications. Nat. Rev. Clin. Oncol. **14**(10):611–629 (2010). Oct 2017. Epub 11 Apr 2017. https://doi.org/10.1038/nrclinonc.2017.44
20. Wang, M., Zhao, J., Zhang, L., Wei, F., Lian, Y., Wu, Y., Gong, Z., Zhang, S., Zhou, J., Cao, K., Li, X., Xiong, W., Li, G., Zeng, Z., Guo, C.: Role of tumor microenvironment in tumorigenesis. J. Cancer **8**(5), 761–773 (2005). 25 Feb 2017. https://doi.org/10.7150/jca.17648
21. Wicha, M.S., Liu, S., Dontu, G.: Cancer stem cells: an old idea–a paradigm shift. Cancer Res. **66**, 1883–1890 (2006)
22. Gener, P., Rafael, D., Seras-Franzoso, J., Perez, A., Pindado, L.A., Casas, G., Arango, D., Fernandez, Y., Diaz-Riascos, Z.V., Abasolo, I., Schwartz, S Jr. (Basel).: Pivotal role of AKT2 during dynamic phenotypic change of breast cancer stem cells. Cancers (Basel) 11(8) 2019 Jul 26. https://doi.org/10.3390/cancers11081058
23. Gupta, P.B., Fillmore, C.M., Jiang, G., Shapira, S.D., Tao, K., Kuperwasser, C., Lander, E.S.: Stochastic state transitions give rise to phenotypic equilibrium in populations of cancer cells. Cell **146**, 633–644 (2011)
24. Cabrera, M.C., FAU., Hollingsworth, R., Hollingsworth, R.E., FAU., Hurt, E., Hurt, E.M.: Cancer stem cell plasticity and tumor hierarchy. World J. Stem Cells **7**(1), 27–36 (2001). Epub 26 Jan 2015. https://doi.org/10.4252/wjsc.v7.i1.27-36
25. Eun, K., Ham, S.W., Kim, H.: Cancer stem cell heterogeneity: origin and new perspectives on CSC targeting. Bmb Rep. **50**(3), 117–125–125 (2003). Mar 2017
26. Sun, Z.A., Wang, L., Dong, L., Wang, X.: Emerging role of exosome signalling in maintaining cancer stem cell dynamic equilibrium. J. Cell Mol. Med. 25 May 2018. https://doi.org/10.1111/jcmm.13676.J
27. Hernandez-Oller, L.A., Seras-Franzoso, J., Andrade, F., Rafael, D.A., Abasolo, I.A., Gener, P.A., Schwartz, S., AUID-ORCID: https: Extracellular vesicles as drug delivery systems in cancer. Pharmaceutics **12**(12) (2012). 26 Nov 2020. https://doi.org/10.3390/pharmaceutics12 121146
28. Jo, D.H., Kim, J.H., Lee, T.G., Kim, J.H.: Size, surface charge, and shape determine therapeutic effects of nanoparticles on brain and retinal diseases. Nanomedicine **7**, 1603–1611 (2015)
29. Meng, H., Leong, W., Leong, K.W., Chen, C., Zhao, Y.: Walking the line: the fate of nanomaterials at biological barriers. Biomaterials **174**, 41–53 (2018). Epub 05 May 2018. https://doi.org/10.1016/jbiomaterials.2018.04.056
30. Rodriguez, P.L., Harada, T., Christian, D.A., Pantano, D.A., Tsai, R.K., Discher, D.E.: Minimal â€œSelfâ€ peptides that inhibit phagocytic clearance and enhance delivery of nanoparticles. Science **339**(6122), 971–975. 22 Feb 2013. https://doi.org/10.1126/science 1229568
31. Mishra, P., Nayak, B., Dey, R.K.: PEGylation in anti-cancer therapy: an overview. AJPS **11**, 337–348 (2020)
32. Libutti, S.K., Tamarkin, L., Nilubol, N.: Targeting the invincible barrier for drug delivery in solid cancers: interstitial fluid pressure. Oncotarget **9**(87), 35723-5 (1987). 6 Nov 2018. https://doi.org/10.18632/oncotarget.26267

33. Miao, L., Newby, J.M., Lin, C., Zhang, L., Xu, F., Kim, W.Y., Forest, M., Lai, S.K., Milowsky, M.I., Wobker, S.E., Huang, L.: The binding site barrier elicited by tumor-associated fibroblasts interferes disposition of nanoparticles in stroma-vessel type tumors. ACS Nano **10**(10), 9243–9258 (2010). 25 Oct 2016. Epub 28 Sep 2016. https://doi.org/10.1021/acsnano.6b02776.
34. Attia, M., Anton, N., Wallyn, J., Omran, Z., Vandamme, T.F.: An overview of active and passive targeting strategies to improve the nanocarriers efficiency to tumour sites. J. Pharm. Pharmacol. 1185–1198 (2019)
35. Murayama, T.F., Gotoh, N.A. Patient-Derived Xenograft Models of Breast Cancer and Their Application. Cells **8**(6) (2006). 20 Jun 2019. https://doi.org/10.3390/cells8060621
36. Dobrovolskaia, M.A., Shurin, M., Shvedova, A.A. Current understanding of interactions between nanoparticles and the immune system. Toxicol. Appl. Pharmacol. **299**, 78–89 15 May 2016. Epub 29 Dec 2015. https://doi.org/10.1016/j.taap.2015.12.022
37. Gazdar, A.F., Hirsch, F.R., Minna, J.D.: From Mice to Men and back â€“ an assessment of preclinical model systems for the study of lung cancers. J. Thorac. Oncol. **11**(3), 287–99 (2003). Mar 2016. Epub 24 Dec 2015. https://doi.org/10.1016/j.jtho.2015.10.009
38. Villaverde, G., Baeza, A.A.: Targeting strategies for improving the efficacy of nanomedicine in oncology. Beilstein J. Nanotechnol. **10**, 168–181 14 Jan 2019. https://doi.org/10.3762/bjnano.10.16-Beilstein
39. Miele, E., Spinelli, G.P., Miele, E., Tomao, F., Tomao, S.: Albumin-bound formulation of paclitaxel (AbraxaneÂ® ABI-007) in the treatment of breast cancer. Int. J. Nanomedicine **4**, 99–105 2009. Epub 20 Apr 2009
40. Xu, X., Ho, W., Zhang, X., Bertrand, N., Farokhzad, O.: Cancer Nanomedicine: from targeted delivery to combination therapy. Trends Mol. Med. **21**(4), 223–32 (2004). Apr 2015 Epub 2 Feb 2015. https://doi.org/10.1016/j.molmed.2015.01.001.32
41. Ventola, C.L.: Progress in nanomedicine: approved and investigational nanodrugs. Pharm. Ther. **42**(12), 742–755 (2012). 2017 Dec
42. Patra, J.K., AUID, Das, G., Fraceto, L.F., Campos, E.V.R., Rodriguez-Torres, M., del, P.A., Acosta-Torres, L.S., AUID-ORCID, Diaz-Torres, L.A., AUID-ORCID, Grillo, R., Swamy, M.K., Sharma, S., Habtemariam, S., Shin, H.S.: Nano based drug delivery systems: recent developments and future prospects. J. Nanobiotechnology **16** 19 Sep 2018. 10 1186/s12951-018-0392-8
43. Anchordoquy, T.J., Barenholz, Y., Boraschi, D., Chorny, M., Decuzzi, P., Dobrovolskaia, M.A., Farhangrazi, Z.S., Farrell, D., Gabizon, A., Ghandehari, H., Godin, B., La-Beck, N.M., Ljubimova, J., Moghimi, S.M., Pagliaro, L., Park, J.H., Peer, D., Ruoslahti, E., Serkova, N.J., Simberg, D.: Mechanisms and barriers in cancer nanomedicine: addressing challenges, looking for solutions. ACS Nano. **11**(1), 12–18 (2001). 24 Jan 2017. Epub 09 Jan 2017. 10 1021/acsnano 6b08244.
44. Cicha, I., Chauvierre, C., Texier, I., Cabella, C., Metselaar, J.M., Szebeni, J., DÃ©zsi, L., Alexiou, C., Rouzet, F., Storm, G., Stroes, E., Bruce, D., MacRitchie, N., Maffia, P., Letourneur, D.: From design to the clinic: practical guidelines for translating cardiovascular nanomedicine. Cardiovasc. Res. **114**(13), 1714–27 (2013). 01 Nov 2018. Epub 27 Aug 2018. https://doi.org/10.1093/cvr/cvy219-27
45. Darmont, F., Rousseau, B.: Translation of nanomedicines from lab to industrial scale synthesis: the case of squalene-adenosine nanoparticles. J. Control. Release **307**, 302–314 (2019)
46. Metselaar, J.M., Lammers, T.: Challenges in nanomedicine clinical translation. Drug Deliv. Transl. Res. **10**(3), 721–725 (2003) (2020). Epub 12 Mar 2020. https://doi.org/10.1007/s13346-020-00740-5

# Swarms: The Next Frontier for Cancer Nanomedicine

**Matimba Swana, Johanna Blee, Namid Stillman, Jonathan Ives, and Sabine Hauert**

**Abstract** Nanomedicine refers to medical products developed using nanotechnology and has the potential to radically change how we diagnose and treat cancer. While the use of nanomedicines has increased in the clinic dramatically, problems persist over the lack of an agreed definition, creating difficulties for safety (including toxicity profiles), governance and transparency. This review assesses the utility of nanomedicines in healthcare, clarifying key concepts in the literature, examining past, present and future nanomedicines and analyzing gaps in current regulations. Advances in nanomedicine offer unique opportunities including programmable and controllable nanoparticles (nanobots) that work cooperatively (nanoswarms), rather than individually, to achieve a targeted, personalized, and intelligent cancer treatment. Swarm behavior can be designed using a systems approach as in silico modelling has now advanced to the point of being a useful tool for selecting nanoparticles that optimize treatment outcomes. We need to understand what the first-in-human clinical trial of nanoswarms should/will look like, and anticipate the associated ethical questions that may arise. To aid clinical adoption of nanoswarms in cancer treatment, a harmonized nanomedicine vocabulary is needed alongside a robust, specific and overarching regulatory framework that can guide researchers, regulators and other key stakeholders.

M. Swana · J. Blee · S. Hauert (✉)
Engineering Mathematics, Faculty of Engineering, University of Bristol, Bristol, UK
e-mail: sabine.hauert@bristol.ac.uk

N. Stillman
Department of Cell and Developmental Biology, University College London, London, UK

J. Ives (✉)
Centre for Ethics in Medicine, Bristol Medical School, University of Bristol, Bristol, UK
e-mail: j.ives@bristol.ac.uk

S. Hauert
Bristol Robotics Laboratory, University of Bristol, Bristol, UK



I. Balaz and A. Adamatzky (eds.), *Cancer, Complexity, Computation*,
Emergence, Complexity and Computation 46,
https://doi.org/10.1007/978-3-031-04379-6_12

**Keywords** Swarm · Nanoswarm · Nanobot · Nanoparticle · Nanomedicine · Cancer · Ethics · Cooperative behavior · Collective behavior · Regulation · Governance

# 1 Cancer Nanomedicine

## *1.1 Cancer*

Cancer is one of the leading causes of death globally, although cancer survival rates differ significantly between different countries. Relatively high (and rising) survival rates in high-income countries are due to accessible early detection, quality treatment and survivorship care, with much lower survival rates in low- and middle-income countries [1]. Human cells multiply through cell division, but when abnormal or damaged cells multiply this can cause a growth or lumps of tissue called a 'solid tumour'. These tumours can be cancerous (malignant) or non-cancerous (benign) and can sometimes spread to other parts of the body causing a secondary tumour (or 'metastasis'), which is the major cause of death from cancer [2, 3]. Current cancer treatments (notably chemotherapy and radiotherapy) cause many side-effects as they cannot differentiate between healthy and cancerous tissues, meaning that healthy tissue is also killed alongside the cancerous. Solutions such as nanomedicines, with triggered release for tailor-made pharmacokinetics, have the potential to offer effective treatments without the deleterious side-effects of current approaches.

## *1.2 Nanomedicine*

Nanomedicine refers to medical products developed using nanotechnology with the aim of preventing, diagnosing, monitoring or treating diseases as well as offering tools for personalized medicine [4]. By providing new ways to deliver drugs to difficult-to-reach target sites, nanomedicines can be used to achieve the same therapeutic effect at smaller doses than their conventional counterparts, for example due to their nano size, which improves their solubility and bioavailability [4, 5]. Advancements of in silico tools that allow multi-scale simulations of nanoparticle transport barriers can facilitate effective exploration of the nanoparticle design space to optimize and select properties of nanoparticles [6]. Nanomedicines are driven by the need to develop therapies that have fewer side-effects, and that are more cost-effective than traditional therapies, in particular for cancer [7, 8]. To achieve this goal, novel nanomaterials, a deep understanding of biology, and computational tools are emerging as the next frontier in nanomedicine drug development [9].

### 1.3 Current State of the Art (Research and the Clinic)

The pharmaceutical industry is responsible for the research, development, production, and distribution of medications. The market has experienced significant growth during the past two decades, and pharma revenues worldwide totaled 1.27 trillion U.S. dollars in 2020 with 99.5 billion pharmaceutical sales in oncology alone [10, 11, 12]. Nanomedicine sales, including cancer nanomedicines from DOXIL® (liposomal doxorubicin) to ONIVYDE® (liposomal irinotecan), reached $16 billion in 2015, and previous reports show the growth of the nanomedicine market size with forecasts of $350 billion by 2025 [7, 13]. Developments in nanomedicine engineering are expected to drive growth, and a rise in funding from government organizations will prime the market [7, 8, 14].

Over the years, various nanoparticles have been developed to assist the delivery of chemotherapy drugs to tumours. They have several routes of administration, as they can be inhaled, swallowed, absorbed through skin or injected. Nanoparticles bring multiple pharmacokinetic, efficacy, safety, and targeting benefits when they are included in drug formulations [15]. Nanomedicines have already entered clinical practice, and even more are being investigated in clinical trials for a wide variety of indications, with over 25 approved for clinical use, either by the Food and Drug Administration (FDA) in the United States, or the European Medicines Agency (EMA) in the European Union [16]. The United States is the global leader for nanomedicines, with the submission trend of drug products containing nanomaterials submitted over the last decade increasing rapidly [17]. Clinicaltrials.gov is the largest clinical trials registry with studies listed in the database conducted in all 50 States and in 220 countries. Within this database a total of 211 active clinical trials involves approved and unapproved nanoparticles, of which 130 are cancer trials (Fig. 1) [18]. These nanoparticle inventions include biological agents (5), medical devices (3), diagnostic tests (4), drugs (100), procedures (5), radiation (1) or in other (12) cases nanoparticles are used as a contrast agent, or in primary/secondary outcome measures [19]. Study types include 112 phase I-III (and 12 where the phase is not applicable) interventional studies (also called clinical trials), 5 observational studies (including patient registries), and 1 expanded access (compassionate use) [19].

Even though more than 50% of clinical trials involving nanoparticles are cancer trials (Fig. 1), the clinical translation of cancer nanomedicines remains low. One reason for this is that only a small percentage of the cancer nanomedicine injection may reach the solid tumour, limiting efficacy [21]. In addition, there is no standard nanomedicines approval process, with approvals for clinical use often carried out on an ad hoc basis, with no standard and specific regulatory or governance processes in place for nanomedicines [22]. Nonetheless, progress in the clinic over the past 20 years (since DOXIL®'s approval) has been made possible by extensive efforts in preclinical, commercial, and clinical studies [23]. Furthermore, the overall outlook of nanoparticle drug delivery systems is promising, as they are also being developed for diagnosing and treating not only cancer, but many other diseases, including the rapid development of lipid nanoparticle mRNA vaccines for COVID-19 [23].

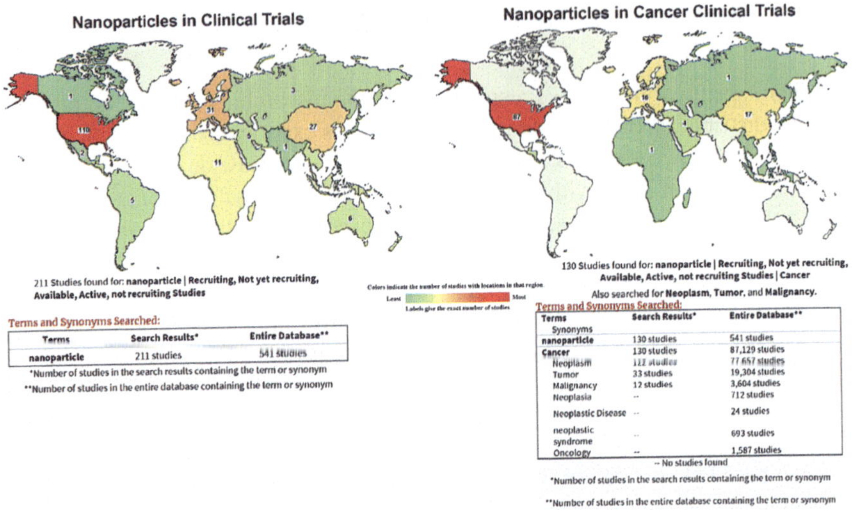


**Fig. 1** Nanoparticles in clinical trials and cancer clinical trials. Data is taken from clinicaltrials.gov (accessed 01 Feb 2022) [20]

## 2 Future Nanomedicines

### *2.1 Nanobots*

The idea of creating an artificial nanoscale world traces its origins to Feynman's *There's plenty of room at the bottom* [24]. An estimated 10,000 different types of biological 'nanomachines' are at work in the human body, performing essential tasks making possible movement and locomotion, energy conversion, transport, and regulating functions. Recognizing this has led to increased research on engineered programmable nanoparticles, or nanobots, able to build and physically manipulate their environment [25–27]. Going forward it is vital that different disciplines, ranging from chemistry to mathematics to robotics, cooperate and learn a common language to improve general problem-solving skills [27–29]. Experiments have shown that nano-sized entities can perform the same kind of logic operations as a silicon-based computer [30–36]. Thus, nanobots might one day be able to dispense drugs after carrying out complex diagnosis and treatment programs with a high level of accuracy and control. Some research groups (e.g. [33, 36, 37]) are already looking to move to animal models, and the next step in the translational pathway will be first-in-human clinical trials. This is likely in the foreseeable future with recent advancements from Los Angeles–based startup Bionaut Labs, who secured a humanitarian use device designation from the FDA to use a microrobot known as BNL-201 to treat Dandy-Walker syndrome, a rare pediatric neurological disorder [38].

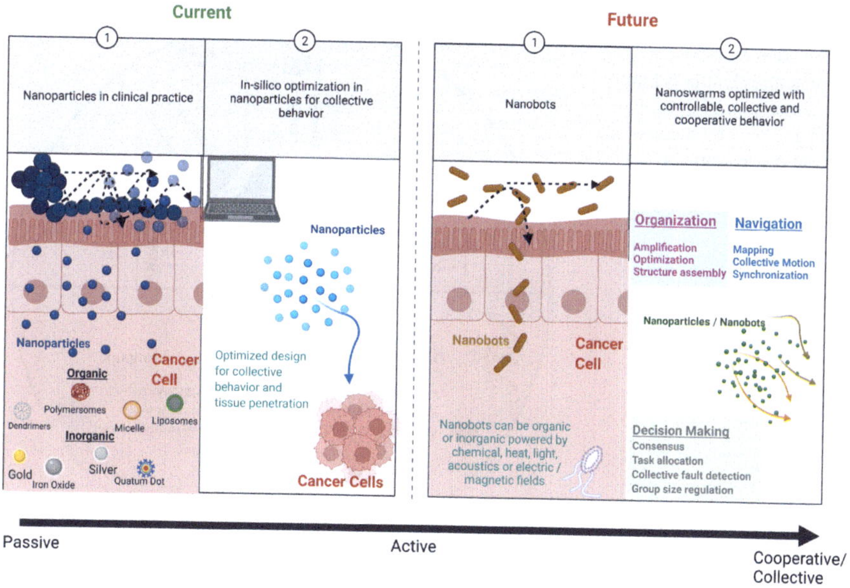


**Fig. 2** Current and future nanoparticles. Nanoparticle designs and interactions determine their behavior in the body

A single nanobot is unable to accomplish a great deal but trillions can. However, trillions of nanobots interacting in a complex tumour environment will require a cooperative systems approach [9], and as such, we need to focus on the next generation of programmable nanoparticles and nanobots that work cooperatively/collaboratively (nanoswarms) (Fig. 2). Nanoswarms build on large numbers of agents (nanoparticles/nanobots) that interact with each other (e.g. through communication) or their environment to achieve a cooperative task (e.g. find and target a tumour, coordinate delivery of a drug, make decisions about a tumour and its treatment, optimise nanoparticle delivery) inspired by swarm behaviors in nature (e.g., in ants, birds, cellular systems, etc.) [39, 40].

## 2.2 *Nanobot Design*

Developing robots on the nanoscale comes with a unique set of engineering challenges, which lead to additional challenges for clinical translation (discussed below in Sect. "Challenges in the clinical pathway"). Figure 3. summarizes some of the driving mechanisms behind, and components currently being explored to develop nanobots that can be controlled in-vivo [41]. Some approaches seek to build entirely new biomaterials, including biochips (small scale devices capable of performing thousands of simultaneous biochemical reactions [42]), while others are harnessing

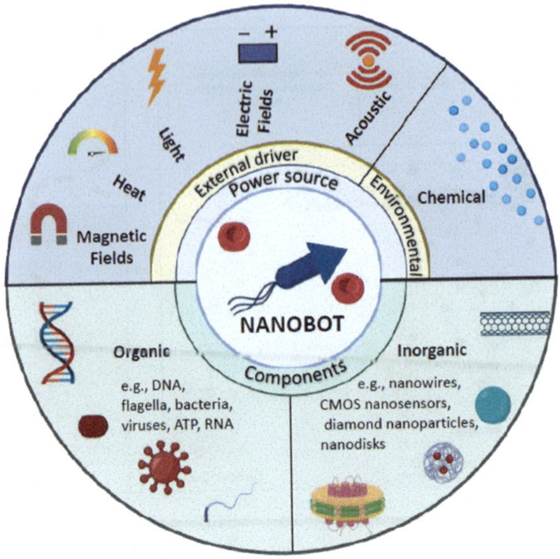


**Fig. 3** Overview of nanobot driving sources and components. Nanobots may be externally, or chemically driven and may be made of a range of organic and/or in-organic components

immune cells and micro-organisms such as bacteria and viruses [43, 44]. Nanobots are being designed with a wide range of different organic (e.g., DNA, flagella) and non-organic (e.g., CMOS nanosensors, nanowires) components with varying biocompatibilities [44–47]. Some nanobots are powered externally by, for example, magnetic fields [48], electricity [46], light [49] and acoustic waves [50], while others are powered chemically [43] (Fig. 3). Computer scientists and biologists have teamed up to make a new class of living robotics that challenge the boundary between synthetic and biological [51–53]. As this technology develops it may be possible to incorporate these into nanoswarms. As we do so, however, we must be attentive to the challenges of nanobot toxicity, which will depend heavily on its specific design, driving method and its application [54].

## 2.3 *Nanoswarms*

Collaborative behaviors among groups of individuals that enable self-organization are present in social animals such as birds, ants, fish and termites [55]. New research is now applying artificial swarm intelligence techniques to controlling and designing nanoswarms [9, 56]. Figure 2 illustrates different nanoswarm behaviors including collective behavior that emerges from interactions between the nanoparticles or nanobots, and between the nanoparticles or nanobots and their environment. Swarm engineering techniques used to design nanobots and their interactions aim to ensure the resultant behaviors are fault tolerant, scalable and flexible. Designing cooperative behaviors requires defining the behavior that should be displayed by the nanoswarm within its environment of action [9]. The challenge is then to understand which individual nanobot design could give rise to a desired collective behavior. In silico tools have been shown to facilitate exploration of nanoswarm design to select nanobots with the potential to deliver their cargo to tumour cells [6]. Moreover, person specific solutions could be engineered to develop targeted therapy, leading to personalized medicine for an individual patient, or to produce sufficient amounts of personalized patient data for machine learning [6]. In silico modelling has now advanced to the point of being an effective tool that can minimize costly trial-and-error design methods, and will be key in cancer modelling and nanoswarm development [6, 57, 58].

One key challenge for developers is that once in silico modelling has run its course, the next step must be first-in-human trials of nano-swarms, for which safety protocols and governance structures are, at best, unclear and, at worst, entirely absent. Although progress is being made to guide developers to focus on functionality and toxicity [59–62], when these technologies are ready for clinical trials, guidelines on trade-offs between nanotoxicity and benefit in terms of patient survival and quality of life need to be in place [21, 59].

### 2.4 *Personalization*

Personalized medical care is based on the patient's genetics and specific disease profile. Personalized cancer medicine can have fewer side effects than other types of treatments because it can target cancer cells without affecting healthy cells, however personalized treatments are not available for many types of cancer and genetic testing can be expensive [63, 64]. Nanoswarms appear to be a promising approach when therapies require different activities to be performed concurrently, when high redundancy and the lack of a single point of failure are desired, and when it is technically infeasible to set up the infrastructure required to control the nanoswarms in a centralized way. Furthermore, researchers could incorporate living systems, maybe even utilizing the patient's own cells, to create personalized therapies. In the future, nanoswarms could be used not just for drug delivery systems or tumour treatment, but for tissue engineering and, eventually, to tailor an increasingly specific and potent immune response, which can result in an effective, and personalized cancer treatment (Fig. 4).

## 3 Challenges in the Clinical Pathway

Alongside their potential, nanomedicines face several challenges, such as ambiguous characterization, possible toxicity issues, a lack of specific regulation, unclear cost–benefit considerations, and waning enthusiasm among some health care professionals

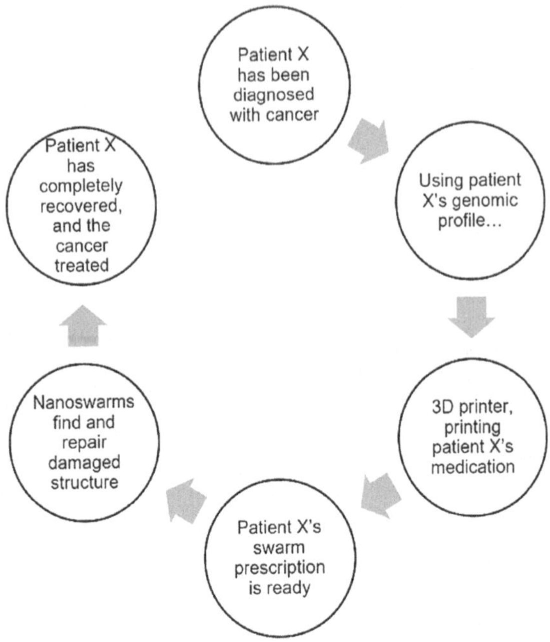


**Fig. 4** Future personalized Swarm Medicine prescription for Patient X's cancer which is treated by injecting or swallowing millions of tiny robots [nanoswarms] that have been created using a 3D bioprinter encoded with the person's specific genome

[15]. Given the sudden expansion of nanoparticles in the clinic and rapid development of nanomedicines, it is crucial to examine these challenges.

## 3.1 Definitions

The International Standards Organization (ISO) have been working towards standardized definitions for the vocabulary of nanotechnology. Although various organizations, including ISO, have proposed several frameworks for the categorization of nanomaterials, a complete internationally agreed terminology is yet to be developed. There are currently two basic classification systems for different types of nanomaterials, based on the structural dimensions of nano-objects and the chemical composition of nanoparticles (Fig. 5), and whilst these offer a useful way to differentiate different aspects of the technology, they fall short of providing the comprehensive vocabulary needed.

Different types of organic and inorganic nanoparticles are used in the healthcare sector and could soon include smart health technologies such as nanobots (Fig. 3). Translational aspects of nanobots are a challenge. One reason is that nanobot designs (as discussed in Sect. "Future Nanomedicines") lack a well-defined and comprehensive design approach with a clear explanation of the requirements for their targeted clinical applications [65, 66]. Design should consider the entire clinical scenario from the point of entry to safe and functional operation within the immune system and to

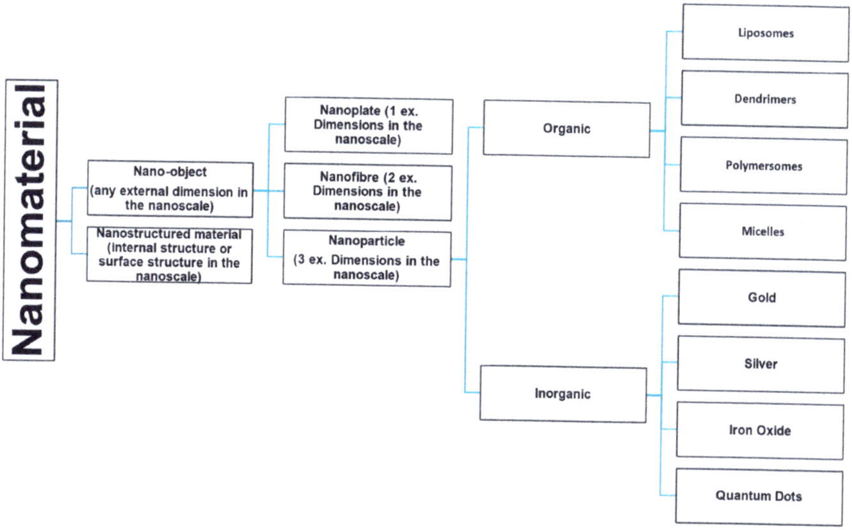


**Fig. 5** Nanomaterials based on the ISO 80004 standards nanotechnologies vocabulary series categorized by structure of nano-objects and chemical composition of nanoparticles

the disposal after completion [65, 67]. In addition, everything must be encoded in the physical design of the nanobot simultaneously [65]. To fully understand the risk posed by a specific nanorobotic application, consideration of its components, how these function in the body, as well as any additional external protocols is required. Currently the wide variety of nanorobotic approaches complicates assessment of the safety of nanobots [41, 68, 69].

The problems arising from a lack of standardized vocabulary are multifaceted. As there are different definitions, across different disciplines and different regulators, the true number of nanomedicines either approved for marketing or under clinical trials is difficult to establish, and this in itself is a cause for concern relating to three main areas: (i) *risk minimization*—without standardized terminology it is harder to locate and aggregate safety data, leading to many nanomedicines being approved on a case-by-case basis, without sufficient assessment of existing data; (ii) *regulation*—a field that lacks definition, and lacks standard vocabulary to describe its activities, risks those activities falling through the regulatory net, with some activity captured by existing policies and guideline, and others not, depending on how they are described. (iii) *communication*—this is key for transparency about ethics, approaches and risks, but is impeded when terminology is inconsistent. Lack of definition can lead to miscommunication, which can in turn lead to mistrust amongst stakeholders [70]. From computation to clinic, a number of safety-related regulatory aspects need to be taken under consideration for current and future nanomedicines including size, toxicity and evaluation as a drug or device [21, 65, 71–73].

## 3.2 Size

One of the most challenging aspects of the lack of a standard vocabulary is that there is not even an agreed definition of what 'nano' means such that in different systems and jurisdictions the technology covered by any specific regulatory systems will likely differ. For example, the US National Nanotechnology Initiative (NNI), considers a structure to be 'nano' if it has dimension of 1–100 nm, whereas the U.K. Royal Society and Royal Academy of Engineering considered as 'nano' anything in the range of 0.2–100 nm. Conversely, the Friends of the Earth Australia recommend defining nanoparticles as anything up to 300 nm in size, whereas it is common in the biological sciences to work with 'nanoparticles' up to 400 nm diameter (as drug carriers), with others considering anything <1000 or <500 nm to be 'nano' [72–75].

The shape and size of nanoparticles is important because increased size can lead to changes in physical, pharmacokinetic and pharmacodynamic properties [76, 77], and biosafety, all of which can affect the impact on the environment and biosystem into which the technology is introduced. Further, different properties of nanoparticles, including size, shape, surface area and charge play a key role in toxicity assessment [59]. There will always be a trade-off for researchers in terms of performance and toxicity when selecting nanoswarms for clinical use, but for these trade-offs to be effectively regulated it is necessary for regulation to effectively capture all relevant activity. Effective oversight will be difficult if it is possible to avoid scrutiny simply by working with a particle size that a specific jurisdiction does not consider 'nano'.

## 3.3 Toxicity

Nanoparticles from natural sources have been used in cosmetics (including sunscreens) for some time, some dating back to over 2000 years ago when ancient Roman and Greek societies used sulfide nano crystals to dye their hair. However, toxicity concerns about nano-sunscreens were first raised in 2008, when a BlueScope Steel report stated that metal oxide nanoparticles in some sunscreens could bleach painted surfaces of coated steel [72] as several sunscreens tested contained nanoparticles that acted as extreme photocatalysts [78–80]. This indicated people with sensitive or damaged skin could be at increased risk of skin penetration by these nanoparticles [78, 81]. Due to public perception that nanoparticles in general had toxic effects, several companies dropped the use of the word "nano" or "nanotechnology" from marketed products that contained nanoparticles and there is still an ongoing policy debate around labeling of nano-enabled consumer products [82–85]. Whilst it goes without saying that toxicity profiles of any nanotechnology will need to be ascertained, and safety established, as the use of nanoparticle-derived applications increases, there will be increasing need for companies to communicate with the public to alleviate their concerns. This could be crucial for cancer treatments, where the benefit of introducing nanomedicine is to reduce the toxic side-effects of current treatments [86, 87].

How nanoparticles behave inside the human body is one of the central issues that need to be considered [21, 59, 60, 62]. For example, in theory they could cause "overload" on phagocytes (cells that ingest and destroy foreign matter), thereby triggering stress reactions that lead to inflammation and weaken the body's defense against other pathogens. Another important issue to consider is the longer-term potential toxic effects of non- or slowly-degradable nanoparticles accumulating in organs over time, which will make continuous monitoring and long-term follow-up after first-in-human trials essential.

### 3.4 Device or Drug

Nanomedicine poses an additional challenge for researchers, industry, and regulators insofar as it is unclear if it should be classified as a medicinal product or a medical or surgical device. The category it fits into determines the regulatory pathway that must be followed for first-in-human trials, and later clinical practice. In many jurisdictions, medicinal/pharmacological interventions are more regulated than medical or surgical devices [21, 88, 89], with the latter sometimes able to be introduced with minimal oversight and little evidence base.

When the FDA first approved DOXIL® they considered making a new category "nano", but decided to market it as a device so that they could expediate the approval process [90]. This raises several questions around how future nanomedicine technology will be approved, and whether regulators will pay sufficient attention to the differences between them and thus differentiate between those that, for example, might merely function as a drug delivery system and those that might be capable of physically manipulating their environment. The former seems more like medicines, and the latter more like surgical instruments, although neither fit neatly into either category. Beyond current nanomedicines, there is a need to classify future nanomedicines. The physical design, composition, medical functions and control methods of a nanobot are application-oriented, so each nanobot should be decided separately for evaluation as a drug or a device [65].

As it stands, many nanoparticle systems have been approved by either the FDA or EMA, and are used in the clinic to either treat or diagnose disease [91]. Most nanoparticles in the clinic fall under the category of Non-Biological Complex Drugs (NBCDs). There are ongoing worldwide discussions on whether current regulatory frameworks for NBCDs are fit for purpose, and there are calls for global harmonization of scientific and technical requirements for NBCDs that benefit stakeholders [92–94].

# 4 Ethical and Regulatory Aspects of Nanoswarms

## 4.1 Ethics

All of the issues discussed above have an ethical component, insofar as our response to the questions they raise will be determined by the extent to which we place value over certain outcomes over others, what risks we are willing to take for what benefits, the extent to which we are prepared to expose vulnerable patients to risk, and the extent to which we believe patients should be allowed to make their own choices. Nanomedicine, including nanoswarms, does not to our mind generate entirely new categories of ethical issue, but it does require us to think carefully about how we apply our current thinking about these kinds of issues to this new frontier of medicine, and consider whether our current theories and approaches can provide the guidance we need. This is not the place to outline and respond to every ethical issue raised. We will, however, make the general observation that, as technology develops it will often challenge our current ethical thinking, and this presents an opportunity, if not a requirement, to re-think our ethical commitments, and possibly either shift our ethical understanding to accommodate the new technology, or to reject it [95]. Nanoswarms in healthcare do present that kind of challenge, as we discuss below.

Nanoswarms have the potential to re-shape our conceptions of what it is to be human. It is unclear whether an advanced technology such as nanoswarm would be a treatment or an enhancement creating 'everyday cyborgs' [96] that appear entirely human but have within them an almost symbiotic and autonomous robotic technology that maintains their health. The very idea of human enhancement is itself controversial [97], and whether a particular technology is considered an enhancement or treatment will likely have significant bearing on its perceived ethical acceptability. Such technologies could take us into a new frontier of risk, with the concurrent challenge of ensuring that patients and research participants sufficiently understand what they are agreeing to put into their bodies, and what risk that poses. This is compounded by the fact that the patients most likely to be in position to trial these treatments are those who are perhaps the most desperate and most vulnerable, in terms of their desire for a specific health outcome.

Whilst this does not rule out conducting human tests, it should give us pause, and highlights the clear obligation on developers to be rigorous and conscientious in their pre-trial safety assessments. Regulations are in-part motivated because we cannot always rely on the rigor and conscientiousness of developers. However, because this technology may blur the lines between our current understanding of pharmacological and surgical interventions, it could present a significant challenge to current regulatory systems, and these will need to adjust accordingly. There are other ethical concerns which we should be attentive to, especially with respect to nanomedicine's role in the wider international community. Concerns have been raised over the potential for nanomedicines to intensify the gap between rich and poor countries, leading to a so-called 'nano-divide' [98]. However, this kind of divide is also foreseeable within individual countries, where new demographics form based on a person's access to technology. This is not a unique problem to nanomedicine but could exacerbate existing health inequalities if some people were to have access to a technology that identifies and manages health risks in real time and others do not.

Additional concerns arise around the volume of data that might need to be recorded and processed in real time for some nanomedicines to function, especially those that are personalized, and the massive health information infrastructure that would be required. For example, there are concerns regarding the security of those data and the possible ways it might be used and abused. Potential military developments raise further concerns. It seems clear that any nanoswarm technology could in principle be deployed to harm, as well as heal. This 'dual use' problem is not unique to this technology, but the potential for this technology to be developed into an invisible and personalized weapon might be as particularly frightening to the public as it might be enticing to those who use, and profit from, arms. The dual use problem cannot be managed by health regulators [98, 99], and so additional legal mechanisms might be required to allay these particular fears.

It is worth noting that all these dystopian concerns do not challenge the substantive ethical status of any nanomedicine technology, but rather concern the way it might be used and the deleterious consequences that *could* follow. These are things it should be possible to manage, and guard against—so long as we have the foresight, and motivation, to do so. Another general ethical concern is that the potential benefits of this technology risk being hindered by an overly restrictive regulatory system. We would certainly not align ourselves with those who call for ethics to 'get out of the way' of research [100], but we are alert to the potential for well-meaning regulation and systems of ethical governance to mire researchers in unnecessary bureaucracy that slows, if not entirely stifles, progress. There is, of course, a middle ground to be sought.

## 4.2 *Towards a Regulatory Framework*

For all their potential medical applications, nanoswarms are still largely in the research and development stage. Clinical trials of nanoswarms in humans have yet to be approved [101], but we ought to start thinking about what the first-in-human clinical trials of nanoswarms could or should look like, and ask how we will regulate the development of this new medical technology. There is significant potential for nanoswarms in medicine, not just for personalized cancer treatment, but many other areas. As and when this technology becomes ready for first-in-human tests, decisions will have to made about how these technologies can and should be safely tested.

As noted above, a harmonized standard nanomedicine vocabulary will be essential, and this is a pre-requisite for an effective regulatory framework. We leave the question of what that framework ought to look like open for now. We intend to return to it, but believe that further research is needed in order to provide evidence, and undertake the necessary consultation needed to begin to draft a framework—this

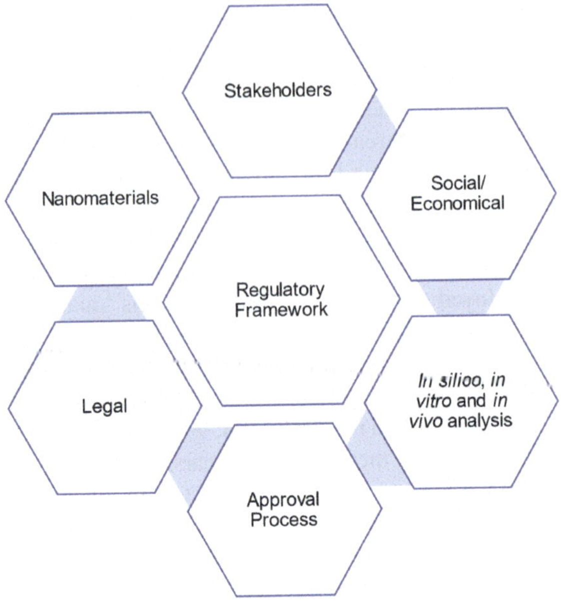


**Fig. 6** Key areas that need to be explored to start creating a framework for past, present and future nanomedicines.

work is underway. We suggest, for now, that there are 6 central domains that need to be explored in order to draft guidance for regulation in this area (Fig. 6).

Here, we add our voice to those of others [4, 102–108], to amplify the critical call for policymakers, researchers, developers, ethicists, patient representatives, healthcare professionals, and the pharmaceutical sector, to cooperate to contribute to the creation of a resilient and sustainable global regulatory landscape, which protects safe and responsible innovation for the benefit of everyone.

## References

1. World Health Organization: Cancer. https://www.who.int/health-topics/cancer#tab=tab_1. Accessed 03 Dec 2021
2. Cancer Research UK: What is cancer?. https://www.cancerresearchuk.org/about-cancer/what-is-cancer. Accessed 03 Dec 2021
3. National Cancer Institute: What Is Cancer?. https://www.cancer.gov/about-cancer/understanding/what-is-cancer#definition. Accessed 03 Dec 2021
4. Hock, S.C., Ying, Y.M., Wah, C.L.: A review of the current scientific and regulatory status of nanomedicines and the challenges ahead. PDA J. Pharm. Sci. Technol. **65**, 177–195 (2011)
5. Lu, W., Yao, J., Zhu, X., Qi, Y.: Nanomedicines: redefining traditional medicine. Biomed. Pharmacother. 134 (2021). https://doi.org/10.1016/J.BIOPHA.2020.111103
6. Stillman, N.R., Kovacevic, M., Balaz, I., Hauert, S.: In silico modelling of cancer nanomedicine, across scales and transport barriers. NPJ Comput. Mater. 6:1. 6, 1–10 (2020). https://doi.org/10.1038/s41524-020-00366-8
7. Grand View Research Inc: Nanomedicine market size worth $350.8 Billion By 2025 CAGR: 11.2%. (2017)

8. Hartshorn, C.M., Grodzinski, P., Farrell, D., Morris, S.A., Fedorova-Abrams, N., Liu, C., Panaro, N., Christ, R.M., Prabhakar, U.: Cancer Nanotechnology Plan 2015. U.S. Department of Health and Human Services, National Institutes of Health (2015)
9. Hauert, S., Bhatia, S.N.: Mechanisms of cooperation in cancer nanomedicine: towards systems nanotechnology. Trends Biotechnol. **32**, 448–455 (2014). https://doi.org/10.1016/J.TIBTECH.2014.06.010
10. Pharmaceutical Market: https://stats.oecd.org/Index.aspx?DataSetCode=HEALTH_PHMC. Accessed 22 Nov 2021
11. Mikulic, M.: Global pharmaceutical industry—statistics and facts. https://www.statista.com/topics/1764/global-pharmaceutical-industry/#dossierKeyfigures. Accessed 22 Nov 2021
12. Association of the British Pharmaceutical Industry: Global pharmaceutical market. https://www.abpi.org.uk/facts-figures-and-industry-data/global-pharmaceutical-market/. Accessed 22 Nov 2021
13. Vasile, C.: Polymeric nanomaterials: recent developments, properties and medical applications. In: Vasile, C. (ed.) Polymeric Nanomaterials in Nanotherapeutics, pp. 1–66. Elsevier (2019). https://doi.org/10.1016/B978-0-12-813932-5.00001-7
14. Boisseau, P., Levy, L., Letourneur, D., Mauberna, P.: Strategic Research and Innovation Nanomedicine Agenda Industry Patient. European Technology Platform for Nanomedicine (2016)
15. Anselmo, A.C., Mitragotri, S., Samir Mitragotri, C.: Nanoparticles in the clinic. Bioeng. Transl. Med. **1**, 10–29 (2016). https://doi.org/10.1002/BTM2.10003
16. Anselmo, A.C., Mitragotri, S.: Nanoparticles in the clinic: An update post COVID-19 vaccines. Bioeng. Transl. Med. 6, (2021). https://doi.org/10.1002/BTM2.10246
17. U.S. Food and Drug Administration: Nanotechnology—Over a Decade of Progress and Innovation (2020)
18. Clinicaltrials.gov: Search of: nanoparticle Recruiting, Not yet recruiting, Available, Active, not recruiting Studies Cancer—List Results—ClinicalTrials.gov. https://clinicaltrials.gov/ct2/results?term=nanoparticle&cond=Cancer&Search=Apply&recrs=b&recrs=a&recrs=d&recrs=c&age_v=&gndr=&type=&rslt=. Accessed 10 Feb 2022
19. Swana, M., Blee, J., Stillman, N., Ives, J., Hauert, S.: Swarms: the next frontier for cancer nanomedicine—Supplementary Material (2022). https://doi.org/10.5281/ZENODO.6077149
20. Search of: nanoparticle Recruiting, Not yet recruiting, Available, Active, not recruiting Studies—List Results—ClinicalTrials.gov, https://clinicaltrials.gov/ct2/results?term=nanoparticle&recrs=abcd. Accessed 20 Nov 2021
21. Wilhelm, S., Tavares, A.J., Dai, Q., Ohta, S., Audet, J., Dvorak, H.F., Chan, W.C.W.: Analysis of nanoparticle delivery to tumours. Nat. Rev. Mater. **1**(5), 1–12 (2016). https://doi.org/10.1038/natrevmats.2016.14
22. Đorđević, S., Gonzalez, M.M., Conejos-Sánchez, I., Carreira, B., Pozzi, S., Acúrcio, R.C., Satchi-Fainaro, R., Florindo, H.F., Vicent, M.J.: Current hurdles to the translation of nanomedicines from bench to the clinic. Drug Deliv. Transl. Res. **12**(3), 500–525 (2021). https://doi.org/10.1007/S13346-021-01024-2
23. Thi, T.T.H., Suys, E.J.A., Lee, J.S., Nguyen, D.H., Park, K.D., Truong, N.P.: Lipid-based nanoparticles in the clinic and clinical trials: from cancer nanomedicine to COVID-19 Vaccines. Vaccines **9**, (2021). https://doi.org/10.3390/VACCINES9040359
24. Feynman, R.P.: There is plenty of room at the bottom. Eng. Sci. **23**, 22–36 (1960)
25. Kay, E.R., Leigh, D.A.: Rise of the molecular machines. Angew. Chem. Int. Ed. **54**, 10080–10088 (2015). https://doi.org/10.1002/ANIE.201503375
26. Drexler, K.: Molecular directions in nanotechnology. Nanotechnology **2**, 113–118 (1991)
27. Baroncini, M., Casimiro, L., de Vet, C., Groppi, J., Silvi, S., Credi, A.: Making and operating molecular machines: a multidisciplinary challenge. Chem. Open **7**, 169–179 (2018). https://doi.org/10.1002/OPEN.201700181
28. Sluysmans, D., Fraser Stoddart, J.: Growing community of artificial molecular machinists. Proc. Natl. Acad. Sci. **115**, 9359–9361 (2018). https://doi.org/10.1073/PNAS.1813973115

29. Kassem, S., van Leeuwen, T., Lubbe, A.S., Wilson, M.R., Feringa, B.L., Leigh, D.A.: Artificial molecular motors. Chem. Soc. Rev. **46**, 2592–2621 (2017)
30. Amir, Y., Ben-Ishay, E., Levner, D., Ittah, S., Abu-Horowitz, A., Bachelet, I.: Universal computing by DNA origami robots in a living animal. Nat. Nanotechnol. **9**, 353–357 (2014). https://doi.org/10.1038/nnano.2014.58
31. Manjunath, A., Kishore, V.: The promising future in medicine: nanorobots. Biomed. Sci. Eng. **2**, 42–47 (2014). https://doi.org/10.12691/BSE-2-2-3
32. Ackerman, E.: Robotic Micro-Scallops Can Swim Through Your Eyeballs—IEEE Spectrum. https://spectrum.ieee.org/robotic-microscallops-can-swim-through-your-eyeballs. Accessed 26 Nov 2021
33. Qiu, T., Lee, T.C., Mark, A.G., Morozov, K.I., Münster, R., Mierka, O., Turek, S., Leshansky, A.M., Fischer, P.: Swimming by reciprocal motion at low Reynolds number. Nat. Commun. **5**, (2014). https://doi.org/10.1038/NCOMMS6119
34. Khalil, I.S.M., Dijkslag, H.C., Abelmann, L., Misra, S.: MagnetoSperm: a microrobot that navigates using weak magnetic fields. Appl. Phys. Lett. **104**, 223701 (2014). https://doi.org/10.1063/1.4880035
35. Jang, B., Gutman, E., Stucki, N., Seitz, B.F., Wendel-García, P.D., Newton, T., Pokki, J., Ergeneman, O., Pané, S., Or, Y., Nelson, B.J.: Undulatory locomotion of magnetic multi-link nanoswimmers. Nano Lett. **15**, 4829–4833 (2015). https://doi.org/10.1021/ACS.NANOLETT.5B01981/SUPPL_FILE/NL5B01981_SI_007.AVI
36. Orozco, C.A., Liu, D., Li, Y., Alemany, L.B., Pal, R., Krishnan, S., Tour, J.M.: Visible-light-activated molecular nanomachines kill pancreatic cancer cells. (2019). https://doi.org/10.1021/acsami.9b21497
37. Jang, J., Lim, D.-H., Choi, I.-H.: The impact of nanomaterials in immune system. Immune Netw. **10**, 85 (2010). https://doi.org/10.4110/IN.2010.10.3.85
38. Bionaut Labs: FDA Grants Humanitarian Use Device Designation to Bionaut Labs for Treatment of Dandy Walker Syndrome—Bionaut Labs. https://bionautlabs.com/fda-grants-humanitarian-use-device-designation-to-bionaut-labs-for-treatment-of-dandy-walker-syndrome/. Accessed 10 Feb 2022
39. Wang, Q., Zhang, L.: External power-driven microrobotic swarm: from fundamental understanding to imaging-guided delivery. ACS Nano **15**, 149–174 (2021). https://doi.org/10.1021/ACSNANO.0C07753
40. Yu, J., Jin, D., Chan, K.F., Wang, Q., Yuan, K., Zhang, L.: Active generation and magnetic actuation of microrobotic swarms in bio-fluids. Nat. Commun. 2019 10:1. 10, 1–12 (2019). https://doi.org/10.1038/s41467-019-13576-6
41. Koleoso, M., Feng, X., Xue, Y., Li, Q., Munshi, T., Chen, X.: Micro/nanoscale magnetic robots for biomedical applications. Mater. Today Bio. **8**, 100085 (2020). https://doi.org/10.1016/J.MTBIO.2020.100085
42. Azizipour, N., Avazpour, R., Rosenzweig, D.H., Sawan, M., Ajji, A.: Evolution of biochip technology: a review from lab-on-a-chip to organ-on-a-chip. Micromachines **11**, 1–15 (2020). https://doi.org/10.3390/MI11060599
43. Ali, J., Cheang, U.K., Martindale, J.D., Jabbarzadeh, M., Fu, H.C., Jun Kim, M.: Bacteria-inspired nanorobots with flagellar polymorphic transformations and bundling. Sci. Rep. **7**(1), 1–10 (2017). https://doi.org/10.1038/s41598-017-14457-y
44. Koudelka, K.J., Pitek, A.S., Manchester, M., Steinmetz, N.F.: Virus-Based Nanoparticles as Versatile Nanomachines. **2**, 379–401 (2015). https://doi.org/10.1146/annurev-virology-100114-055141
45. Chen, A.Y., Deng, Z., Billings, A.N., Seker, U.O.S., Lu, M.Y., Citorik, R.J., Zakeri, B., Lu, T.K.: Synthesis and patterning of tunable multiscale materials with engineered cells. Nat. Mater. **13**, 515–523 (2014). https://doi.org/10.1038/nmat3912
46. Shin, S.R., Migliori, B., Miccoli, B., Li, Y.C., Mostafalu, P., Seo, J., Mandla, S., Enrico, A., Antona, S., Sabarish, R., Zheng, T., Pirrami, L., Zhang, K., Zhang, Y.S., Wan, K.T., Demarchi, D., Dokmeci, M.R., Khademhosseini, A.: Electrically driven microengineered bio-inspired soft robots. Adv. Mater. (Deerfield Beach, Fla.). **30**, (2018). https://doi.org/10.1002/ADMA.201704189

47. Hu, M., Ge, X., Chen, X., Mao, W., Qian, X., Yuan, W.E.: Micro/nanorobot: a promising targeted drug delivery system. Pharmaceutics **12**, 1–18 (2020). https://doi.org/10.3390/PHARMACEUTICS12070665
48. Zhou, H., Mayorga-Martinez, C.C., Pané, S., Zhang, L., Pumera, M.: Magnetically driven micro and nanorobots. Chem Rev **121**, 4999–5041 (2021). https://doi.org/10.1021/ACS.CHEMREV.0C01234
49. Wang, J., Xiong, Z., Tang, J.: The encoding of light-driven micro/nanorobots: from single to swarming systems. Adv. Intell. Syst. **3**, 2000170 (2021). https://doi.org/10.1002/AISY.202000170
50. Aghakhani, A., Yasa, O., Wrede, P., Sitti, M.: Acoustically powered surface-slipping mobile microrobots. Proc. Natl. Acad. Sci. U.S.A. **117**, 3469–3477 (2020). https://doi.org/10.1073/PNAS.1920099117/VIDEO-10
51. Kriegman, S., Blackiston, D., Levin, M., Bongard, J.: Kinematic self-replication in reconfigurable organisms. In: Proceedings of the National Academy of Sciences of the United States of America, vol. 118. https://doi.org/10.1073/PNAS.2112672118/-/DCSUPPLEMENTAL (2021)
52. Blackiston, D., Lederer, E., Kriegman, S., Garnier, S., Bongard, J., Levin, M.: A cellular platform for the development of synthetic living machines. Sci. Robot. **6**, (2021). https://doi.org/10.1126/SCIROBOTICS.ABF1571
53. Kriegman, S., Blackiston, D., Levin, M., Bongard, J.: A scalable pipeline for designing reconfigurable organisms. Proc. Natl. Acad. Sci. U.S.A. **117**, 1853–1859 (2020). https://doi.org/10.1073/PNAS.1910837117/-/DCSUPPLEMENTAL
54. Saadeh, Y., Vyas, D.: Nanorobotic applications in medicine: current proposals and designs. Am J Robot. Surg. **1**, 4 (2014). https://doi.org/10.1166/AJRS.2014.1010
55. Boonrong, P., Kaewkamnerdpong, B.: Canonical PSO based nanorobot control for blood vessel repair. Int. J. Biomed. Biol Eng **5**, 428–478 (2011)
56. Alhafnawi, M., Hauert, S., O'Dowd, P.: Self-Organised saliency detection and representation in robot swarms. IEEE Robot. Autom. Lett. **6**, 1487–1494 (2021). https://doi.org/10.1109/LRA.2021.3057567
57. Molins, P., Stillman, N., Hauert, S.: Trail formation using large swarms of minimal robots. **50**, 693–710 (2019). https://doi.org/10.1080/01969722.2019.1677336
58. Stillman, N.R., Balaz, I., Tsompanas, M.A., Kovacevic, M., Azimi, S., Lafond, S., Adamatzky, A., Hauert, S.: Evolutionary computational platform for the automatic discovery of nanocarriers for cancer treatment. NPJ Comput. Mater. **7**, 150 (2021). https://doi.org/10.1038/S41524-021-00614-5
59. Akçan, R., Aydogan, H.C., Yildirim, M.Ş, Taştekin, B., Sağlam, N.: Nanotoxicity: a challenge for future medicine. Turk. J. Med. Sci. **50**, 1180 (2020). https://doi.org/10.3906/SAG-1912-209
60. Lewinski, N., Colvin, V., Drezek, R.: Cytotoxicity of nanopartides. Small **4**, 26–49 (2008). https://doi.org/10.1002/smll.200700595
61. Arora, S., Rajwade, J.M., Paknikar, K.M.: Nanotoxicology and in vitro studies: the need of the hour. Toxicol. Appl. Pharmacol. **258**, 151–165 (2012). https://doi.org/10.1016/J.TAAP.2011.11.010
62. Xue, H.Y., Liu, S., Wong, H.L.: Nanotoxicity: a key obstacle to clinical translation of siRNA-based nanomedicine. Nanomedicine (London, England). **9**, 295 (2014). https://doi.org/10.2217/NNM.13.204
63. Jackson, S.E., Chester, J.D.: Personalised cancer medicine. Int. J. Cancer **137**, 262–266 (2015). https://doi.org/10.1002/IJC.28940
64. Krzyszczyk, P., Acevedo, A., Davidoff, E.J., Timmins, L.M., Marrero-Berrios, I., Patel, M., White, C., Lowe, C., Sherba, J.J., Hartmanshenn, C., O'Neill, K.M., Balter, M.L., Fritz, Z.R., Androulakis, I.P., Schloss, R.S., Yarmush, M.L.: The growing role of precision and personalized medicine for cancer treatment. Technology **6**, 79 (2018). https://doi.org/10.1142/S2339547818300020

65. Ceylan, H., Yasa, I.C., Kilic, U., Hu, W., Sitti, M.: Translational prospects of untethered medical microrobots. Prog. Biomed. Eng. **1**, 012002 (2019). https://doi.org/10.1088/2516-1091/AB22D5
66. Schmidt, C.K., Medina-Sánchez, M., Edmondson, R.J., Schmidt, O.G.: Engineering microrobots for targeted cancer therapies from a medical perspective. Nat. Commun. **11**, 1–18 (2020). https://doi.org/10.1038/s41467-020-19322-7
67. Liu, D., Wang, T., Lu, Y.: Untethered microrobots for active drug delivery: from rational design to clinical settings. Adv. Healthcare Mater. **11**, 2102253 (2022). https://doi.org/10.1002/ADHM.202102253
68. Dixit, S.S., Luqman, N.: Nanobots: development and future. Int. J. Biosens. Bioelectron. **2**, (2017). https://doi.org/10.15406/IJBSBE.2017.02.00037
69. Novotný, F., Wang, H., Pumera, M.: Nanorobots: machines squeezed between molecular motors and micromotors. Chem. **6**, 867–884 (2020). https://doi.org/10.1016/J.CHEMPR.2019.12.028
70. Birchley, G., Ives, J., Huxtable, R., Blazeby, J.: Conceptualising surgical innovation: an eliminativist proposal. HCA J. Health Philos. Policy **28**, (2020). https://doi.org/10.1007/S10728-019-00380-Y
71. Hua, S., de Matos, M.B.C., Metselaar, J.M., Storm, G.: Current trends and challenges in the clinical translation of nanoparticulate nanomedicines: Pathways for translational development and commercialization. Front Pharmacol **9**, (2018). https://doi.org/10.3389/FPHAR.2018.00790/FPHAR_09_00790_PDF.PDF
72. Murday, J.S., Siegel, R.W., Stein, J., Wright, J.F.: Translational nanomedicine: Status assessment and opportunities. Nanomed. Nanotechnol. Biol. Med. **5**, 251–273 (2009). https://doi.org/10.1016/j.nano.2009.06.001
73. Satalkar, P., Elger, B.S., Hunziker, P., Shaw, D.: Challenges of clinical translation in nanomedicine: a qualitative study. Nanomed. Nanotechnol. Biol. Med. **12**, 893–900 (2016). https://doi.org/10.1016/j.nano.2015.12.376
74. Mudshinge, S.R., Deore, A.B., Patil, S., Bhalgat, C.M.: Nanoparticles: emerging carriers for drug delivery. Saudi Pharm. J. **19**, 129–141 (2011). https://doi.org/10.1016/J.JSPS.2011.04.001
75. Pudlarz, A., Szemraj, J.: Nanoparticles as carriers of proteins, peptides and other therapeutic molecules. Open Life Sci. **13**, 285 (2018). https://doi.org/10.1515/BIOL-2018-0035
76. Zoubari, G., Staufenbiel, S., Volz, P., Alexiev, U., Bodmeier, R.: Effect of drug solubility and lipid carrier on drug release from lipid nanoparticles for dermal delivery. Eur. J. Pharm. Biopharm. : Off. J. Arbeitsgemeinschaft fur Pharmazeutische Verfahrenstechnik e.V. **110**, 39–46 (2017). https://doi.org/10.1016/J.EJPB.2016.10.021
77. Kim, J.H., Kim, Y.S., Kim, S., Park, J.H., Kim, K., Choi, K., Chung, H., Jeong, S.Y., Park, R.W., Kim, I.S., Kwon, I.C.: Hydrophobically modified glycol chitosan nanoparticles as carriers for paclitaxel. J. Control. Release: Off. J. Control. Release Soc. **111**, 228–234 (2006). https://doi.org/10.1016/J.JCONREL.2005.12.013
78. Barker, P.J., Branch, A.: The interaction of modern sunscreen formulations with surface coatings. Prog. Org. Coat. **62**, 313–320 (2008). https://doi.org/10.1016/J.PORGCOAT.2008.01.008
79. Steel Direct: Prevention of sunscreen damage. https://cdn.dcs.bluescope.com.au/download/technical-bulletin-tb-37-prevention-of-sunscreen-damage. Accessed 11 Dec 2021
80. Tran, D.T., Salmon, R.: Potential photocarcinogenic effects of nanoparticle sunscreens. Australas. J. Dermatol. **52**, 1–6 (2011). https://doi.org/10.1111/J.1440-0960.2010.00677.X
81. Jacobs, J.F., van de Poel, I., Osseweijer, P.: Sunscreens with titanium dioxide (tio2) nanoparticles: a societal experiment. NanoEthics **4**, 103 (2010). https://doi.org/10.1007/S11569-010-0090-Y
82. Carbonell, R.: Fresh concern over nano-particles hidden in sunscreen. https://www.abc.net.au/news/2013-03-05/fresh-concern-over-nano-particles-in-sunscreen/4552522. (2013)
83. D'Silva, J., Bowman, D.M.: To label or not to label?—it's more than a nano-sized question. Eur. J. Risk Regul. **1**, 420–427 (2010). https://doi.org/10.1017/S1867299X00000891

84. Gruére, G.P.: Labeling nano-enabled consumer products. Nano Today **6**, 117–121 (2011). https://doi.org/10.1016/J.NANTOD.2011.02.005
85. Akin, H., Yeo, S.K., Wirz, C.D., Scheufele, D.A., Brossard, D., Xenos, M.A., Corley, E.A.: Are attitudes toward labeling nano products linked to attitudes toward GMO? Exploring a potential 'spillover' effect for attitudes toward controversial technologies **6**, 50–74 (2018). https://doi.org/10.1080/23299460.2018.1495026
86. Shi, J., Kantoff, P.W., Wooster, R., Farokhzad, O.C.: Cancer nanomedicine: progress, challenges and opportunities. Nat. Rev. Cancer **17**, 20–37 (2017). https://doi.org/10.1038/NRC.2016.108
87. Patil, R.M.: Nanomedicine for early diagnosis of breast cancer. Nanomedicines Breast Cancer Theranostics 153–173 (2020). https://doi.org/10.1016/B978-0-12-820016-2.00008-2
88. Lytton-Jean, A.K.R., Kauffman, K.J., Kaczmarek, J.C., Langer, R.: Cancer nanotherapeutics in clinical trials. Cancer Treat. Res. **166**, 293–322 (2015). https://doi.org/10.1007/978-3-319-16555-4_13
89. US Food and Drug Administration: Classification of Products as Drugs and Devices and Additional Product Classification Issues FDA, https://www.fda.gov/regulatory-information/search-fda-guidance-documents/classification-products-drugs-and-devices-and-additional-product-classification-issues. Accessed 03 Feb 2022
90. Min, Y., Caster, J.M., Eblan, M.J., Wang, A.Z.: Clinical translation of nanomedicine. Chem. Rev. **115**, 11147 (2015). https://doi.org/10.1021/ACS.CHEMREV.5B00116
91. Muthu, M.S., Leong, D.T., Mei, L., Feng, S.S.: Nanotheranostics—application and further development of nanomedicine strategies for advanced theranostics. Theranostics **4**, 660 (2014). https://doi.org/10.7150/THNO.8698
92. Klein, K., Stolk, P., de Bruin, M.L., Leufkens, H.G.M., Crommelin, D.J.A., de Vlieger, J.S.B.: The EU regulatory landscape of non-biological complex drugs (NBCDs) follow-on products: observations and recommendations. Eur. J. Pharm. Sci. **133**, 228–235 (2019). https://doi.org/10.1016/J.EJPS.2019.03.029
93. Gaspar, R.S., Silva-Lima, B., Magro, F., Alcobia, A., da Costa, F.L., Feio, J.: Non-biological complex drugs (NBCDs): complex pharmaceuticals in need of individual robust clinical assessment before any therapeutic equivalence decision. Front. Med. **7**, 590527 (2020). https://doi.org/10.3389/FMED.2020.590527
94. Schellekens, H., Stegemann, S., Weinstein, V., de Vlieger, J.S.B., Flühmann, B., Mühlebach, S., Gaspar, R., Shah, V.P., Crommelin, D.J.A.: How to regulate nonbiological complex drugs (NBCD) and their follow-on versions: points to consider. AAPS J. **16**, 15–21 (2014). https://doi.org/10.1208/S12248-013-9533-Z
95. Ives, J.: A method of reflexive balancing in a pragmatic, interdisciplinary and reflexive bioethics. Bioethics **28**, 302–312 (2014). https://doi.org/10.1111/BIOE.12018
96. Quigley, M., Ayihongbe, S.: Everyday cyborgs: on integrated persons and integrated goods. Med. Law Rev. **26**, 276–308 (2018). https://doi.org/10.1093/MEDLAW/FWY003
97. Harrison, P., Wolyniak, J.: The history of 'transhumanism.' Notes Queries **62**, 465–467 (2015). https://doi.org/10.1093/NOTESJ/GJV080
98. Royal Academy of Engineering: Nanoscience and nanotechnologies: opportunities and uncertainties (2004)
99. Fischer, S.: Regulating nanomedicine: new nano tools offer great promise for the future?if regulators can solve the difficulties that hold development back. IEEE Pulse **5**, 21–24 (2014). https://doi.org/10.1109/MPUL.2013.2296797
100. Pinker, S.: The moral imperative for bioethics. https://www.bostonglobe.com/opinion/2015/07/31/the-moral-imperative-for-bioethics/JmEkoyzlTAu9oQV76JrK9N/story.html (2015)
101. Yu, J., Wang, B., Du, X., Wang, Q., Zhang, L.: Ultra-extensible ribbon-like magnetic microswarm. Nat. Commun. **9**, 1–9 (2018). https://doi.org/10.1038/s41467-018-05749-6
102. Soares, S., Sousa, J., Pais, A., Vitorino, C.: Nanomedicine: principles, properties, and regulatory issues. Front. Chem. **6**, 360 (2018). https://doi.org/10.3389/FCHEM.2018.00360
103. Bawa, R., Johnson, S.: The ethical dimensions of nanomedicine. Med. Clin. North Am. **91**, 881–887 (2007). https://doi.org/10.1016/J.MCNA.2007.05.007

104. Demetzos, C.: Regulatory framework for nanomedicines. Pharm. Nanotechnol. 189–203 (2016). https://doi.org/10.1007/978-981-10-0791-0_7
105. Tinkle, S., Mcneil, S.E., Mühlebach, S., Bawa, R., Borchard, G., Barenholz, Y.C., Tamarkin, L., Desai, N.: Nanomedicines: addressing the scientific and regulatory gap. Ann. N. Y. Acad. Sci. **1313**, 35–56 (2014). https://doi.org/10.1111/NYAS.12403
106. Chan, V.S.W.: Nanomedicine: an unresolved regulatory issue. Regul. Toxicol. Pharmacol. **46**, 218–224 (2006). https://doi.org/10.1016/J.YRTPH.2006.04.009
107. Vishakha Tambe, Maheshwari, R., Chourasiya, Y., Choudhury, H., Gorain, B., Tekade, R.K.: Chapter 18. clinical aspects and regulatory requirements for nanomedicines. In: Tekade, R.K. (ed.) In Advances in Pharmaceutical Product Development and Research, Basic Fundamentals of Drug Delivery, pp. 733–752. Elsevier (2019)
108. Mühlebach, S., Borchard, G., Yildiz, S.: Regulatory challenges and approaches to characterize nanomedicines and their follow-on similars. Nanomedicine **10**, 659–674 (2015). https://doi.org/10.2217/NNM.14.189

# Study of Tumour Induced Vessel Displacement in the Tumour Progression Rate with Advanced Bioinspired Computational Tools

**Ioannis Karafyllidis, Georgios Ch. Sirakoulis, and Raphael Sandaltzopoulos**

**Abstract** In this chapter, a high level, namely tumour level, model and a corresponding simulation algorithm are introduced for the study of the effect of tumour-induced vessel displacement, on tumour progression rate prior to the onset of angiogenesis. The proposed model attributes its successful characteristics to a well known parallel bio-inspired computational tool, i.e. Cellular Automata (CAs) and solves diffusion differential equations to compute oxygen and glucose distribution into the tumour mass. CAs have been proven as efficient alternative to differential equations computational models that, despite their simplicity, exhibit complex dynamical behavior and can describe successfully the underlying phenomena for various physical, chemical and biological systems. More specifically, the studied phenomenon arrives from the displacement of the existing vessels, caused by mechanical forces owing to tumour growth, that heavily affect the aforementioned distribution of glucose and oxygen and, consequently and in turn, affect the tumour growth itself. The growth of a large number of tumours that initiated at various distances from a vessel, for the same number of time steps has been successfully modeled enabling us to further investigate and understand the underlying dynamics of early tumour growth. Simulation results showed that the properties of the physical diffusion processes in the case of moving vessel-tumour boundaries, affect directly tumour progression in the avascular progression phase. As such, the proposed model can be further utilized to explore various hypotheses of tumour growth relevant to drug delivery in chemotherapy, as well as to study access of growth factors and other plasma factors to the tumour.

---

I. Karafyllidis · G. Ch. Sirakoulis (✉)
Department of Electrical and Computer Engineering, Democritus University of Thrace, 67100 Xanthi, GR, Greece
e-mail: gsirak@ee.duth.gr

I. Karafyllidis
e-mail: ykar@ee.duth.gr

R. Sandaltzopoulos
Department of Molecular Biology and Genetics, Democritus University of Thrace, 68100 Alexandoupoli, GR, Greece
e-mail: rmsandal@mbg.duth.gr



I. Balaz and A. Adamatzky (eds.), *Cancer, Complexity, Computation*, Emergence, Complexity and Computation 46,
https://doi.org/10.1007/978-3-031-04379-6_13

## 1 Introduction

Cancer is not a single disease, but rather a highly complex and heterogeneous set of diseases that can adapt in an opportunistic manner, even under a variety of stresses [39]. In particular, cancer describes a group of genetic and epigenetic diseases, characterized by uncontrolled growth of cells, leading to a variety of pathological consequences and frequently death [36]. During the last decades, a great number of researchers have focused on modeling of cancer and tumour growth as a multifactorial process. Hence, over the last decades the development of numerous models that attempt to simulate the mechanisms controlling solid tumours morphogenesis [77]. Data and observations of tumour growth, evolution and adaptability has attracted the attention of theoreticians from many different fields, becoming one of the most important areas of active research in the theoretical biology community. Such data and observations strongly indicate that tumours do not behave randomly, but rather constitute self-organized versatile systems able to collect information from their surroundings, alter and exploit their microenvironment and evolve in an organized manner [10, 33, 34]. As already stated, tumour growth is a complex phenomenon that integrates genetic, biochemical, chemical and mechanical processes. Although recent progress has been made in understanding certain aspects of the complex tumour-host interactions that may be responsible for invasive cancer behaviors, many mechanisms are either not fully understood or remain completely elusive at the moment. Towards this direction, several aspects of tumour growth and tumour induced angiogenesis have been extensively studied and modeled [4, 8, 9, 23, 25, 28, 63, 75, 77, 86]. In this chapter, we focus mainly on the modeling of avascular tumour growth.

The number of models analysing the different stages in tumour development, from its initial avascular phase to invasion and metastasis through vascularization via tumour-induced angiogenesis, is huge [5]. Models for tumour growth can be continuous, discrete or hybrid. Continuous models are based on the use of partial differential equations or stochastic methods and treat the tumour as a (homogeneous or inhomogeneous) continuum surrounded by advancing boundaries. These models actually calculate the velocity and direction of the various tumour boundary points under conditions imposed by the flow of oxygen, nutrients and drugs [18, 28, 45, 86]. In discrete models, the area occupied by the tumour and its environment is divided into small equal areas, thus forming a discrete lattice, and the growth process follows local rules [20, 53, 59, 66]. In this case, tumour growth emerges as a result of the collective behavior of the processes that take place in each of these areas. Discrete models of tumour growth can easily be decomposed in biochemical pathways and provide a useful bridge between the macroscopic tumour behavior and the molecular processes that drive it [48]. Hybrid models combine the discrete models with the solution of partial differential equations on the discrete lattice of these models. Cellular Automata (CAs) models have been widely acknowledged as an appropriate alternative to partial differential equations [12, 15, 17, 19, 57, 76]. The main advantage of utilizing CA in cancer modeling is their inherent ability to formalize experimentally observable single-cell kinetics and observe emerging population level dynamics without a-priori knowledge of tumour behavior [60]. Because of their apparent resemblance of in vitro cell culture models, CA may be referred to as in silico experiments [60].

Tumour growth and development comprises three general phases: first, the tumour should escape from the immune system surveillance. Then, it needs to reduce the extracellular matrix. Finally, after sufficient growing, it should assist blood vessels to survive in its tissue [50]. Avascular growth phase is the common phase between both benign and malignant tumours and can be better experimentally simulated using in vitro platforms. In this phase, the growth rate is limited to diffusion and consumption of the nutrients. Studies in avascular tumour growth by means of diffusion have been extensively conducted over the past decades, and many models have been presented [75]. Nevertheless, a shortcoming of most of the aforementioned tumour growth models is that they do not consider the displacement of the surrounding tissue caused by tumour growth as a basic model parameter [43]. Even prior to the neo-angiogenesis phase, tumour growth exerts mechanical forces that push any vessels in its vicinity away from the tumour volume, thus inhibiting oxygen and nutrients to reach all tumour cells [49]. This effect enhances hypoxia and possible necrosis in tumour areas. Hence, there is a cycle of events: Tumour growth and the concomitant tissue displacement cause displacement of nearby blood vessels; displacement of vessels alters the distribution of $O_2$ and nutrients, which in turn affects tumour growth, which affects vessel distances from the tumour and so on. Later on, tumour induced angiogenesis enters this cycle by the production of vascular growth factors such as vascular endothelial growth factor from tumour cells [85] rendering the tumour growth process more complex. It is therefore essential to develop an accurate model of early tumour growth which can serve as a basis and can be extended to incorporate new processes such as angiogenesis and the dynamics of metastasis.

In this chapter, we present our efforts towards the development of a high level, i.e. tumour level, model and a simulation algorithm based on it, for the study of the effect of tumour growth and tumour-induced vessel displacement. Although our simulated environment is still a highly idealized version of the situation in vivo, we have used this algorithm to simulate the growth of a number of tumours that initiated at various distances from a vessel, for the same number of time steps. Note here that similarly to any mathematical model of a physical system, simplifications and multiple levels of abstractions are incorporated in the proposed model. However, simulation results showed that the properties of the physical diffusion processes in the case of moving vessel-tumour boundaries, affect directly tumour progression in the avascular progression phase.

In Sect. 2, the necessary background of Cellular Automata (CAs) are presented. In Sect. 3, the model of tumour growth and tumour-induced vessel displacement is proposed. Finally, in Sect. 4 the simulation results of the tumour growth and tissue developments are introduced, while in Sect. 5, conclusions are finally drawn.

## 2 Cellular Automata and Modeling Aspects

Several systems in physics, biology, chemistry, finance, as well as social sciences are characterized by emerging features that are not easy to guess from the elementary interactions of their microscopic individual components. In the past, such macroscopic behavior of the aforementioned systems was modeled by assuming that the collective dynamics of microscopic components can be effectively described collectively by equations acting on spatially continuous density distributions. It turns out that, to the contrary, taking into account the actual individual/discrete character of the microscopic components of these systems is crucial for explaining their macroscopic behavior [66]. Towards this direction and for the enhancement of modeling of such systems, the application of appropriate bio-inspired inherently parallel computational models like Cellular Automata (CAs) that can capture the essential features of systems in which global behavior emerges from the collective effect of simple components, which interact locally [69]. During the last decades, CAs have been extensively used for mimicking several natural processes and systems to find fine solutions in many complex hard to solve computer science and engineering problems.

A CA is a computational model originally proposed by John von Neumann and Stanislav Ulam in the 1940s [83]. Despite that, the interest around them mostly grew in the 1970s, when the Scientific American published a 2-D CA called Game of Life by the British mathematician, John Conway [29] and they were more widely studied in the next decades by S. Wolfram [89]. Their popularity is due to the fact that they allow the emergence of complex phenomena employing simple structures and resulting to inherent emergent computation and self-organization [2, 88].

A CA configuration can be defined using specific attributes, like a finite size $N$ dimensional grid of CA cells [2]. In other words, a CA consists of $N \in \mathbb{N}$ dynamical modules, namely the cells, in regular spatial formation that are locally interconnected. The whole configuration constitutes a discrete dynamical system, whose temporal evolution is governed by the inter-cell interactions and the inter-cell dynamics. For every CA cell, the CA's neighborhood is described as a set of attached cells connected to the CA cell, affecting its time evolution. In particular, each individual cell $C_i\big|_{i \in [1,N]}$, is defined by a state $s_i$ from a predefined set $S$ of states that is evolving in discrete time intervals according to the transition rule $f$, which takes into account the cell's and its neighbors' current states and as such it depends on the states $s_j\big|_{j \in [1,N]}$, of the neighboring cells $Nb_i$, i.e. the cells that are interconnected with $C_i$.

In particular, Elementary Cellular Automata (ECAs) describe an 1-D CA with binary state cells, $s_i \in \{0, 1\}$, and each cell is interconnected only with the two adjacent cells, $Nb_i = j\big|_{|i-j| \leq 1}$, thus the transition rule is:

$$s_i' = f(s_{i-1}, s_i, s_{i+1}) \tag{1}$$

where $s_{i-1}$, $s_i$, and $s_{i+1}$ are the current states of cells $C_{i-1}$, $C_i$, and $C_{i+1}$, respectively, and $s_i'$ is $C_i$'s next state.

In more detail, the transition rule maps all the current state combinations $LCR = \{s_{i-1}, s_i, s_{i+1}\}$ to the next state $s_i'$, i.e. $f : \{0, 1\}^3 \rightarrow \{0, 1\}$, thus the transition rule is defined as the truth table for the 8 ($2^3$) possible input combinations $LCR$ (Left, Central and Right cell). According to Wolfram's notation [87], the transition rule $f$ can be effectively coded to numbers from 0 to 255 ($2^{2^3} - 1$), in relation to the rule's outcome set. For example, the ECA transition rule $f_k$, where $k = 45_{10}$ equals to $k = 00101101_2$, which is presented for the better comprehension of the rule notation and while the Big-endian format is considered for the binary number representation, i.e. the most significant bit is in the leftmost position.

Furthermore, the current state of all CA cells $S_B = \{s_1, s_2, ..., s_N\}$ can be also coded to a binary number representation, named as the global state [55]. In specific, the global state is a $N$-bit binary number where each bit is the current state of a single CA cell, thus, in decimal representation, the global state is coded to a single decimal number in the range $[0, 2^N)$. If, for example, $N = 4$ and the current CA cells' states are $S_B = \{s_1, s_2, s_3, s_4\} = \{1, 0, 0, 1\}$ then the global state in decimal form is $S_D = 9$. In addition, a global transition rule $F_k$ can be described by a $2^N \times 2^N$ matrix such as:

$$F_k(i, j) = \begin{cases} 1, & S_D = (j+1) \ \& \ S_D' = (i+1) \\ 0, & elsewhere \end{cases} \tag{2}$$

where, the transition $S_D \rightarrow S_D'$ is the result of the local transition rule $f$ on the whole CA cell grid.

In the case of two-dimensional $(2-D)$ CA, the most well known neighborhoods are (i) the von Neumann neighborhood, where the state transition rule $F$ can be defined as follows, practically introducing the involvement of the cell states of the most nearest adjacent neighbors:

$$S_C^{\tau+1} = F\left(S_C^\tau, S_N^\tau, S_E^\tau, S_S^\tau, S_W^\tau\right), \tag{3}$$

while in case (ii) of the Moore neighborhood, beyond the aforementioned adjacent cells, the state transition rule successfully involves also the states of the diagonal neighboring cells as presented in the following Eq. 4:

$$S_C^{\tau+1} = F\left(S_C^\tau, S_N^\tau, S_E^\tau, S_S^\tau, S_W^\tau, S_{NE}^\tau, S_{NW}^\tau, S_{SE}^\tau, S_{SW}^\tau\right). \tag{4}$$

Regarding the boundary cells ($i = \{1, N\}$) that are missing one adjacent neighbor, there are various ways to handle their neighborhood. One may consider the boundary cells as non-dynamical or mirror the existing neighboring cell to the missing one. However, the most commonly used when compared to adiabatic, reflective and fixed boundaries are the periodic/cyclic ones, where the other boundary cell is used as the missing neighbor. Nevertheless, the selection of the boundary conditions depends on the specific system under examination as well as the neighborhood connections in every specific application.

Having an identical rule throughout the lattice and simultaneously applying it, leads to synchronous dynamics. However, one can introduce spatial or, even, temporal inhomogeneities [16, 70, 84]. Modeling physical systems with CA is in harmony with the contemporary notion of unified space-time and can lead to the state-of-the-art concept of in-memory computing [6]. The CA cell state is representative of the memory, while the CA local rule is appropriate of a processing unit. Moreover, CA have been widely acknowledged as an alternative to partial differential equations [57, 76]. More specifically, there are limitations related with the approximation of differential equations in serial computers. Partial differential equations contain much more information than is usually needed, because variables may take an infinite number of values in a continuous space. Moreover, the value of a physical quantity cannot be measured at a point, but instead it is measured over a finite volume [76]. Thus, to alleviate these limitations, several well-known partial differential equations have been successfully solved on CA lattices, namely the Diffusion equation [17], the Laplace equation [19], the Poisson equation [15], and the Weyl, Dirac, and Maxwell equations [12], just to name a few. It has been also proven that CA can trivially manage complex initializations, boundary geometries, anisotropies and inhomogeneities [38, 69]. The CA approach is consistent with the modern notion of unified space-time. In computer science, space corresponds to memory and time to processing unit. In CAs, memory (CA cell state) and processing unit (CA local rule) are inseparably related to a CA cell . Models based on CA methodology have several advantages, like inherent parallel nature that can be taken advantage when implemented on specialized hardware [21, 44, 46, 56, 67, 74, 78, 81]. The applicability and robustness of CAs in real world problems have been proven numerous times in the past [26, 82]. For instance spanning from the prediction of the epidemic fronts spreading [72] and the modeling of the underlying mechanisms of phenomena arising from combining light and chemical oscillations and their macroscopic behaviour successfully simulated by the Oregonator equations [3, 80] up to non probabilistic enhanced stereo vision simultaneous localization and mapping [54]. Also, the implementation of CAs was studied in the projection and simulation of the dynamic crowd movement [30, 40] also coupled with robotic guidance [14], the study of modern urban transport networks [1] and those of the past [24]. The motivation behind using CAs is based on the fact that they are proven to be efficient tools for fast prototyping and in-depth analysis of waves dynamics in chemical systems [35, 51, 79]. The CA models allow for studying a curvature and dispersion of wave patterns [31], dynamics of waves in anisotropic medium [64], turbulence [35].

Regarding biological and chemical systems, CAs models have been widely used in the past decades owing to almost all of the aforementioned reasons and their prominent and inherent characteristics, not to mention the intrinsic similarity to them as bio-inspired computational tools [22, 52, 68, 71]. More specifically, and concerning the modeling of cancer with the help of CAs, several attempts to model tumour growth [5, 20, 32, 36, 37, 41, 58, 59, 65], tumour cell invasion [39, 58], and tumour interactions with various environmental factors [50, 61] have been reported to literature so far. A previous extensive review of the main methodologies for CA models describing tumour growth has been delivered by [13], emphasizing mainly to the fact that by utilizing CAs, researchers have managed to consider a microscopic scale to describe the macroscopic characteristics of tumour morphology as well as in [61] where a comparison of relative strengths and weaknesses of various cell based models also referring to the CA-based ones is provided.

## 3 Cellular Automata Model of Tumour Growth and Tumour-Induced Vessel Displacement

Normal cells proliferate much slower than tumour cells and the volume of the normal tissue is maintained almost constant because of the balance between cell proliferation and cell death. In our tumour growth model we assume that normal cell proliferation around the tumour has no significant effect on tumour growth. The tumour and the tissue around the tumour are modeled as a two-dimensional cellular automaton (CA) lattice. We set each lattice site to correspond to a square with side length equal to $10\mu m$, similar to the size of a typical cell. Obviously, if required, lattice dimensions may be adjusted according to our preferences without affecting the mechanics of the algorithm. The state, $TC$, of each lattice site rectangle, say the $(i, j)$ site, at a time $t$, is the probability that the cell covered by this lattice site rectangle is a tumour cell. The state of a lattice site covering a normal cell is 0 and the state of a lattice site covering a tumour cell is 1. All the lattice sites located at the border between normal and tumour cells have states between 0 and 1.

We follow up with the assumption that during tumour growth the state of a lattice site that is adjacent to the tumour is initially at state 0 and as the tumour grows the state also grows till the state 1 is reached. As a consequence, the tumour growth phenomenon is modeled by the corresponding growth of the states of the CA lattice sites. It must be noticed that the state of a lattice site grows at discrete time steps, allowing for the modeller to adopt any time evolution requested [42, 72, 73]. The state at time step $t + 1$, $TC_{i,j}^{t+1}$ is directly dependent of all the neighboring sites at time step $t$, and is calculated according to the following equation:

$$
\begin{aligned}
TC_{i,j}^{t+1} = {} & \left(\alpha \times o_{i,j}^{t} + \beta \times g_{i,j}^{t}\right) \times TC_{i,j}^{t} + \left(\alpha \times o_{i-1,j}^{t} + \beta \times g_{i-1,j}^{t}\right) \times TC_{i-1,j}^{t} + \\
& \left(\alpha \times o_{i,j-1}^{t} + \beta \times g_{i,j-1}^{t}\right) \times TC_{i,j-1}^{t} + \left(\alpha \times o_{i+1,j}^{t} + \beta \times g_{i+1,j}^{t}\right) \times TC_{i+1,j}^{t} + \\
& \left(\alpha \times o_{i,j+1}^{t} + \beta \times g_{i,j+1}^{t}\right) \times TC_{i,j+1}^{t} + 0.18 \times \Big[ \left(\alpha \times o_{i-1,j-1}^{t} + \beta \times g_{i-1,j-1}^{t}\right) \times TC_{i-1,j-1}^{t} + \\
& \left(\alpha \times o_{i+1,j-1}^{t} + \beta \times g_{i+1,j-1}^{t}\right) \times TC_{i+1,j-1}^{t} + \left(\alpha \times o_{i-1,j+1}^{t} + \beta \times g_{i-1,j+1}^{t}\right) \times TC_{i-1,j+1}^{t} + \\
& \left(\alpha \times o_{i+1,j+1}^{t} + \beta \times g_{i+1,j+1}^{t}\right) \times TC_{i+1,j+1}^{t} \Big]
\end{aligned}
\tag{5}
$$

where $o_{i,j}^{t}$ is considered to be the normalized oxygen concentration and $g_{i,j}^{t}$ is considered to be the normalized nutrient (for example glucose) concentration at time step $t$ at lattice site $(i, j)$. As described later on in this section, these concentra-

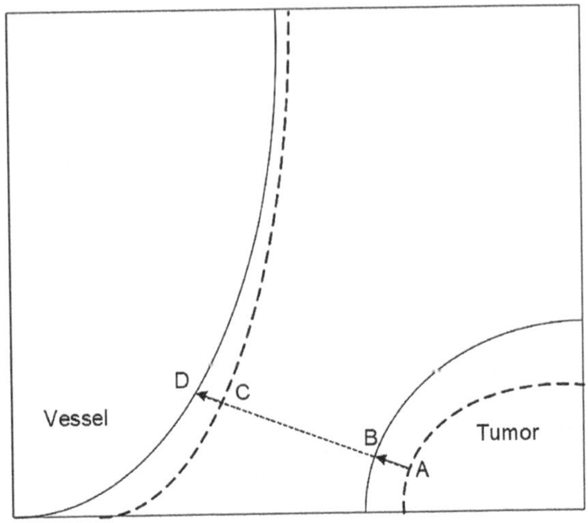


**Fig. 1** Schematic representation of vessel displacement due to tumour growth. Solid lines represent tumour and vessel borders at present time and dashed lines the borders at a previous time

tions are computed at each time step by solving the diffusion equations for oxygen and glucose on the CA lattice, with boundary conditions that are different for each time step because of the displacement of blood vessels caused by tumour growth. The parameters $\alpha$ and $\beta$ are user defined calibration parameters, which determine the effect of oxygen and glucose concentration on tumour growth. Indeed, there are experimental ways to define these parameters, but this is not the purpose of our work.

During tumour growth, mechanical forces develop and cause mechanical displacement of the surrounding tissue and blood vessels [11]. In areas relatively away from blood vessels, the concentration of oxygen and nutrients drops almost exponentially with distance. Therefore, the effect of even a minor displacement, which drives vessels away from the tumour, may not be trivial. To model vessel displacement due to tumour growth, we assume that the tissue between the tumour and the vessel is homogeneous and isotropic. We further assume that during the initial phase of tumour growth the forces are not strong enough to cause tissue cell damage, i.e. they do not exceed the elastic limit of the cells, hence the Hooke's law can be applied [47]. Figure 1 depicts vessel displacement caused by tumour growth. Solid lines represent the tumour and vessel borders at present and dashed lines the borders at a previous time point. At the tumour's surface, the forces caused by tumour growth are within normal limits.

In Fig. 1 the force caused by the displacement $AB$, acts in the direction of the line $AD$ and causes a vessel displacement $CD$. According to Hooke's law [47]:

$$\mathbf{F} = k_e q \times \mathbf{x} \tag{6}$$

where $\mathbf{F}$ is the force, $k_{eq}$ is the equivalent elastic constant of tumour and surrounding tissue and $x$ is the total displacement. This displacement comprises the line segments AB and CD. Assuming that the tumour and tissue have different elastic constants, $k_c$

and $k_t$ respectively, the equivalent constant is given by the following equation:

$$\frac{1}{k_{eq}} = \frac{1}{k_c} + \frac{1}{k_t} \tag{7}$$

and the displacement ratio by:

$$\frac{AB}{CD} = \frac{k_t}{k_c} \tag{8}$$

Since different solid tumours and tissues are expected to have diverse elastic constants, these constants are user- defined parameters in the simulation algorithm which will be described later on.

The results of vessel displacement simulation are shown on Fig. 2. Initially, as shown in Fig. 2a, the tumour is relatively small and the forces are not large enough to displace the nearby vessel, but as the tumour grows the vessel is displaced (Fig. 2b).

Equation (5) relates tumour growth rate with the normalized concentrations of oxygen and glucose, $o$ and $g$. Oxygen and glucose diffuse from the vessels and are distributed in the surrounding tissue. Normal tissue and tumour cells uptake oxygen and glucose and act as sinks in the diffusion process. In our model, we choose to use normalized concentrations by setting the values of oxygen and glucose concentrations at lattice sites that are adjacent to vessels equal to 1. These concentrations are two-dimensional functions that are solutions of the diffusion equations:

$$\frac{\partial o(x,y,t)}{\partial t} = D_{ox} \times \left(\frac{\partial^2 o(x,y,t)}{\partial x^2} + \frac{\partial^2 o(x,y,t)}{\partial y^2}\right) - U_{ox}(x,y) \times o(x,y,t) \tag{9}$$

and

$$\frac{\partial g(x,y,t)}{\partial t} = D_{gl} \times \left(\frac{\partial^2 g(x,y,t)}{\partial x^2} + \frac{\partial^2 g(x,y,t)}{\partial y^2}\right) - U_{dg}(x,y) \times g(x,y,t) \tag{10}$$

Equation (9) describes the diffusion of oxygen, where $D_{ox}$ is the diffusivity (or diffusion coefficient) of oxygen and $U_{ox}$ is the oxygen uptake rate with units $(1/s)$. Equation (10) describes the diffusion of glucose. $D_{gl}$ is the diffusivity of glucose and $U_{gl}$ is the glucose uptake rate with units $(1/s)$.

We discretize equations (9) and (10) on the CA lattice and calculate the values of oxygen and glucose normalized concentrations at points that are located at the center of the squares that correspond to lattice sites [7]. Both $o$ and $g$ are functions of $x$, $y$ and $t$ and their discrete counterparts used in Eq. (5) are:

$$o(x,y,t) \rightarrow o^t_{i,j} and g(x,y,t) \rightarrow g^t_{i,j} \tag{11}$$

**Fig. 2** **a** Vessel location and shape at the onset of the tumour growth **b** Vessel displacement because of the tumour growth

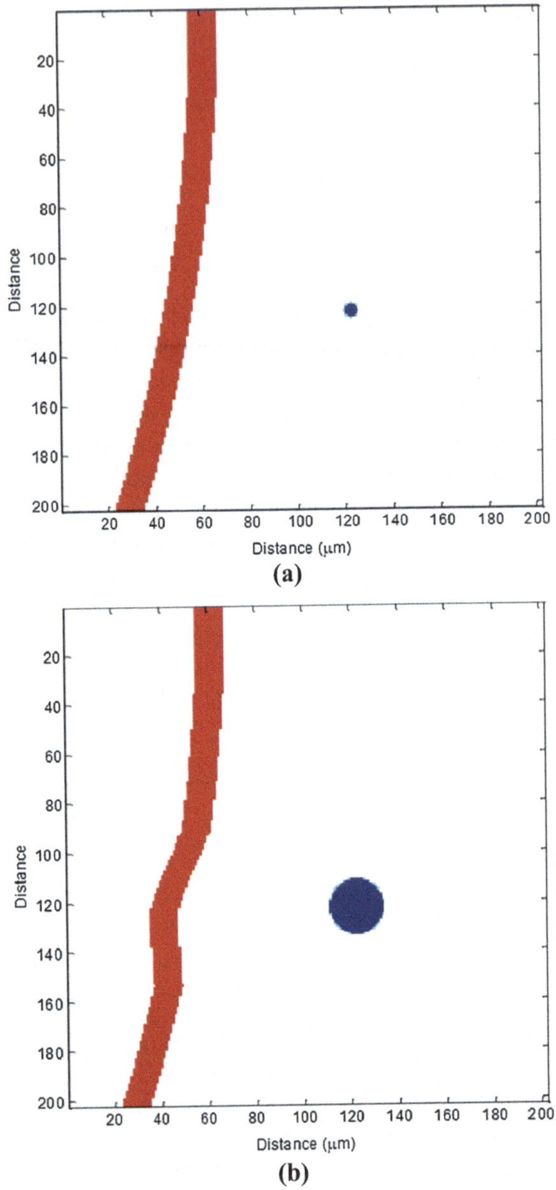


Figures 3 and 4 show the distribution of oxygen and glucose, respectively, for the vessel-tumour relative positions depicted in Fig. 2b. The boundary conditions were imposed by setting the values of both normalized concentrations equal to 1 in lattice sites that are adjacent to vessels and equal to 0 in lattice sites that are located more than $200\mu m$ away from the vessel.

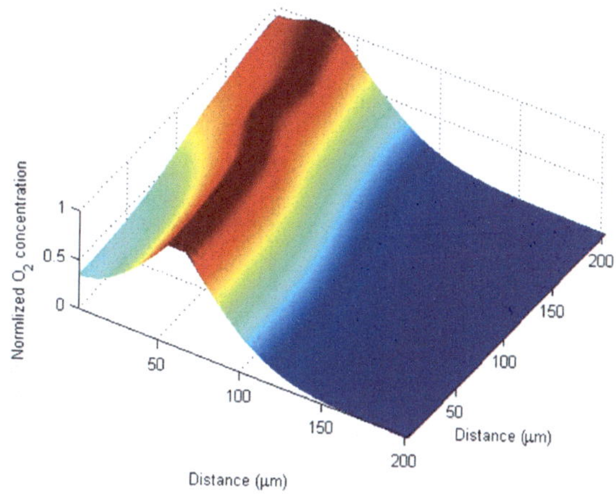


**Fig. 3** Distribution of oxygen for the vessel-tumour relative positions depicted in Fig. 2b

The diffusivity values are taken to be $D_{ox} = 10^{-5} cm^2/s$ and $D_{gl} = 10^{-7} cm^2/s$ [27]. The uptake rate for oxygen is $U_{ox} = 10^{-1} s^{-1}$ and the uptake rates of glucose for normal and tumour areas are $U_{gl} = 10^{-4} s^{-1}$ and $U_{gl} = 10^{-3} s^{-1}$, respectively [85]. In normal tissue lattice sites, where the state $TC = 0$, the uptake rate is $U_{gl} = 10^{-4} s^{-1}$ and in tumour lattice sites where the state $TC = 1$, the uptake rate is $U_{gl} = 10^{-3} s^{-1}$. In lattice sites that contain both normal and tumour cells the uptake rate is a linear function of the site state $TC$. As shown in Fig. 4, this results in increasing glucose consumption as we move from normal tissue into the tumour area.

## 4 Simulation Results of Tumour Growth and Tissue Displacement

Based on the model described in the previous section we developed an algorithm for the simulation of the effect of tissue displacement and tumour growth on tumour progression. The flowchart of this algorithm is shown in Fig. 5. The user can either draw a hypothetical vessel (or vessels) or import a digitized image of a tissue, in which she wishes to study the dynamics of avascular tumour growth. Then, the user selects a small tissue area (as small as a pixel in the case of an imported image) where the tumour is initiated. If there is already a tumour in the imported image, the user determines the tumour area as the tumour initiation area. Vessel shape and location and tumour initiation area are the inputs to the algorithm. These inputs also determine the initial and boundary conditions for the tumour growth simulation during the first time step.

After input introduction the algorithm takes the first time step ($t = 1$). Diffusion equations (9) and (10) are discretized and solved on the CA lattice. Thus the distribution of the normalized oxygen and glucose concentrations, $o$ and $g$, and their values in all lattice sites are determined. These values are used to compute tumour growth using Eq. (5). Tumour growth causes vessel displacement. This displacement is com-

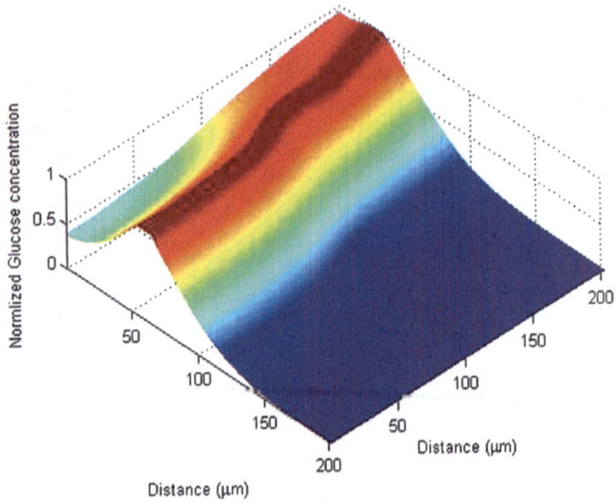


**(a)**

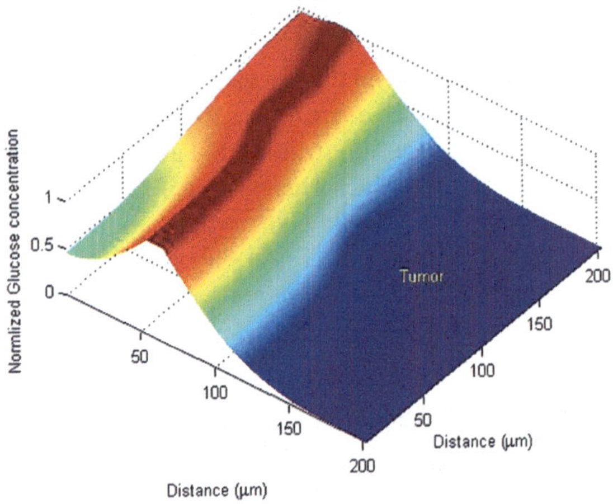


**(b)**

**Fig. 4** **a** Distribution of glucose, originally and **b** for the vessel-tumour relative positions depicted in Fig. 2b

puted and the new vessel location and shape are obtained. The new vessel location and shape and the new size of the tumour determine the new boundary conditions for the two diffusion equations (9) and (10). The next time step is taken and the diffusion equations are solved with the new boundary conditions determined in the previous time step. The tumour growth, and the new vessel displacement and shape are computed. These will be the new boundary conditions that will be imposed in the next time step and so on until the user determined maximum number of time steps, $T_{max}$, is reached. $T_{max}$ is the time when the avascular phase is supposed to end and the neo-angiogenesis phase begin. Tumour growth causes vessel displacement, which changes the distributions of oxygen and glucose, which in turn affect tumour growth, which causes vessel displacement and this process goes on until neo-angiogenesis is triggered.

**Fig. 5** Flowchart of the modeling algorithm

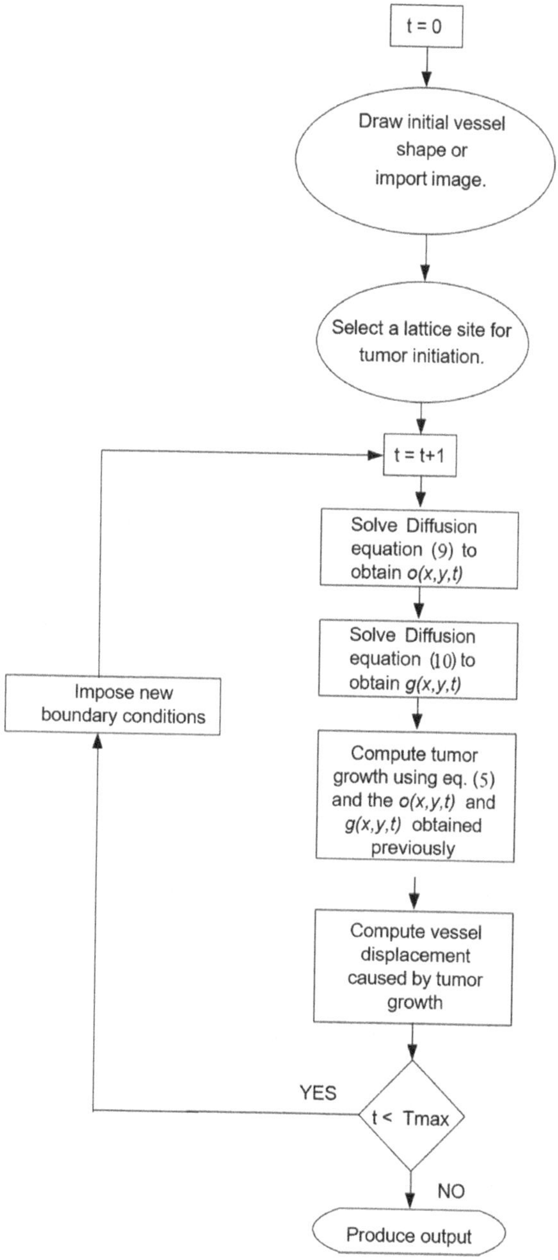

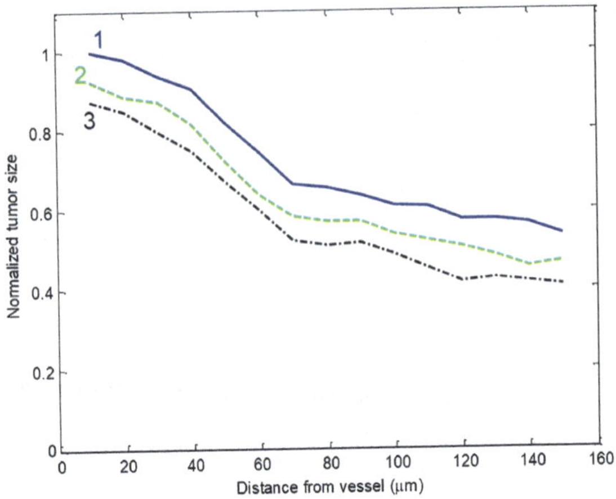


**Fig. 6** Normalized tumour sizes as a function of the distance of the tumour initiation lattice site from the vessel. Line (1) corresponds to $\alpha = 0.01$, $\beta = 0.05$, line (2) to $\alpha = 0.005$, $\beta = 0.03$ and line (3) to $\alpha = 0.001$, $\beta = 0.002$

We used this algorithm to simulate the growth of a number of tumours that are initiated at various distances from a vessel, for the same number of time steps. We did this for three different pairs of values of the user defined parameters $\alpha$ and $\beta$ of Eq. (5). The results are shown in Fig. 6. Line (1) corresponds to $\alpha = 0.01$ and $\beta = 0.05$. The first tumour initiation lattice site is located $10\mu m$ from the vessel. In our series of digital experiments, this tumour reached the maximum area (maximum size), $TA_{max}$, in the given time steps. If the size reached by another tumour, say tumour $j$, located further from the vessel in the same number of time steps is $TA_j$, the normalized size of this tumour, $NS_j$, is:

$$NS_j = \frac{TA_j}{TA_{max}} \tag{12}$$

The normalized size of the tumour that reached the maximum size is 1. Line (2) corresponds to $\alpha = 0.005$ and $\beta = 0.03$. Line (3) corresponds to $\alpha = 0.001$ and $\beta = 0.002$. In each case the growth of fifteen tumours was simulated. The first one is located at a distance of $10\mu m$ from the vessel and the rest fourteen are located at $10\mu m$ intervals at distances ranging from $20\mu m$ up to $150\mu m$ from the vessel. In all three cases, tumour normalized sizes are large at distances from $10\mu m$ to $40\mu m$ from the vessel. The sizes are reduced significantly at distances from 40 to $70\mu m$ from the vessel and are further slightly reduced at distances from $70\mu m$ to $150\mu m$ from the vessel.

We attribute these digital experimental results to the nature of the diffusion process. Concentrations of oxygen and glucose are relative high near the vessel, then drop almost exponentially and finally reach a nearly constant minimum value. Our simulation results strongly suggest that the properties of the physical diffusion process in the case of moving vessel-tumour boundaries, affect directly tumour progression during the avascular phase. The values of parameters $\alpha$ and $\beta$ may vary greatly among different tissues. Therefore, it was not necessary to define the exact parameter values in order to support our main conclusion. Our simulation analysis showed that given any set of parameter values, the vessel displacement affected tumour growth. We should like to stress that it is the displacement, not the nature of these parameters that is evaluated by our approach. Therefore, we chose to make our model as general as possible and offer the option to define these parameters according to the properties of the tissue under study.

Based on our analysis, we propose that our model may prove useful in the study of tumour dynamics and consequently the simulation algorithm presented herein may find applications in the study of processes influenced by them, such as optimization of drug delivery strategies [62]. Obviously, the progression of tumour development as a function of time may not be observed experimentally because the dissection of a tumour, at any given time step, requires the sacrifice of the specimen. In addition, technically it is quite difficult to monitor vessel displacement on a tissue section; the plane of the section must pass from the longitudinal axis of the deformed vessel as well as from the solid tumour itself. Analysis is further complicated by the fact that

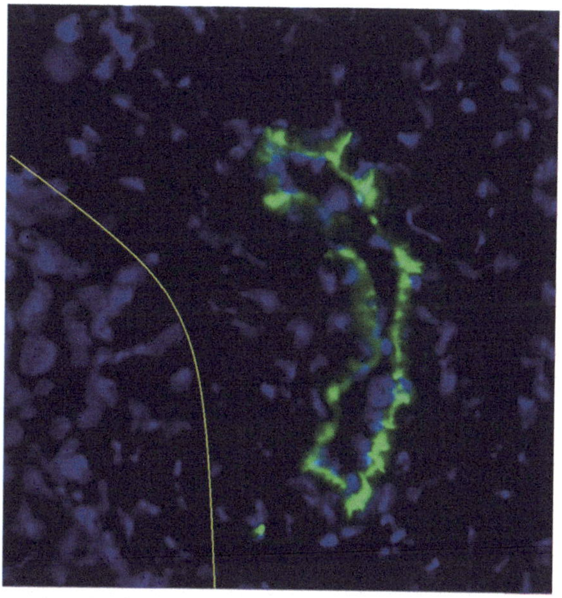

**Fig. 7** Vessel displacement due to avascular growth of a neighboring solid tumour. Sections (thickness: 8$\mu$m) of a frozen, early human ovarian tumour were fixed in cold acetone for 10min and washed in phosphate buffered saline (PBS). After blocking of the slides with 10% normal goat serum for 30min and washing in 1xPBS, the slides were incubated with anti-CD31 FITC (BD) (1:50) at 4 °C overnight. The following day the slides were washed 3 times in 1xPBS and mounted with DAPI containing medium. The CD31 antigen (bright green) is a specific marker present on the surface of endothelial cells. DAPI (blue dye) stains cell nuclei. The yellow line is an approximate delineation of the border of the solid tumour at the left bottom corner. The concave side of the deformed vessel reflects the contour of the tumour border, indicating that it is caused by the growth of the nearby tumour. This tumour is still in the avascular phase as no endothelial cells may be observed within its mass. Such observations are consistent with our hypothesis that tumour growth may cause vessel displacement during the avascular phase

vessels are usually not straight and appear branched. However, we have been able to document vessel distortions and deformations induced by tumour growth during the avascular phase (Fig. 7). Such observations are consistent with our simulation analysis indicating that tumour growth causes vessel displacement during the avascular phase.

## 5 Conclusions

We developed a model and a simulation algorithm for the study of the interplay between tumour growth and tumour induced vessel displacement. Our digital experiments strongly indicated that tumour induced vessel displacement may affect tumour progression. Therefore, our model highlights the importance of taking into account a factor that was overlooked in previous models and urges for experimental validation in a biological system. Furthermore, the model and algorithm can be extended to model tumour growth after the angiogenesis onset by incorporating the diffusion of tumour-derived angiogenetic factors, such as VEGF, towards the blood vessel and the formation of new vessels. The model presented here can be expanded to three dimensions by considering a 3-D CA lattice, where the lattice sites will be cubes, and by solving the 3-D diffusion equations for oxygen and glucose on this lattice. This will include the 3-D motion of the molecules and may elucidate further the dynamics of tumour growth and vessel displacement.

Our model and algorithm can be used for the study of various tumour growth scenarios with applications such as understanding early tumour growth dynamics and mechanics, drug and nanoparticles delivery in chemotherapy, access of growth factors and other plasma factors to the tumour, etc.

## References

1. Adamatzky, A.: Bioevaluation of World Transport Networks. World Scientific (2012)
2. Adamatzky, A.: Cellular Automata: A Volume in the Encyclopedia of Complexity and Systems Science, 2nd edn. Springer (2018)
3. Adamatzky, A., Phillips, N., Weerasekera, R., Tsompanas, M.A., Sirakoulis, G.C.: Street map analysis with excitable chemical medium. Phys. Rev. E **98**(1), 012306 (2018)
4. Ghazani, M., Saghafian, M., Jalali, P., Soltani, M.: Mathematical simulation and prediction of tumor volume using rbf artificial neural network at different circumstances in the tumor microenvironment. Proc. Instit. Mech. Eng. Part H J. Eng. Med. **235**(11), 1335–1355 (2021). PMID: 34247529
5. Alarcón, T., Byrne, H.M., Maini, P.K.: A cellular automaton model for tumour growth in inhomogeneous environment. J. Theor. Biol. **225**(2), 257–274 (2003)
6. Alonso-Sanz, R.: Cellular Automata with Memory, vol. 3. Archives contemporaines(2008)
7. Ames, W.F.: Numerical Methods for Partial Differential Equations, 3rd edn, 380p. Academic Press (2014)
8. Amoddeo, A.: Modeling avascular tumor growth: approach with an adaptive grid numerical technique. J. Multiscale Model. **09**(03), 1840002 (2018)

9. Araujo, R.P., Sean McElwain, D.L.: A history of the study of solid tumour growth: the contribution of mathematical modelling. Bull. Math. Biol. **66**(5), 1039–1091 (2004)
10. Axelrod, R., Axelrod, D.E., Pienta, K.J.: Evolution of cooperation among tumor cells. Proc. Natl. Acad. Sci. **103**(36), 13474–13479 (2006)
11. Balaguera, M.I., Briceño, J.C., Glazier, J.A.: An object-oriented modelling framework for the arterial wall. Comput. Methods Biomech. Biomed. Eng. **13**(1), 135–142 (2010). PMID: 19603305
12. Bialynicki-Birula, I.: Weyl, Dirac, and Maxwell equations on a lattice as unitary cellular automata. Phys. Rev. D **49**(12), 6920 (1994)
13. Boondirek, A., Triampo, W., Nuttavut, N.: A review of cellular automata models of tumor growth. Int. Math. Forum **5**(61), 3023–3029 (2010)
14. Boukas, E., Kostavelis, I., Gasteratos, A., Sirakoulis, G.C.: Robot guided crowd evacuation. IEEE Trans. Autom. Sci. Eng. **12**(2), 739–751 (2014)
15. Chen, H., Matthaeus, W.H., Klein, L.W.: Theory of multicolor lattice gas: a cellular automaton poisson solver. J. Comput. Phys. **88**(2), 433–466 (1990)
16. Chopard, B., Droz, M.: Cellular Automata, vol. 1. Springer (1998)
17. Chopard, B., Droz, M.: Cellular automata model for the diffusion equation. J. Stat. Phys. **64**(3), 859–892 (1991)
18. Cui, S., Escher, J.: Asymptotic behaviour of solutions of a multidimensional moving boundary problem modeling tumor growth. Commun. Partial Differ. Eqn. **33**, 636–655 (2008)
19. Danikas, M.G., Karafyllidis, I., Thanailakis, A., Bruning, A.M.: Simulation of electrical tree growth in solid dielectrics containing voids of arbitrary shape. Model. Simul. Mater. Sci. Eng. **4**(6), 535–552 (1996)
20. Dormann, S., Deutsch, A.: Modeling of self-organized avascular tumor growth with a hybrid cellular automaton. In Silico Biol. **2**, 393–406 (2002)
21. Dourvas, N., Tsompanas, M.A., Sirakoulis, G.C., Tsalides, P.: Hardware acceleration of cellular automata physarum polycephalum model. Parallel Process. Lett. **25**(01), 1540006 (2015)
22. Dourvas, N.I., Sirakoulis, G.C., Adamatzky, A.: Cellular automaton belousov–zhabotinsky model for binary full adder. Int. J. Bifurcat. Chaos **27**(06), 1750089 (2017)
23. Enderling, H., Chaplain, M.: Mathematical modeling of tumor growth and treatment. Curr. Pharm. Des. **20**, 11 (2013)
24. Evangelidis, V., Jones, J., Dourvas, N., Tsompanas, M.A., Sirakoulis, G.C., Adamatzky, A.: Physarum machines imitating a roman road network: the 3d approach. Sci. Rep. **7**(1), 1–14 (2017)
25. Fasano, A., Bertuzzi, A., Gandolfi, A.: Mathematical modeling of tumour growth and treatment. Complex Syst. Biomed. 71–108 (2007)
26. Feynman, R.P.: Simulating physics with computers. Int. J. Theor. Phys. **21**(6/7), 467–488 (1982)
27. Frieboes, H.B., Edgerton, M.E., Fruehauf, J.P., Rose, F.R.A.J., Worrall, L.K., Gatenby, R.A., Ferrari, M., Cristini, V.: Prediction of drug response in breast cancer using integrative experimental/computational modeling. Cancer Res. **69**(10), 4484–4492 (2009)
28. Friedman, A.: Cancer Models and Their Mathematical Analysis, vol. 1872, pp. 223–246. Springer Berlin, Heidelberg (2005)
29. Gardner, M.: Mathematical games: The fantastic combinations of john conway's new solitaire game "life". Sci. Am. **223**(4), 120–123 (1970)
30. Gerakakis, I., Gavriilidis, P., Dourvas, N.I., Georgoudas, I.G., Trunfio, G.A., Sirakoulis, G.C.: Accelerating fuzzy cellular automata for modeling crowd dynamics. J. Comput. Sci. **32**, 125–140 (2018)
31. Gerhardt, M., Schuster, H., Tyson, J.J.: A cellular automation model of excitable media including curvature and dispersion. Science **247**(4950), 1563–1566 (1990)
32. Gerlee, P., Anderson, A.R.A.: An evolutionary hybrid cellular automaton model of solid tumour growth. J. Theor. Biol. **246**(4), 583–603 (2007)
33. Hanahan, D., Weinberg, R.A.: The hallmarks of cancer. Cell **100**(1), 57–70 (2000)
34. Harrington, L.A., Tannock, I.F., Hill, R.P., Cescon, D.W.: The basic Science of Oncology, 6th edn. McGraw Hill, New York (2021)

35. Hartman, H., Tamayo, P.: Reversible cellular automata and chemical turbulence. Physica D **45**(1–3), 293–306 (1990)
36. Hatzikirou, H., Brusch, L., Schaller, K.L., Simon, M., Deutsch, A.: Prediction of traveling front behavior in a lattice-gas cellular automaton model for tumor invasion. Comput. Math. Appl. **59**(7), 2326–2339 (2010) (Mesoscopic Methods in Engineering and Science)
37. Interian, R., Rodríguez-Ramos, R., Valdés-Ravelo, F., Ramírez-Torres, A., Ribeiro, C.C., Conci, A.: Tumor growth modelling by cellular automata. Math. Mech. Complex Syst. **5**(3–4) (2017)
38. Ioannidis, K., Sirakoulis, G.C., Andreadis, I.: A path planning method based on cellular automata for cooperative robots. Appl. Artif. Intell. **25**(8), 721–745 (2011)
39. Jiao, Y., Torquato, S.: Emergent behaviors from a cellular automaton model for invasive tumor growth in heterogeneous microenvironments. PLOS Comput. Biol. **7**(12), 1–14 (2011)
40. Kalogeiton, V.S., Papadopoulos, D.P., Georgilas, I.P., Sirakoulis, G.C., Adamatzky, A.I.: Cellular automaton model of crowd evacuation inspired by slime mould. Int. J. General Syst. **44**(3), 354–391 (2015)
41. Kansal, A.R., Torquato, S., Harsh, G.R., Chiocca, E.A., Deisboeck, T.S.: Simulated brain tumor growth dynamics using a three-dimensional cellular automaton. J. Theor. Biol. **203**(4), 367–382 (2000)
42. Karafyllidis, I., Thanailakis, A.: A model for predicting forest fire spreading using cellular automata. Ecol. Model. **99**(1), 87–97 (1997)
43. Karafyllidis, I.G., Sasaroli, D., Karapetsas, A., Sandaltzopoulos, R.: Modeling the effect of tissue displacement during avascular tumor growth on tumor progression. Asian J. Fuzzy Appl. Math. **2**(1) (2014)
44. Karamani, R.E., Fyrigos, I.A., Tsakalos, K.A., Ntinas, V., Tsompanas, M.A., Sirakoulis, G.C.: Memristive learning cellular automata for edge detection. Chaos Solitons Fractals **145**, 110700 (2021)
45. Koch, G., Walz, A., Lahu, G., Schropp, J.: Modeling of tumor growth and anticancer effects of combination therapy. J. Pharmacokinet. Pharmacodyn. **36**, 179–97 (2009)
46. Kotoulas, L., Tsarouchis, D., Sirakoulis, G.C., Andreadis, I.: 1-d cellular automaton for pseudorandom number generation and its reconfigurable hardware implementation. In: 2006 IEEE International Symposium on Circuits and Systems, pp. 4. IEEE (2006)
47. Landau, L.D., Lifshitz, E.M.: Mechanics, 3rd, vol. 1 (Course of Theoretical Physics). Butterworth-Heinemann (1976)
48. Lowengrub, J.S., Frieboes, H.B., Jin, F., Chuang, Y.L., Li, X., Macklin, P., Wise, S.M., Cristini, V.: Nonlinear modelling of cancer: bridging the gap between cells and tumours. Nonlinearity **23**(1), R1–R91 (2009)
49. Macklin, P., Lowengrub, J.: Nonlinear simulation of the effect of microenvironment on tumor growth. J. Theor. Biol. **245**(4), 677–704 (2007)
50. Mallet, D.G., De Pillis, L.G.: A cellular automata model of tumor-immune system interactions. J. Theor. Biol. **239**(3), 334–350 (2006)
51. Markus, M., Hess, B.: Isotropic cellular automaton for modelling excitable media. Nature **347**(6288), 56–58 (1990)
52. Mizas, C., Sirakoulis, G.C., Mardiris, V., Karafyllidis, I., Glykos, N., Sandaltzopoulos, R.: Reconstruction of dna sequences using genetic algorithms and cellular automata: towards mutation prediction? Biosystems **92**(1), 61–68 (2008)
53. Moreira, J., Deutsch, A.: Cellular automaton models of tumor development: a critical review. Adv. Complex Syst. **05**(02n03), 247–267 (2002)
54. Nalpantidis, L., Sirakoulis, G.C., Gasteratos, A.: Non-probabilistic cellular automata-enhanced stereo vision simultaneous localization and mapping. Meas. Sci. Technol. **22**(11), 114027 (2011)
55. Ntinas, V., Sirakoulis, G.C., Rubio, A.: Memristor-based probabilistic cellular automata. In: 2021 IEEE International Midwest Symposium on Circuits and Systems (MWSCAS), pp. 792–795. IEEE (2021)

56. Ntinas, V.G., Moutafis, B.E., Trunfio, G.A., Sirakoulis, G.C.: Parallel fuzzy cellular automata for data-driven simulation of wildfire spreading. J. Comput. Sci. **21**, 469–485 (2017)
57. Omohundro, S.: Modelling cellular automata with partial differential equations. Physica D **10**(1), 128–134 (1984)
58. Patel, A.A., Gawlinski, E.T., Lemieux, S.K., Gatenby, R.A.: A cellular automaton model of early tumor growth and invasion: The effects of native tissue vascularity and increased anaerobic tumor metabolism. J. Theor. Biol. **213**(3), 315–331 (2001)
59. Piotrowska, M.J., Angus, S.D.: A quantitative cellular automaton model of in vitro multicellular spheroid tumour growth. J. Theor. Biol. **258**(2), 165–178 (2009)
60. Poleszczuk, J., Enderling, H.: A high-performance cellular automaton model of tumor growth with dynamically growing domains. Appl. Math. **5**, 09 (2013)
61. Rejniak, K.A., McCawley, L.J.: Current trends in mathematical modeling of tumor-microenvironment interactions: a survey of tools and applications. Exp. Biol. Med. **235**(4), 411–423 (2010)
62. Ribba, B., You, B., Tod, M., Girard, P., Tranchand, B., Trillet-Lenoir, V., Freyer, G.: Chemotherapy may be delivered based on an integrated view of tumour dynamics. IET Syst. Biol. **3**, 180–190 (2009)
63. Roose, T., Chapman, S.J., Maini, P.K.: Mathematical models of avascular tumor growth. SIAM Rev. **49**(2), 179–208 (2007)
64. Schönfisch, B.: Anisotropy in cellular automata. Biosystems **41**(1), 29–41 (1997)
65. Shahmoradi, S., Rahatabad, F.N., Maghooli, K.: A stochastic cellular automata model of growth of avascular tumor with immune response and immunotherapy. Inf. Med. Unlocked **12**, 81–87 (2018)
66. Shnerb, N.M., Louzoun, Y., Bettelheim, E., Solomon, S.: The importance of being discrete: Life always wins on the surface. Proc. Natl. Acad. Sci. **97**(19), 10322–10324 (2000)
67. Sirakoulis, G.C.: A TCAD system for VLSI implementation of the CVD process using VHDL. Integr. VLSI J. **37**(1), 63–81 (2003)
68. Sirakoulis, G.C.: Parallel application of hybrid dna cellular automata for pseudorandom number generation. J. Cell. Automata **11**(1), 63–89 (2016)
69. Sirakoulis, G.C., Adamatzky, A. (eds.) Robots and Lattice Automata, vol. 13, 1st edn. Emergence, Complexity and Computation. Springer International Publishing Cham (2014)
70. Sirakoulis, G.C., Bandini, S.: Cellular Automata. In: Proceedings of the 10th International Conference on Cellular Automata for Research and Industry, ACRI 2012, vol. 7495, Santorini Island, Greece, 24–27 September 2012. Springer (2012)
71. Sirakoulis, G.C., Karafyllidis, I., Mizas, C., Mardiris, V., Thanailakis, A., Tsalides, P.: A cellular automaton model for the study of dna sequence evolution. Comput. Biol. Med. **33**(5), 439–453 (2003)
72. Sirakoulis, G.C., Karafyllidis, I., Thanailakis, A.: A cellular automaton model for the effects of population movement and vaccination on epidemic propagation. Ecol. Model. **133**(3), 209–223 (2000)
73. Sirakoulis, G.C., Karafyllidis, I., Thanailakis, A.: A cellular automaton for the propagation of circular fronts and its applications. Eng. Appl. Artif. Intell. **18**(6), 731–744 (2005)
74. Sirakoulis, G.C., Karafyllidis, I., Thanailakis, A., Mardiris, V.: A methodology for VLSI implementation of cellular automata algorithms using VHDL. Adv. Eng. Softw. **32**(3), 189–202 (2001)
75. Taghibakhshi, A., Barisam, M., Saidi, M.S., Kashaninejad, N., Nguyen, N.T.: Three-dimensional modeling of avascular tumor growth in both static and dynamic culture platforms. Micromachines **10**(9) (2019)
76. Toffoli, T.: Cellular automata as an alternative to (rather than an approximation of) differential equations in modeling physics. Physica D **10**(1–2), 117–127 (1984)
77. Tracqui, P.: Biophysical models of tumour growth. Rep. Prog. Phys. **72**(5), 056701 (2009)
78. Tsiftsis, A., Georgoudas, I.G., Sirakoulis, G.C.: Real data evaluation of a crowd supervising system for stadium evacuation and its hardware implementation. IEEE Syst. J. **10**(2), 649–660 (2016)

79. Tsompanas, M.A., Fyrigos, I.A., Ntinas, A.V., Sirakoulis, G.C.: Cellular automata implementation of oregonator simulating light-sensitive belousov–zhabotinsky medium. Nonlinear Dyn. **104**, 4103–4115 (2021)
80. Tsompanas, M.A., Fyrigos, I.A., Ntinas, V., Adamatzky, A., Sirakoulis, G.C.: Light sensitive belousov–zhabotinsky medium accommodates multiple logic gates. Biosystems **206**, 104447 (2021)
81. Tsompanas, M.A.I., Sirakoulis, G.C.: Modeling and hardware implementation of an amoeba-like cellular automaton. Bioinspiration Biomimetics **7**(3), 036013 (2012)
82. Vichniac, G.Y.: Simulating physics with cellular automata. Physica D **10**(1), 96–116 (1984)
83. von Neumann, J.: Theory of Self-reproducing Automata. University of Illinois Press (1966)
84. Was, J., Sirakoulis, G.C., Bandini, S.: Cellular Automata. In: Proceedings of the 11th International Conference on Cellular Automata for Research and Industry, ACRI 2014, Krakow, Poland, 22–25 September 2014, vol 8751. Springer, Cham (2014)
85. Weinberg, R.A.: The Biology of Cancer. W.W. Norton (2013)
86. Wodarz, D., Komarova, N.: Computational Biology Of Cancer. Lecture Notes And Mathematical Modeling. World Scientific Publishing Company (2005)
87. Wolfram, S.: Statistical mechanics of cellular automata. Rev. Mod. Phys. **55**(3), 601 (1983)
88. Wolfram, S.: A New Kind of Science. Wolfram Media Inc., Champaign, Ilinois, USA (2002)
89. Wolfram, S.: Cellular Automata and Complexity. Collected Papers. CRC Press (2018)

# Complexities of Drug Resistance in Cancer: An Overview of Strategies and Mathematical Models

**Benedetta Casadei, Marta Giacosa, Alessandro Maula, Sara Plos, Letizia Zappulla, Camilla Viotto, Marco A. Deriu, and Jack A. Tuszynski**

**Abstract** There are several standard modalities of cancer therapy, among which chemotherapy is the most commonly employed. Drugs used in chemotherapy could reduce the tumour mass or cause cancer cell death but, they are almost always associated with some complications and side effects causing damage to healthy cells. Multidrug-resistance is the main reason of cancer treatment failures, and it involves three elements: macroscopic, microscopic and mesoscopic resistance. The heterogeneity of tumour masses contributes significantly to the development of drug resistance. The mechanisms behind drug-resistance can be classified into four groups: drug efflux, alterations in target molecules, blocked apoptosis and tumour microenvironment. These mechanisms may include alterations in target molecules, altered membrane transport, decreased drug activation, altered expression of drug-metabolism enzymes, drug inactivation, enhanced DNA repair, failure of apoptosis and modifications to tumour microenvironment. Several mathematical models have been proposed to describe the cellular heterogeneity of cancer in order to optimize the development of anti-cancer treatment. These models are based on differential equations or integro-differential equations depending on the complexity of the model. Many strategies have been developed to overcome multidrug resistance in cancer, in particular, the RNAi and three different generations of MDR inhibitors have been generated thus far, although problems still persist. Other methods designed to overcome multidrug resistance include: nanomedicine therapeutic approaches based on different types of nanoparticles or a combination of nanoparticles, anticancer drugs and MDR modulators, and antiangiogenic therapy.

B. Casadei · M. Giacosa · A. Maula · S. Plos · L. Zappulla · C. Viotto · M. A. Deriu · J. A. Tuszynski (✉)
DIMEAS, Politecnico Di Torino, Corso Duca degli Abruzzi 24, 10129 Turin, Italy
e-mail: jack.tuszynski@gmail.com



I. Balaz and A. Adamatzky (eds.), *Cancer, Complexity, Computation*, Emergence, Complexity and Computation 46,
https://doi.org/10.1007/978-3-031-04379-6_14

# 1 Introduction

Treatments for patients suffering from cancer aim at destroying cancer cells while striving to reduce damage to the healthy cells or tissues. Each treatment depends on the type of pathology, the molecular features of the cancer, its location in the body, the general health status of the patient, and the presence or absence of metastases. The modalities generally employed to treat cancer include surgery, radiotherapy and chemotherapy. These procedures can be used alone, in combination, simultaneously or in sequence. Among the many treatments, one of the most frequently employed is chemotherapy (CTX), which consists of a large group of cytotoxic drugs which are intended to reduce the tumour mass by inducing cancer cell death. Since chemotherapy usually involves systemic administration, it also affects healthy cells due to its non-specific effect on the patient. Chemotherapeutic drugs can only be used for a limited period of time due to their sometimes debilitating side effects. Moreover, the patient undergoes a significant amount of psychological and physical stress [1].

CTX drugs are antineoplastic (or antiblastic) drugs which act by means of two main mechanisms. First, their direct interaction with DNA, mainly employing anti-alkylating agents, prevents DNA from duplicating. Second involves their interactions with the biosynthetic pathway of DNA and RNA precursors (for instance, antimetabolites). Additionally, other pharmacological agents in combination therapies target microtubules and another group of drugs interrupt signaling pathways. In the first case, the effect of the drug is independent of the time of exposure to it, although it depends strongly on the drug concentration. In the second case, the therapeutic effect depends on the time of exposure, because the longer the exposure time, the greater the number of the cells going through the metabolic pathway are being blocked by the drug. These drugs can enter the cells along the concentration gradient by the ABC transporter molecules (passive transport), but the absorption of high concentrations of the drug only occurs through active transport [2].

Most of the CTX drugs are natural compounds extracted from plants, while others are synthetically generated small molecules. They can be classified into four major groups, according to their mode and site of actions (Fig. 1):

- Antimetabolites: they include folate, pyrimidine and purine antagonists. Folate antagonists inhibit the dihydrofolate reductase (DHFR), which is an enzyme of the nucleotide metabolism. Pyrimidine antagonists, especially 5-fluorouracil and arabinosylcytosine, are inhibitors of pyrimidine nucleotide formation. In addition, 5-fluorouracil affects some of the processes involving the synthesis of thymine from uracil. Purine antagonists are another type of antimetabolites that hinder the formation of adenine and guanine [3].
- Genotoxic agents: they affect enzymes involved in the apoptotic process. They can be divided into several clusters, among which are alkylating agents, which are genotoxic and may cause DNA mutations, intercalating agents, which interfere with the polymerase activity, and also enzyme inhibitors. An adverse effect caused by genotoxic agents is the possibility of the growth of secondary cancer cells.

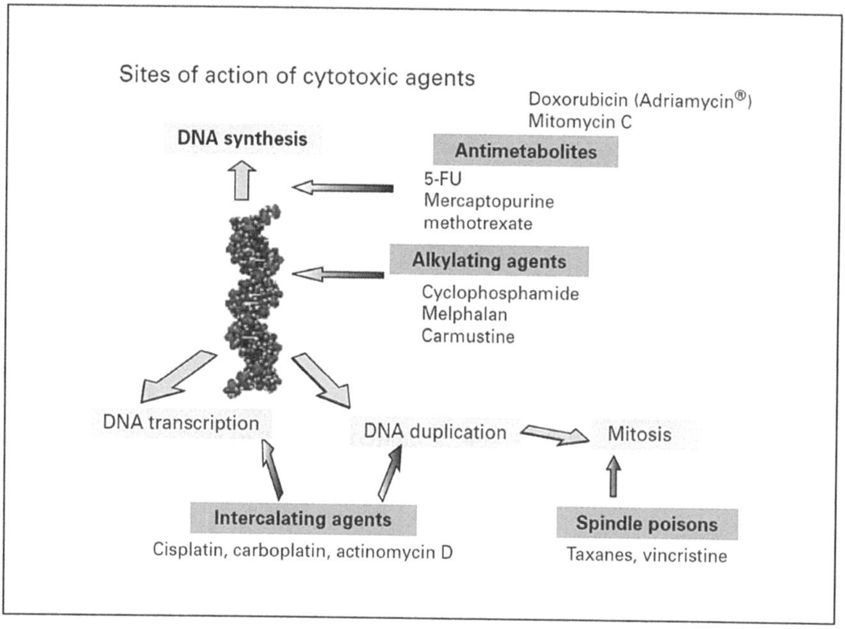


**Fig. 1** Sites of action of cytotoxic agents

- Mitotic spindle inhibitors: during the mitotic process, they affect the generation of microtubule containing mitotic spindles, which are indispensable in the chromosome alignment and segregation. They may also cause adverse side effects on healthy cells, especially those characterized by rapid replication [1, 4].
- Agents used to target tumour vasculature in order to interrupt the tumour progression, since the tumour cells require a steady blood supply for the provision of oxygen and nutrients. Angiogenesis inhibitors obstruct the formation of blood vessels in the area in and around the tumour site. These agents, however, might also limit the perfusion of cytotoxic drugs [5].

## 2 The Problem of Drug Resistance

In the case of chemotherapy-based drug treatments, the tumour often develops a complex mechanism of multidrug-resistance, meaning a resistance to the action of several drugs. Drug resistance can be divided into three main categories: pharmacological, physiological and cellular resistance (see Fig. 2). Taken together, they lead to the failure of the treatment [2]. These three categories are also called respectively macroscopic (systemic) resistance, mesoscopic (physical, mechanical) resistance and microscopic (local) resistance. Macroscopic resistance is affected by pharmacokinetics, which, in turn, consists of four stages: absorption, distribution, metabolism

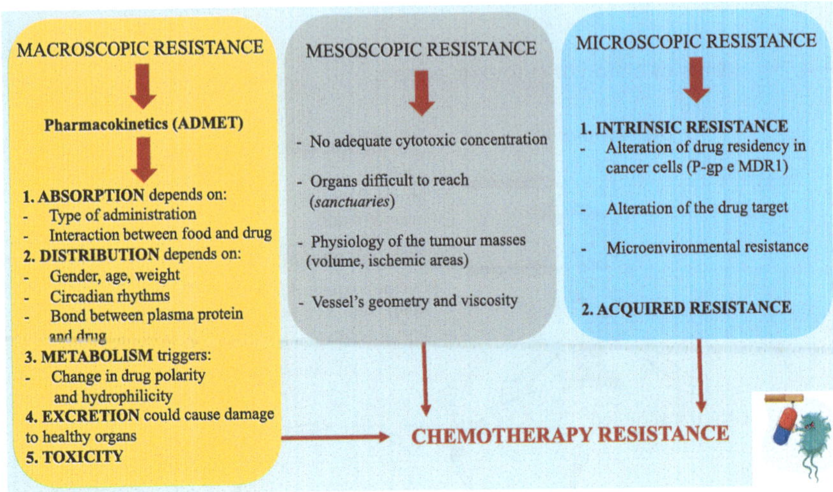


**Fig. 2** Types of drug resistance

and excretion (ADME). Absorption is influenced by the type of administration of the drug and by the interaction between food and antineoplastic drugs. Distribution depends on gender (women are more likely subject to pharmacokinetic variability during their menstrual cycle), age and weight of the patient (during the treatment the patients usually lose weight, so that a dose adjustment is needed). Moreover, drug distribution is also affected by the circadian rhythms and by the bonding between the drug and the plasma protein (albumin and alpha-1-acid glycoprotein). As regards drug metabolism, it mainly triggers a change in the drug polarity and its hydrophilicity. Finally, the excretion of the drug may be controlled and could change during therapy causing damage to healthy organs, for examples kidneys [5].

Resistance to a specific drug occurs when the adequate cytotoxic concentration, for the required amount of time, is not achieved. This mechanism takes place when the neoplastic cells are confined to organs which are difficult for drugs to reach—the so-called *sanctuaries*, such as the central nervous system and the testicles, which are separated by blood barriers. Complications in the therapeutic strategy may add to this process, e.g. an excessively low dose [3]. Also, the physiology of the tumourous masses can induce drug resistance, for instance a tumour of large volume produces vast ischemic areas that can hinder the transport of the drug and the transport of oxygen near the cancer cells. Moreover, the presence of ischemic areas induces the selection of cells with P53 mutations. These cells are not able to undergo apoptosis, which is why they are resistant to many drugs. The geometry of the vessels, the blood viscosity in the vessels near the tumour and the role of the extracellular matrix, which causes impairments in drug diffusion, are important factors involved in the mesoscopic resistance [5]. Cellular resistance can be divided into two clusters: the intrinsic resistance, i.e. resistance developed during the first drug administration, and the acquired or secondary resistance, which is the resistance to a drug that at the beginning of the cure was successful and which may show up if relapse occurs [6, 7]. The microscopic resistance is due to the alteration of drug residency in cancer cells by proteins such as P-glycoprotein 1 (P-gp) and multidrug resistance-associated protein 1 (MDR1), but it is also caused by the alteration of the drug target and by the so-called microenvironmental resistance. Furthermore, this kind of resistance is featured by the concentration of oxygen and glucose, the pH gradient, which is more acidic extracellularly and more alkaline intracellularly, the tumour-tumour or tumour-stromal communication processes and the activity of cytokines and growth factors [2, 5].

A vast majority (approximately 90%) of pharmacological cancer therapies eventually meet resistance to the drug. In particular, chemo-resistance, often called tachyphylaxis, takes place when a cancer which has begun reducing its size due to chemotherapy drugs, starts to grow again. The main cause of the onset of chemo-resistance is that the tumour is a heterogeneous agglomeration of different types of cells derived from one mutated stem cell which has replicated over a number of generations. Two same kinds of tumours will never be molecularly identical, since they are extremely heterogenous. Genetic, epigenetic, transcriptomic and proteomic properties cause the heterogeneity of the tumour. In particular, genotypic factors include mutations, gene amplification, deletions, chromosomal rearrangements, transposition of genetic elements, translocation and microRNA alteration, while epigenetic factors, which are often consequences of genotypic factors, may lead to stochastic variations between cells or to a hierarchical organization of the cell [2]. This makes it hard for pharmacological treatments to affect tumours, because one drug can attack or eliminate only one or few types of cells within the tumour mass [3, 4].

Many solutions have been attempted in an effort to overcome the phenomenon of drug resistance. Usually, patients are given several antineoplastic drugs, following a specific procedure called polychemotherapy. The main aim of a polychemotherapeutic regimen is to avoid the selection of the neoplastic population which is resistant to the drug. Moreover, the clinical outcome of the combination of drugs is generally better than the sum of the effects of the agents employed individually in the monotherapy. We refer to this as synergistic action of drug combinations. Consequently, polychemotherapeutic treatment can provide the same results with lower dosage, thus causing less toxic effects in the organism. However, the main problem with this method is that one drug may counteract or neutralize the effects of another [5] (Fig. 3).

## 3 Mechanisms of Drug Resistance

As stated above, multidrug resistance (MDR) is a phenomenon in which cancer cells exhibit a cross-resistant phenotype against multiple unrelated drugs that are structurally and/or functionally different and may also have different molecular targets [8]. Several host factors are involved in the development of both intrinsic and acquired MDR, including those that impair the delivery of anticancer drugs to the cancer

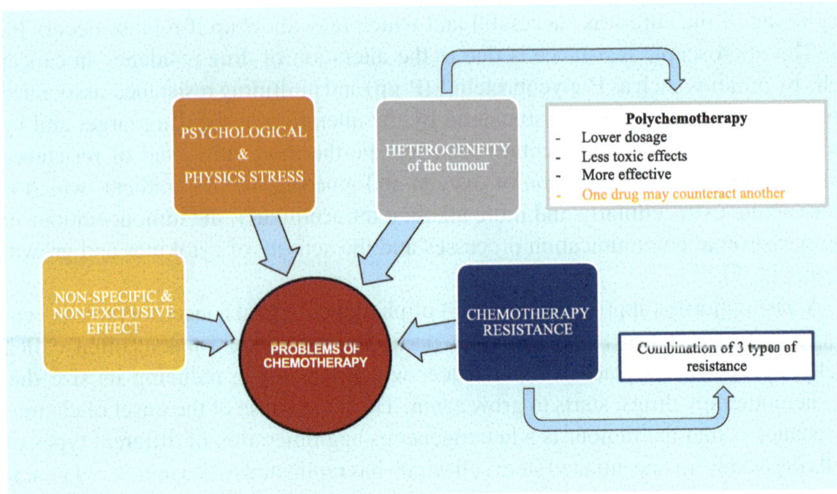


**Fig. 3** Causes of chemotherapy failures

cells and nullify their cytotoxic effects, and those that alter the genetic or epigenetic factors of cancer cells, which lead to drug insensitivity [8]. It is possible to identify four main mechanisms, which are responsible for the drug resistance progress and the possible failure of a particular anti-cancer treatment (see Fig. 4).

The drug efflux mechanism is an over-expression of the biological process whose purpose is to prevent over-accumulation of toxins within the cell. It is probably one

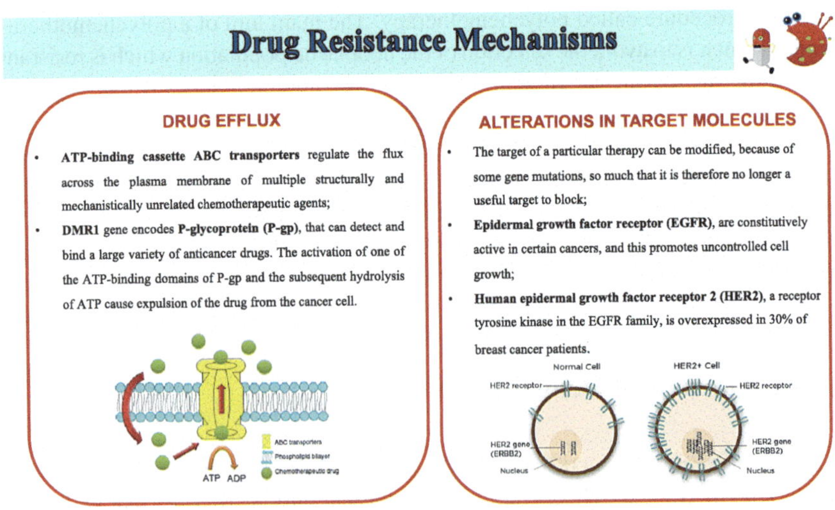


**Fig. 4** Upregulation of ABC transporters on cancer cell membranes effectively removes drugs and cytotoxic agents as a means of drug resistance

of the most significant forms of resistance which reduces the efficacy of the variety of antineoplastic agents currently used, caused by a group of membrane proteins, which extrude cytotoxic molecules, keeping intracellular drug concentration below a cell-killing threshold [1]. These transmembrane proteins are called ATP-binding cassette (ABC) transporters, comprising a family of 49 members in humans, which regulate the flux across the plasma membrane of multiple structurally and mechanistically unrelated chemotherapeutic agents (see Fig. 4).

These transporters have broad substrate specificity and are able to cause efflux from cells of many xenobiotics, including vinca alkaloids, epipodophyllotoxins, anthracyclines, taxanes, and kinase inhibitors. Hence, they protect cancer cells from many chemotherapeutic drugs. An example is MDR1 gene (Multi Drug Resistance 1) which is normally expressed in almost all tissues at low levels, but it becomes overexpressed in many tumours (causing intrinsic drug resistance) and, sometimes, the expression of MDR1 can be induced by chemotherapy (also resulting in the acquired development of MDR) [9]. The MDR1 gene encodes P-glycoprotein (P-gp), a plasma membrane protein, which consists of two ATP binding cassettes and two transmembrane regions. P-gp can detect and bind a large variety of anticancer drugs and other hydrophobic compounds, including anthracyclines, epipodophyllotoxins, vinca alkaloids, and taxanes [8]. This drug binding activity results in the activation of one of the ATP-binding domains of P-gp and the subsequent hydrolysis of ATP, leading to a major change in the shape of P-gp, which causes expulsion of the drug from the cancer cell [8].

In several cases, the target of a particular therapy can be modified to the point that it becomes no longer useful. The cause of these processes is linked to gene mutations, which are very common in cancer cells and can be treated by cytotoxic drugs. These drugs are able to disable a component, whose continuous function is necessary for the cell survival. There are some cells that, surviving the treatment, can carry a gene for that target, which has mutated in such a way that it produces a protein that retains its activity but no longer binds to the drug for stereo-chemical reasons, and is therefore not inhibited by it [1]. This mechanism produces drug resistance. For example, several kinases, such as members of the epidermal growth factor receptor (EGFR) family, are constitutively active in certain cancers, and this leads to uncontrolled cell growth. In most circumstances, mutations cause the over-activation of kinases. However, the same effect may sometimes result from gene over-expression. Human epidermal growth factor receptor 2 (HER2), a receptor tyrosine kinase in the EGFR family, is overexpressed in 30% of breast cancer patients, and drug resistance can result after long term use of inhibitors targeting this kinase. Another example of drug target alteration has been observed in the androgen receptor. In almost 30% of prostate cancers, the androgen receptor is genomically-amplified, thus enabling these cancers to become resistant to androgen deprivation therapy using the drugs leuprolide and bicalutamide. However, these drugs cannot inhibit all molecular targets [3].

The aim of most anticancer drugs is to activate cell cycle arrest or to induce the cellular death pathways inside the cell (see Fig. 5). Apoptosis has two established pathways: one is an intrinsic pathway mediated by the mitochondria, which involves B-cell lymphoma 2 (BCL-2) family proteins, caspase-9 and Akt. The other is an

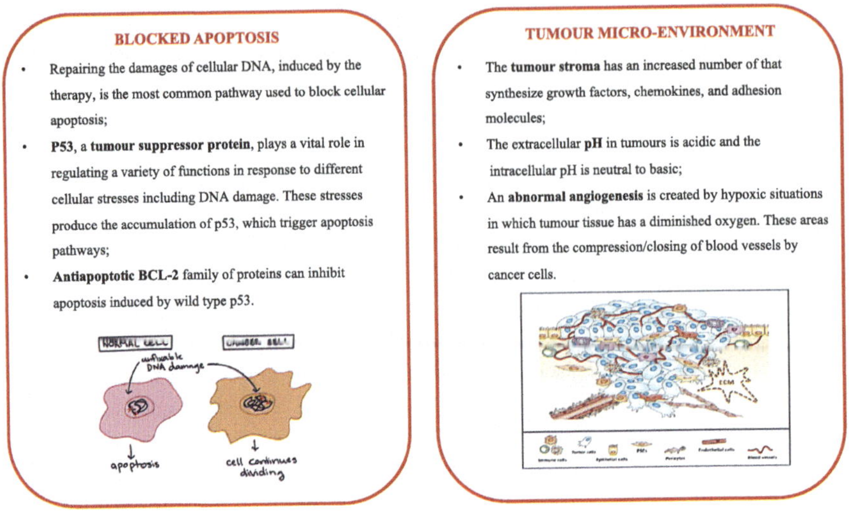


**Fig. 5** Unfixable DNA damage in a normal cell and in a cancer one (left panel) and an illustration of the tumour microenvironment (right panel)

extrinsic pathway that involves the death receptors on the cell surface. The intrinsic and extrinsic pathways merge through the activation of down-stream caspase-3, which eventually causes apoptosis. However, there is also additional cross-talk between these pathways [3]. One of the most common pathways aims to repair the damage of cellular DNA, induced by the therapy.

Many chemotherapeutic drugs induce DNA damage either directly, in case of platinum-based drugs, or indirectly, for example, using topoisomerase inhibitors. Unfortunately, dysregulation or impairment of certain DDR genes and mechanisms either by mutations or epigenetic silencing are common in many cancers. DNA damage induces cell cycle arrest, which has evolved to allow cells time to repair the damage.

In some cancers, the regulation of cell cycle arrest is disrupted owing to gain-of-function alterations to oncogenes and/or loss-of-function alterations to tumour suppressor genes [9]. In this respect, P53, a tumour suppressor protein, plays a crucial role in regulating a variety of functions in response to different cellular stresses, including DNA damage and oncogene activation. Stress causes accumulation of P53, which is responsible for cell cycle arrest in the G1 phase or can trigger apoptosis; the intensity of these processes depends on the extent of DNA damage. Recent studies show how the loss of P53 function in cells with mutated P53 genes induces MDR. An example of this aspect is found in lymphoblastic leukemia, melanoma, osteosarcoma, breast, ovarian, and testicular cancers, which have been reported to express mutant P53. Mutations, amplifications, chromosomal translocations and overexpression of the genes encoding particular proteins have been linked to various malignancies and to the resistance to chemotherapy and targeted therapies. An example of these proteins is given by the anti-apoptotic BCL-2 family of proteins, which can inhibit apoptosis induced by wild type P53. For this reason, an overexpression of BCL-2 proteins results in the resistance of cells to different drugs such as paclitaxel, etoposide, mitoxantrone and doxorubicin [8].

The microenvironment of a solid tumour is composed of cancer cells and stromal cells, including fibroblasts and immune cells, embedded in an extracellular matrix (see Fig. 5). The tumour stroma has a large number of fibroblasts, which synthesize growth factors, chemokines, and adhesion molecules. The interactions between cancer cells and these factors can affect the sensitivity of the cells to apoptosis signals and their response to chemotherapy, and is known as cell adhesion-mediated drug resistance (CAM-DR). Tumour cells can also form polarized three-dimensional structures by means of interactions with the basement membrane and ligation of β4 integrins, which regulate polarity and NF-κB activation. Another important factor is related to the pH of the tumour microenvironment, which can influence the efficacy of cytotoxic drugs and may inhibit their active transport. The extracellular pH in tumours is acidic and the intracellular pH is neutral to basic. Thus, weakly basic drugs, such as doxorubicin, are protonated and have reduced cellular uptake while weakly acidic drugs, such as cyclophosphamide, tend to concentrate in neutral extracellular space. Drug distribution is also affected by the composition and organization of the extracellular matrix. In tumours with a well-organized collagen network, high-molecular weight drugs are not able to penetrate when compared to a poorly organized collagen structure. Further, the tumour microenvironment can create hypoxic situations in which tumour tissue has a diminished oxygen supply contributing to MDR. These areas result from abnormal angiogenesis or from the compression/closing of blood vessels by cancer cells. This reduced blood flow may lower the concentrations of chemotherapeutics in hypoxic cells. In addition, hypoxia can lead to the activation of genes associated with angiogenesis, survival, and glycolysis through the transcription factor hypoxia-inducible factor 1 (HIF-1) and may contribute to the emergence of a drug-resistant phenotype [10].

## 4 Differential Equation Models Used to Represent Cellular Heterogeneity

An ordinary differential equation (ODE) contains one or more functions of an independent variable and its derivatives. It is important to underline the presence of a single independent variable because it justifies the fact that if these models are used to describe a cell population, spatial discretization must be introduced. Specifically, the heterogeneity of a cell population varies over time and according to a phenotypic variable, if in the model the time is assumed as an independent variable, the only way to introduce heterogeneity is through spatial compartments within the phenotypic variable, which remains constant. In this way models with spatially-structured population are introduced [11]. The use of this compartmentalized space does not

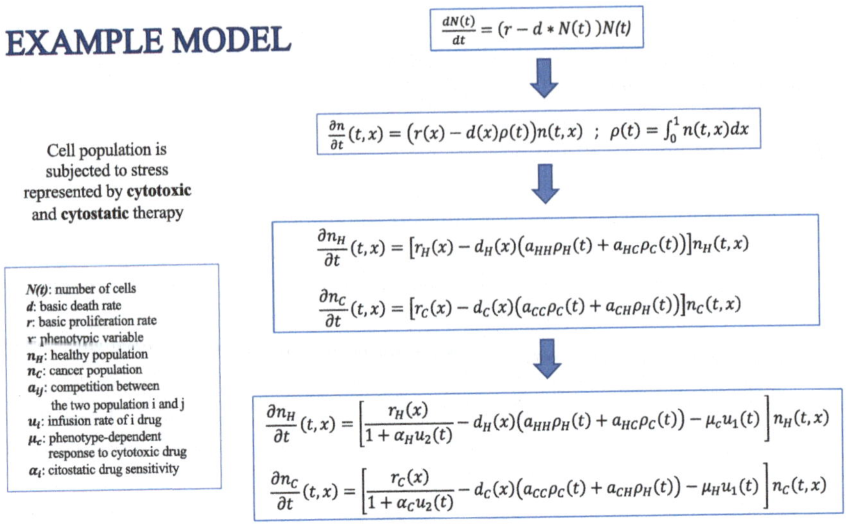


**Fig. 6** The model shown here is based on ODE equations and represents a cell population with phenotypic heterogeneity and genetic homogeneity [13]

best represent the physiological reality but can be applied to represent the resistance to cancer cells, as is described below in the analysis of ODE models (see Fig. 6).

The integro-differential equation (IDE) models introduce a phenotypic variable that describes the phenotypic heterogeneity in a continuous way, and no longer constant in each spatial compartment as in the ODE models. Differential terms differ from the ODE models only in the integral term that represents non-local interactions, for example the interaction of a cell with the other cells of the population, whatever their phenotype [11]. Such models represent a population of cancer cells exposed to cytotoxic drugs with the goal to investigate how intra-tumour heterogeneity affects multi-drug resistance [12]. Cancer cells were assumed to experience uniform exposure to a cytotoxic drug and drug resistance of a cell was assumed to represent small changes between cell generations through genetic or epigenetic mechanisms. The net growth rate of the cell during the therapy was represented by the level of resistance and the total size of the cell population. The numerical simulations performed made it possible to understand that therapy acts as a selection process that causes the expansion of resistant clones and the increase heterogeneity in cancer cell population. The authors thus proposed that treatment which reduces phenotype alteration rates may improve targeted therapy [12].

Evolution of an ODE model to an IDE model is often used to study drug resistance. The main objective is to start from a simple model and gradually progress by better representing the physiological reality of drug resistance in cancer cells.

The cancer cell population is subjected to stress represented by cytotoxic and cytostatic therapy. The evolution over time of the number of cancer cells is described by the function N(t) with an initial condition $N_0$. The factor *r* is the basic proliferation rate, also called *net selection rate* [13]. The term d*N(t), called a *logistic term* [13], indicates an additional death rate directly proportional to the number of cells. A fundamental hypothesis is the non-exponential proliferation of cancer cells due to competition for space and nutrients in the area of solid tumour development. Now, by applying changes to the previous model it is possible introducing a phenotype-structured model with a phenotypic variable representing phenotypic heterogeneity. This leads to a passage from an ODE model to an IDE model. The number of cells is represented by $n(t, x)$ at time $t$ and phenotype expression level $x$ that varies between 0 and 1 and defines the resistance to a given drug. The model also introduces $\rho(t) = \int^1 n(t, x)x$ where $\rho(t)$ sums all cells with different phenotypes. Net selection rate is a function of the phenotype variable and the logistic term depends both on time and phenotype x. The model is often used in the adaptive dynamics of the cell population to describe the selection phenomenon, so only cells with certain phenotypes can survive. This elementary model is based on two main concepts: the convergence of $\rho(t)$ to maximum value and the density of cells $n(t, x)$ with a given phenotype x. The areas of maximum concentration are the zones of the phenotypic variable that resistance to drug is found. This reasoning can be interpreted as the convergence of $n(t, x)$ to a sum of Dirac delta functions located on resistant phenotypes.

To better represent the physiological reality, it is good to modify the model by introducing two populations of cells, one healthy and one cancer [13]:

$$\frac{\partial n_H}{\partial t}(t, x) = [r_H(x) - d_H(x)(a_{HH}\rho_H(t) + a_{HC}\rho_C(t))]n_H(t, x)$$

$$\frac{\partial n_C}{\partial t}(t, x) = [r_C(x) - d_C(x)(a_{CC}\rho_C(t) + a_{CH}\rho_H(t))]n_C(t, x)$$

where:

$$\rho_H(t) = \int_0^1 n_H(t, x)dx, \ \rho_C(t) = \int_0^1 n_C(t, x)dx$$

These equations incorporate terms, which quantify the competition between the two populations for proliferation and survival ($a_{HC}$, $a_{CH}$). The competition exists both between cells of different types and between cells of the same type since we find the terms: $a_{HH}$ and $a_{CC}$. The cells compete more with those of the same type rather than with the cells of the other type, so we assume: $a_{HC} < a_{HH}$ and $a_{CH} < a_{CC}$ [13].

The current model describes exhaustively the time course of the healthy and cancerous populations, but to better represent the effect of the drugs to which they are subjected it is useful to make new changes to the model. Hence infusion rates of the cytotoxic drug $u_1$ and cytostatic drug $u_2$ are therefore introduced [13].

$$\frac{\partial n_H}{\partial t}(t,x) = \left[\frac{r_H(x)}{1+\alpha_H u_2(t)} - d_H(x)(a_{HH}\rho_H(t) + a_{HC}\rho_C(t)) - \mu_C u_1(t)\right] n_H(t,x)$$

$$\frac{\partial n_C}{\partial t}(t,x) = \left[\frac{r_C(x)}{1+\alpha_C u_2(t)} - d_C(x)(a_{CC}\rho_C(t) + a_{CH}\rho_H(t)) - \mu_H u_1(t)\right] n_C(t,x)$$

where $\mu_{[H,C]}$ represents the phenotype-dependent response to the cytotoxic drug, with concentration $u_1(t)$, and $\alpha_{[H,C]}$ represents the sensitivities to drug $u_2$.The cytotoxic drug kills the cell by increasing the death term, while the cytostatic drug only slows down proliferation, but does not arrest it. The goal of this model is to minimize $\rho_C(t)$ in a therapeutic time-window [0, T]. The model must simultaneously meet these requirements:

1. Do not exceed the maximum tolerable dose $0 \le u_1(t) \le u_1$, $0 \le u_2(t) \le u_2$
2. Prevent the growth of too large a solid tumour $\rho_H(t)/[\ \rho H(t) + \rho_C(t)] \ge \theta_{HC}$ where $\theta_{HC}$ [14] denotes the $\rho_H(t) + \rho_C(t)$ proportion of division with mutations.
3. Avoid side effects on the healthy cell population $\rho_H(t) \ge \theta_H\ \rho_H$.

The model described here allows to study the effects of different drug doses on cell populations. For example, if we consider a high dose administration for a long time, we do not respect the last two constraints of the model and so we do not arrive at an effective treatment.

## 5 Stochastic Models

In order to account for rapid mutations occurring in realistic analysis of tumour heterogeneity, stochastic models involving probabilities instead of continuous variables have been developed. This includes probabilities of cell division, cell death and mutation events. Figure 7 shows a schematic description of such stochastic models. Figure 8 demonstrates difference between single- and multiple-drug resistance models. In Fig. 9 we show possible mutation states of cell phenotype depending on the drug resistance effect.

The probability that an event happens depends on the treatment applied. We denote two parameters: the growth rate of cancer cells (*L*) and their death rate in the absence of the drugs (*D*). Hence, it is possible to observe a clonal expansion in the absence of treatment when L > D. In general, the *turnover* of cancer cells is the ratio of the natural death rate and the proliferation rate in the absence of treatment, that is $0 \le D/L < 1$. If $D = 0$ or $L \gg D$, a low-turnover cancer case occurs that defines a low-death cancer (almost no cell death), while $D \approx L$ describes high-turnover, such as slow-growth cancers. Also introducing the effect of drug-induced death rate, we obtain $D + H > L$ that causes an imbalance that leads to the colony reduction (with the net cell death rate being larger than the birth rate). This parameter is important to estimate the drug resistance, as well as the intensity of the therapy. In fact, if a cell is resistant to all drugs applied, then its drug-induced death rate is simply

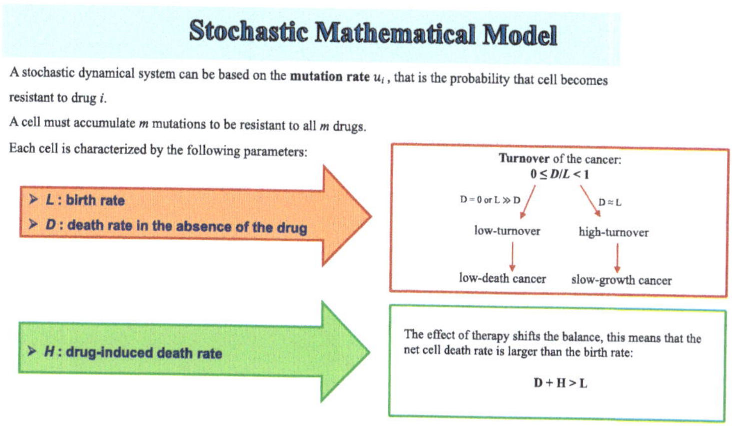


**Fig. 7** A summary of the assumptions used in stochastic models of drug resistance in tumours

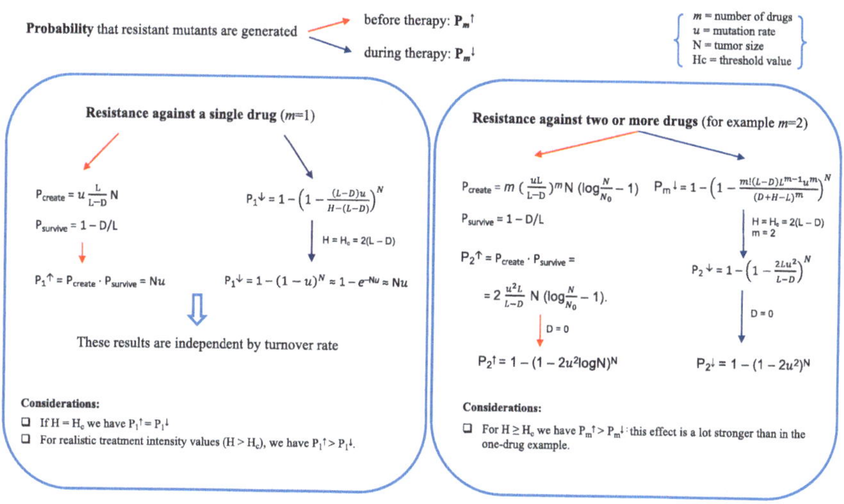


**Fig. 8** Differences between mathematical models of single drug and multiple drug resistance

zero. On the other hand, if a cell is susceptible to at least one of the drugs, then its drug-induced death rate is equal to a constant, $H$, which quantifies the intensity of the therapy [15]. In the case in which the cell is resistant to a single drug ($m = 1$) and not to all others used. We now focus on mutant's creation: it is possible to distinguish the probability that this happens before the start of treatment (P$m\uparrow$) and during therapy (P$m\downarrow$). Another important parameter is the size of the tumour. It is difficult to estimate the age of cancer at the beginning of therapy but is possible to measure the size. In particular, the probability that resistant mutants pre-exist before treating a tumour of size $N$, defined as "treatment size", has been investigated in Ref. [16]. Figure 10 plots the tumour size N as a function of the death rate D for various

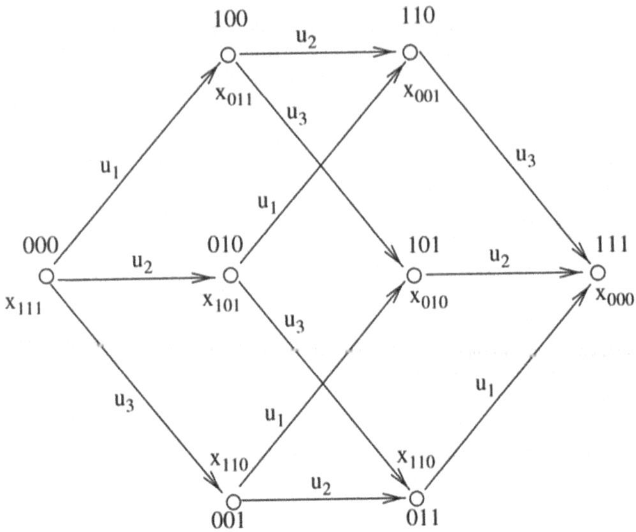


**Fig. 9** Mutation diagram corresponding to three drugs [16]. Each node is a cellular type (phenotype) identifies by a binary number of length *m*. It describes the resistance properties: "0" means *susceptible* to the drug corresponding to its position and "1" means *resistant*. An example is 101: this phenotype is resistant to drug 1 and 3 but not to drug 2

**Fig. 10** Plot of the probability of producing resistant mutants before treatment, depending on the death rate of tumour cells, D [16]

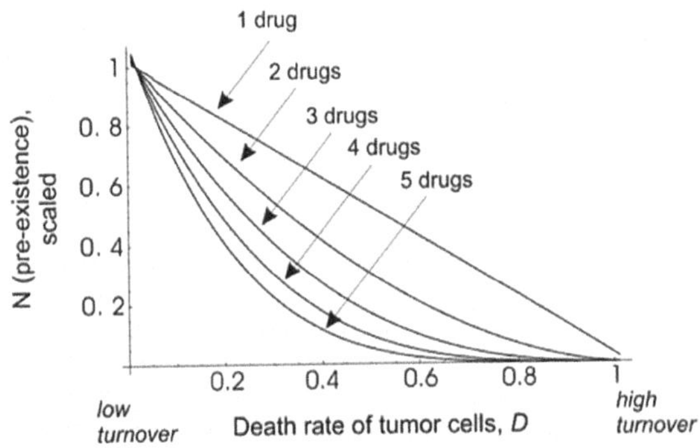


numbers of chemotherapy drugs used, m. For $m = 1$ this is linear, instead for $m = 2$ and higher, this dependence becomes increasingly strongly nonlinear. This means that, if the number of drugs increases, the higher natural death rate of tumour cells causes a higher production of the resistant mutants and thus the treatment will likely fail.

Below we list our final considerations in this area:

- The dependence on the turnover rate is very strong in the pre-treatment phase: for high values of the ratio D/L, the probability that the resistant mutants are generated before the therapy begins is very high.

- For H = Hc and H > Hc we have Pm↑ > Pm↓ again, which indicates the role of the pre-existing resistance, in addition to the fact that this effect is a lot stronger than in the single-drug case. It indicates that the pre-treatment phase always plays an important role in treatment failure.

In conclusion, resistance arises mainly before the start of treatment and, for cancers with high turnover rates, combination therapy is less advantageous over single-drug therapy. In contrast, for low-turnover rates, increasing the number of drugs will increase the success of therapy.

The methods described above can be applied to cancers treated with targeted therapy. An example of this is CML (chronic myeloid leukemia) because blast crisis corresponds to the clonal expansion processes described by the above model, and drug activity, as well as resistance mechanisms, are well defined [16]. The treatment is developed in three phases: the first is the chronic phase, characterized by expansion of terminally differentiated cells; the second is the accelerated phase, in which there is an increase of undifferentiated cells; finally, the blast crisis, where undifferentiated cancer cells undergo large expansion caused by genomic instability. The drug used to treat CML, imatinib mesylate (Gleevec), is not mutagenic and it is a small-molecule inhibitor of the BCR-ABL kinase [16]. The resistance to non-mutagenic drugs is mostly generated before treatment. In fact, treatment of blast crisis often fails because of drug resistance. Instead, using Gleevec at the start of cancer progression is usually successful. Simulations invovling biological parameters of the phenomenon of genetic instability [25] were based on experiments with susceptible CML cell lines with a measured viability at about 90% [26]. Komarova and Wodarz [16] started from the range of D/L = 0.1 – 0.5 and considered that the size of advanced cancers was $<10^{13}$ cells [27]. They proposed the use of different values of *D/L* and *u* to analyze various treatment options and found that a combination of three drugs should prevent resistance and ensure successful therapy even for advanced cancers. As long as the point mutation rate is elevated less than 100-fold by BCR-ABL, triple drug therapy should prevent resistance [16].

## 6 Methods to Overcome Multidrug Resistance

Many efforts have been made during the past several decades to enhance the efficacy of chemotherapy by suppressing or evading MDR mechanisms. Several strategies are now available (see Fig. 11) including the use of new anticancer drugs that could escape from the efflux reaction, MDR inhibitors, multifunctional nanocarriers, and RNA interference (RNAi) therapy [8].

**MDR Inhibitors**: A method to overcome resistance to anticancer drugs is to administer compounds that would not be toxic themselves but would inhibit ABC transporters [17]. These MDR modulators have been classified as first, second and third generation MDR modulators based on their affinity for transporter proteins and their

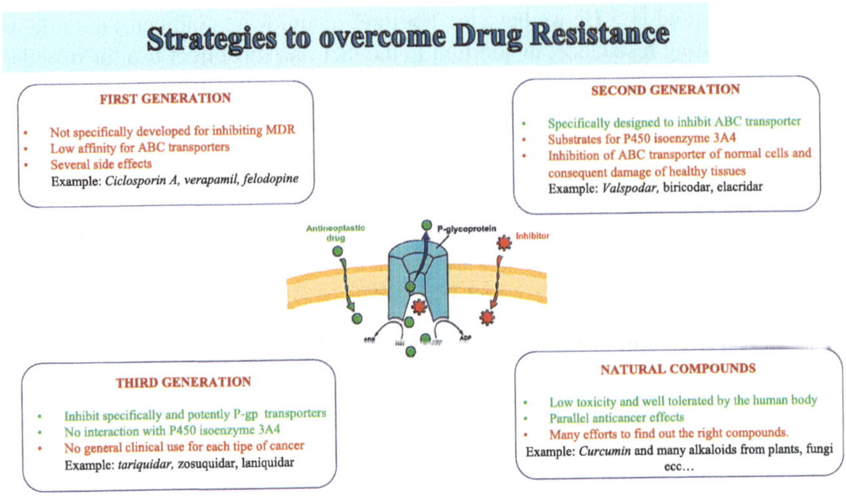


**Fig. 11** A summary of the strategies to overcome drug resistance

relative side effects [8]. First-generation MDR drugs were not specifically developed for inhibiting MDR and their affinity was low for ABC transporters and, consequently, they led to several side effects and a high dose of the cancer drug was needed to produce sufficient intracellular concentrations [17]. Among them, cyclosporin A, a commonly used immunosuppressant, remains one of the most effective first generation MDR modulators [17]. Figure 12 shows 2D and 3D structures of the compounds discussed in this section.

Second-generation chemosensitizers were designed to reduce the side effects of first- generation drugs. However, they still retain some characteristics that limit their clinical usefulness [17]. Many of the anticancer drugs are metabolized by the cytochrome P450 isoenzyme 3A4, but most of the second-generation MDR chemosensitizers are also substrates for cytochrome P450 3A4 and this could inhibit the normal metabolism of the anticancer drug and result in unpredictable pharmacokinetic interactions [8]. Indeed, co-administration of an MDR modulator usually elevates plasma concentrations of an anticancer drug by interfering in its metabolism and excretion, thus leading to unacceptable toxicity that necessitates chemotherapy dose reductions in the clinical setting down to pharmacologically ineffective levels. Another problem with second-generation MDR modulators is their specificity. In fact, they do not just inhibit some of the most common ABC transporters present in cancer cells, but they are substrates for almost the entire ABC transporter family [17]. However, ABC transporters have well- defined physiological roles, often involving the elimination of xenobiotics, in regulating the permeability of the central nervous system (blood–brain barrier) and the placenta, thus preventing these systems from being exposed to cytotoxic agents circulating in the blood. Inhibition of these transporters could lessen the ability of normal cells and tissues [17]. Valspodar (PSC 833)

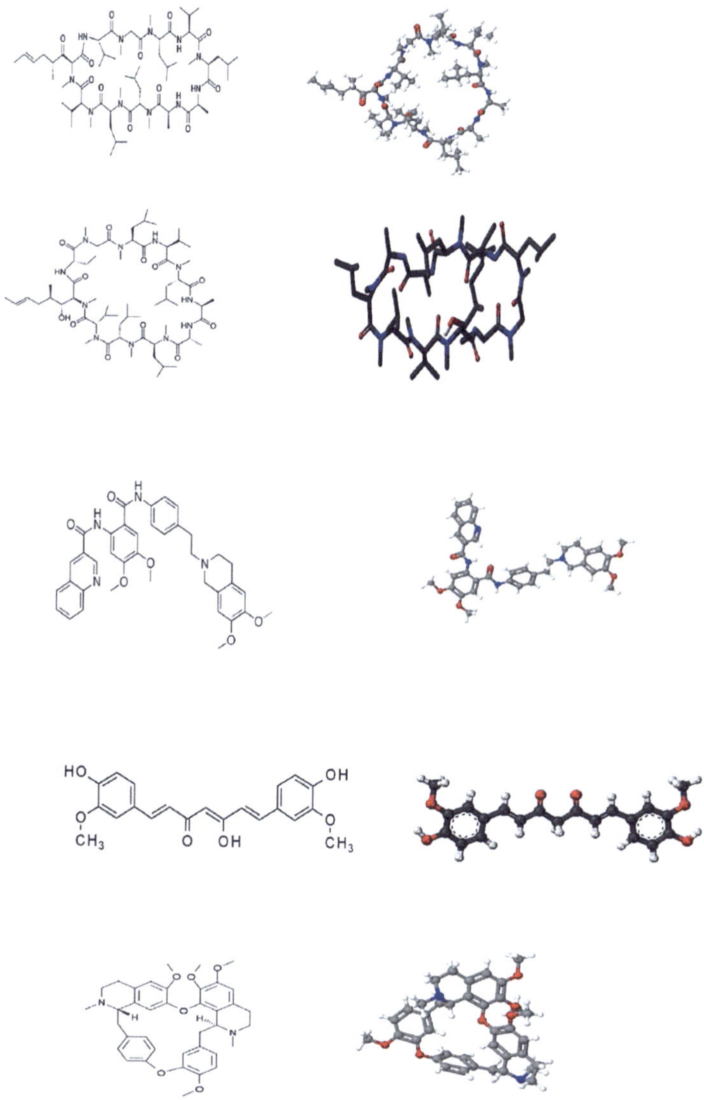

**Fig. 12.** 2D and 3D chemical structures of (from top to bottom): cyclosporin, valspodar, tariquidar, curcumin, and tetrandine

is an example of second generation chemosensitizer. It is a derivative cyclosporin D and it is approximately 5-to 30-fold more potent than cyclosporine A in vitro [8].

Third-generation molecules have been developed to overcome the limitations of the second generation MDR modulators. They are not metabolized by cytochrome P450 3A4 and they do not alter the plasma pharmacokinetics of anticancer drugs.

Third-generation agents specifically inhibit P-gp and do not inhibit other ABC transporters. Because of their specificity for P-gp transporters and lack of interaction with cytochrome P450 3A4, third-generation P-gp inhibitors offer significant improvements in chemotherapy without a need for chemotherapy dose reductions [18]. Clinical trials with these new third-generation agents are ongoing with the aim for survival prolongation in cancer patients. None of them has found a general clinical use so far [17]. One of the most promising third-generation P-gp inhibitors is tariquidar, which binds specifically and non- competitively to the P-gp pump with a high affinity and potently inhibits the activity of the P-gp transporter [17]. Recently, many promising strategies have emerged using natural compounds to overcome MDR. These modulators are low in toxicity and are well tolerated in the human body [8].

For example studies on curcumin revealed that it has many anticancer properties mainly due to its ability to inhibit the transcriptional nuclear factor kappa beta (NFκβ), which is a master regulator of inflammation, cell proliferation, apoptosis, and multidrug resistance in cancer cells [19]. There are also many natural alkaloids possessing potent inhibition of P-gp efflux pump and other related pumps responsible for the development of resistance. For example, recently a natural alkaloid, tetrandrine (CBT-01) has advanced to the clinical phase [20] and provided an impetus for the discovery of more of such small molecule natural products as fourth generation P-gp inhibitors [20].

**RNA interference therapy (RNAi)** RNAi is a biological process that cells use to inhibit or silence specific gene expression through the destruction of specific mRNA filaments triggered by RNA molecules [8]. The process starts when a double-stranded RNA molecule (dsRNA) is recognized and cleaved by the enzyme into shorter fragments of 21–23 nucleotides, called siRNAs. These siRNAs assemble with protein components forming RNA-induced silencing complexes (RISC) and here the dsRNA unwinds, removing the sense strand (see Fig. 13). The activated RISC containing only a single-stranded (antisense) siRNA binds to the target mRNA. A component of RISC called Argo cleaves the target mRNA that is subsequently destroyed by the cell, thereby preventing it from being used as a translational template and silencing the expression of the gene from which the mRNA was transcribed [8]. This natural process can be used for regulation of genes of choice and in the case of cancer to downregulating the expression of P-gp or alternative ABC transporters in various tumours [21]. RNAi can be triggered by two different pathways: (1) an RNA-based approach where the effector siRNAs are delivered to target cells or (2) a DNA-based strategy in which the siRNA effectors are produced by intracellular processing of longer RNA hairpin transcripts [21]. The DNA-based approach is mainly involved in the nuclear synthesis of short hairpin RNAs (shRNAs). Expression of shRNAs in cells can be achieved via the delivery of plasmids or through viral or bacterial vectors. Once the vector has integrated into the host genome, the shRNAs are transcribed in the nucleus, which is transported to the cytoplasm and are processed into siRNAs by Dicer [8]. Although RNAi therapy offers a very promising strategy for minimizing the MDR effect, it is not easy to deliver siRNAs to tumour cells via systemic administration. In fact, naked siRNAs is rapidly degraded by serum

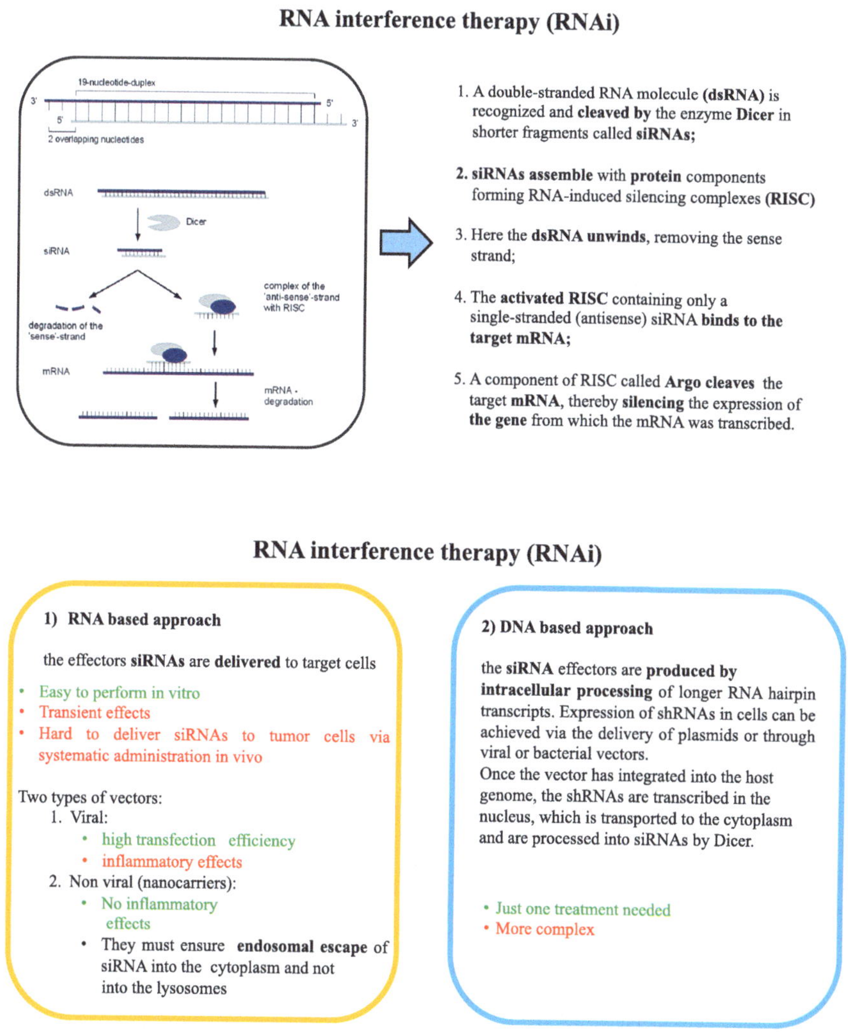


**Fig. 13** The mechanism of RNA interference in mammalian cells

ribonucleases and moreover siRNAs hardly cross the cell membrane because of their structure. For these reasons different vectors have been developed in order to enhance the efficacy of RNAi therapy in vivo [22]. Some of them are based on the use of viral carrier that have a high transfection efficiency, but they enter into the cytoplasm and not into the lysosomes where the environment is too acidic [8].

**Nanoparticles as Drug Delivery System**: Strategies for circumventing MDR by using a Drug Delivery System (DDS) have the potential for serving as an innovative and promising alternative to conventional small-molecule chemotherapeutics, by encapsulating and conjugating drug molecules within a nanocarrier [23].

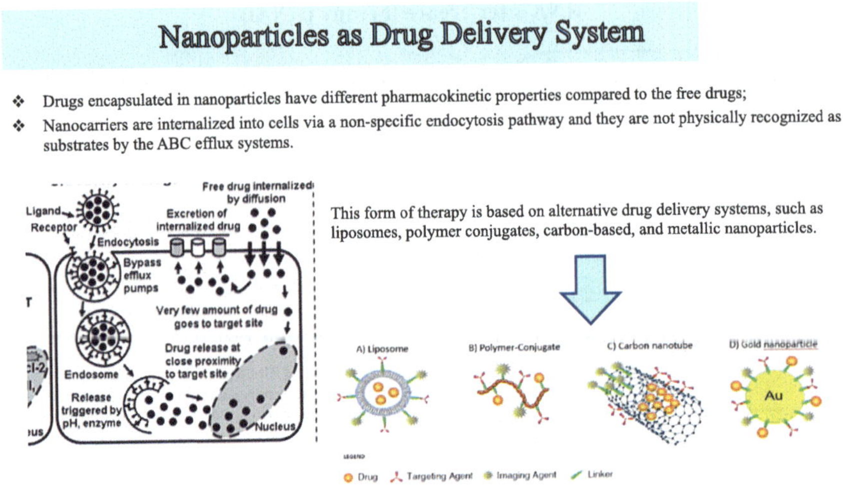


**Fig. 14** Schematic illustration of the endocytosis of nanoparticles loaded with an active drug [23]

Drugs encapsulated in nanoparticles (Fig. 14) have different pharmacokinetic properties compared to free drugs. Generally, the drug efflux pumps present on the cell membrane can sense free drug molecules as they cross the membrane and prevent them from entering the cell, with their subsequent ejection [23].

To overcome this problem due to the drug efflux pumps and to increase the drugs' efficacy, the latter can be loaded or encapsulated in nanocarriers and then delivered to the cellular internal organelles. Nanocarriers are internalized into cells via a non-specific endocytosis pathway and cross the cell membrane in an 'invisible' form. This is because the encapsulated drugs are not physically recognized as substrates by the ABC efflux systems. After crossing the efflux containing membranes on endosomal delivery pathway, the free chemotherapeutics are released into the perinuclear region of the cytoplasm and can unfold their activities. As an example, a novel polymeric micelle consisting of doxorubicin and two block copolymers, one conjugated to TAT, has been produced. The micelle surface hides the TAT during circulation and only exposes it at a slightly acidic tumour extracellular pH to allow for TAT-induced internalization into cancerous cells. The micelle core then disintegrates in the early endosomal pH of the cells to release doxorubicin. Further, the ionization of the block copolymers aids in disrupting the endosomal membrane, allowing the drug to accumulate in the cytosol [10].

Thus, nanomedicine is an emerging form of therapy aimed at improving the treatment efficacy and reducing the adverse side effects to normal tissues. Currently, the knowledge base of nanoparticles is still expanding rapidly with an emphasis on safety and efficacy [10]. This form of therapy is based on alternative drug delivery systems, such as liposomes, polymer conjugates, carbon-based, and metallic nanoparticles.

1. Liposomes are lipid-based vesicles that have the ability to carry payloads in either an aqueous compartment or embedded in the lipid bilayer; generally

based on passive targeting especially on the enhanced permeability and retention (EPR) effect, for which a leaky tumour vasculature is necessary. One example of liposomes, modified with monoclonal antibodies, as suitable carrier for targeted delivery of chemotherapeutic drugs is mAb 2C5 with doxorubicin (Doxil®), which is in the preclinical phase, whereas others are in clinical trials [10].

2. Polymeric nanoparticles can either covalently attach to or encapsulate therapeutic payloads. Their solid cores, ideal for hydrophobic drugs, are highly stable, have a relatively uniform size and are able to control the drug release. Currently, two polymers, polylactide (PLA) and poly(lactideco-glycolide) (PLGA), are polymeric biodegradable nanoplatforms that are used for synthesis of FDA-approved nanomedicines, whereas many others are undergoing clinical trials [10].
3. Carbon nanotubes have the ability to enter cells using "needle-like penetration" [10] and transfer molecules into the cytoplasm. However, toxicity concerns remain a major challenge for their clinical development.
4. Gold nanoparticles can be used to deliver small molecules such as proteins, DNA, or RNA. Drugs can easily be attached through ionic or covalent bonds, or through adhesion. To gain stability and raise circulation time PEG can be attached to the surface of metallic nanoparticles. Magnetic fields can also be used to guide the drug to the intended target area within the body. Unfortunately, their potential clinical use is not presently possible due to the acute in vivo toxicity.

There are two types of targeting: the nanoparticles that depend on the characteristics of the tumour for drug accumulation are passively targeted, and those conjugated to therapeutic antibodies, as well as targeting antibodies such as 2C5 are considered to be actively targeted. Both passive and active tumour targeting abilities of nanoparticles can reduce systemic toxicity and potentially circumvent the problem of drug resistance [8].

Nanodrugs have a better accumulation profile within a tumour than free drugs, including paclitaxel, doxorubicin, and many others [10]. Active tumour targeting is achieved by conjugating nanocarrier systems with various types of ligands that are specific to receptors overexpressed in cancer cells as compared to normal cells. The actively targeted nanoparticles improve efficacy, increase binding affinity to the cancer cells, reach higher accumulation of drug within the tumour rather than other vital organs, avoid toxicity to normal tissue, and let the drug overcome MDR. To this end, different agents can be conjugated with the nanoparticles, such as monoclonal antibodies (mAbs), e.g. rituximab (Rituxan) for non-Hodgkin's lymphoma, the anti-HER2 trastuzumab (Herceptin), the anti-VEGF bevacizumab (Avastin) to inhibit angiogenesis, and the anti-EGFR cetuximab [10]. However, they may also bind to the receptors on normal cells causing an activated signaling cascade that may result in increased immunogenicity.

Aptamers are considered to have high affinity and low immunogenicity but lack flexibility and in vivo nuclease stability, hence are still in preclinical phase.

Receptor ligands (peptides) have a small size, ease of synthesis, and typical non-immunogenicity. Tumour homing peptides include those with an RGD sequence motif, a binding motif for integrins, which is specifically expressed on tumour endothelia, and those that have a form of aminopeptidase N (CD13) that binds peptides with the NGR motif. Delivery of TNFα using both RGD and NGR peptides has shown to decrease the effective dose by up to 1000-fold [10]. Further, these targeting agents play a role in tumour penetration, as is the case with iRGD, because of the exit from blood vessels, and this represents a major advantage in treating resistant tumours. However, peptides such as RGD can bind to other integrins on normal tissue making specificity of the peptide a crucial consideration.

**Anti-angiogenic therapy**: Without the formation of new blood vessels, a tumour cannot grow larger than about 1–2 $mm^3$ [23]. Through a complex process called angiogenesis, endothelial cells are able to divide and grow in order to create new blood vessels. Anti-angiogenesis is the process of stopping the formation of new blood vessels, and, consequently, of stopping the tumour's growth or spread. Currently, there are more than 20 angiogenesis inhibitors being tested on a variety of cancers in clinical trials. Some of them are available commercially and approved by the FDA for other uses. Other anti-angiogenesis drugs are not yet approved by the FDA and can only be given to patients enrolled in clinical trials. However, it is too early to tell if anti-angiogenesis drugs will damage healthy blood vessels that may be needed elsewhere in the body. The benefits and risks of anti- angiogenesis drugs will be determined through clinical trials over the next several years.

## 7 Conclusions

As stated above, the different approaches to overcoming drug resistance have both advantages and limitations. Among them, nanocarriers offer particular advantages such as the ability to bypass drug efflux mechanisms via ABC transporters and can also be conjugated with ligands that can specifically target cancer cells with overexpressed receptors, presenting a powerful and versatile therapeutic platform. Most efforts in this field have focused on P-gp. Such studies have increased the awareness about the complexity of drug resistance in patients, which is consistent with our knowledge that multiple ABC transporters and MDR mechanisms exist, and may contribute to this failure. Despite many clinical setbacks in reversing MDR there are reasons to be optimistic. Indeed, there are already available drugs which could act to inhibit multiple ABC transporters and to deliver specific MDR reversing agents to target sites.

RNAi has evolved from a powerful laboratory tool to elucidate the function of novel genes, to a potential new therapeutic modality against cancer. It provides the ability to silence virtually any gene with artificial triggers of RNAi, producing favorable effects in vitro and in preclinical animal models, although the translation of such findings to the clinical setting remains a serious challenge. As we begin to better understand the mechanisms by which RNAi regulates gene expression, we can hope to exploit this powerful tool as an adjunct in the multimodal therapy. Moreover, through nanomedicine it has been possible to increase circulation time, refine multiple targeting mechanisms, enhance drug accumulation at the tumour site, deliver into the cytoplasm or nuclei of cancer cells, and to carry combinations of therapeutic payloads. For these reasons this can be considered an attractive treatment option in overcoming MDR. Numerous unique nanodrugs have been created and researched extensively, and are already in clinical development. Computational modeling efforts have allowed for treatment optimization and can further assist in improving clinical outcomes in the future [28].

## References

1. Luqmani, Y.A.: Mechanisms of drug resistance in cancer chemotherapy. Med. Princ. Pract. **14**(Suppl. 1), 35–48 (2005)
2. Mansoori, B., Mohammadi, A., Davudian, S., Shirjang, S., Baradaran, B.: The different mechanisms of cancer drug resistance: a brief review. Adv. Pharm. Bull. **7**(3), 339 (2017)
3. Housman, G., Byler, S., Heerboth, S., Lapinska, K., Longacre, M., Snyder, N., Sarkar, S.: Drug resistance in cancer: an overview. Cancers **6**(3), 1769–1792 (2014)
4. Takimoto, C.H., Calvo, E.: Principles of oncologic pharmacotherapy. Cancer management: a multidisciplinary approach **11**, 1–9 (2008)
5. Alfarouk, K.O., Stock, C.M., Taylor, S., Walsh, M., Muddathir, A.K., Verduzco, D., Rauch, C.: Resistance to cancer chemotherapy: failure in drug response from ADME to P-gp. Cancer Cell Int. **15**(1), 1–13 (2015)
6. Raguz, S., Yagüe, E.: Resistance to chemotherapy: new treatments and novel insights into an old problem. Br. J. Cancer **99**(3), 387–391 (2008)
7. Rebucci, M., Michiels, C.: Molecular aspects of cancer cell resistance to chemotherapy. Biochem. Pharmacol. **85**(9), 1219–1226 (2013)
8. Saraswathy, M., Gong, S.: Different strategies to overcome multidrug resistance in cancer. Biotechnol. Adv. **31**(8), 1397–1407 (2013)
9. Holohan, C., Van Schaeybroeck, S., Longley, D.B., Johnston, P.G.: Cancer drug resistance: an evolving paradigm. Nat. Rev. Cancer **13**(10), 714–726 (2013)
10. Markman, J.L., Rekechenetskiy, A., Holler, E., Ljubimova, J.Y.: Nanomedicine therapeutic approaches to overcome cancer drug resistance. Adv. Drug Deliv. Rev. **65**(13–14), 1866–1879 (2013)
11. Chisholm, R.H., Lorenzi, T., Clairambault, J.: Cell population heterogeneity and evolution towards drug resistance in cancer: biological and mathematical assessment, theoretical treatment optimisation. *Biochimica et Biophysica Acta (BBA)-General Subjects, 1860*(11), 2627–2645 (2016)
12. Lavi, O., Greene, J.M., Levy, D., Gottesman, M.M.: The role of cell density and intratumoral heterogeneity in multidrug resistance. Can. Res. **73**(24), 7168–7175 (2013)
13. Almeida, L., Chisholm, R.H., Clairambault, J., Lorenzi, T., Lorz, A., Pouchol, C., Trélat, E.: Why is evolution important in cancer and what mathematics should be used to treat cancer? Focus on drug resistance. Trends Biomath. Model. Optim. Comput. Prob. 107–120 (2018)
14. Lorz, A., Lorenzi, T., Hochberg, M.E., Clairambault, J., Perthame, B.: Populational adaptive evolution, chemotherapeutic resistance and multiple anti-cancer therapies. ESAIM: Math. Modell. Numer. Anal. 47(2), 377–399 (2013)
15. Komarova, N.: Stochastic modeling of drug resistance in cancer. J. Theor. Biol. **239**(3), 351–366 (2006)

16. Komarova, N.L., Wodarz, D.: Drug resistance in cancer: principles of emergence and prevention. Proc. Natl. Acad. Sci. **102**(27), 9714–9719 (2005)
17. Ozben, T.: Mechanisms and strategies to overcome multiple drug resistance in cancer. FEBS Lett. **580**(12), 2903–2909 (2006)
18. Lee, C.H.: Reversing agents for ATP-binding cassette (ABC) transporters: application in modulating multidrug resistance (MDR). Curr. Med. Chem. (2004)
19. Tomita, M., Kawakami, H., Uchihara, J.N., Okudaira, T., Masuda, M., Takasu, N., Mori, N.: Retracted: Curcumin (diferuloylmethane) inhibits constitutive active NF-κB, leading to suppression of cell growth of human T-cell leukemia virus type I-infected T-cell lines and primary adult T-cell leukemia cells. Int. J. Cancer **118**(3), 765–772 (2006)
20. Joshi, P., Vishwakarma, R.A., Bharate, S.B.: Natural alkaloids as P-gp inhibitors for multidrug resistance reversal in cancer. Eur. J. Med. Chem. **138**, 273–292 (2017)
21. Aagaard, L., Rossi, J.J.: RNAi therapeutics: principles, prospects and challenges. Adv. Drug Deliv. Rev. **59**(2–3), 75–86 (2007)
22. Pai, S.I., Lin, Y.Y., Macaes, B., Meneshian, A., Hung, C.F., Wu, T.C.: Prospects of RNA interference therapy for cancer. Gene Ther. **13**(6), 464–477 (2006)
23. Kibria, G., Hatakeyama, H., Akiyama, K., Hida, K., Harashima, H.: Comparative study of the sensitivities of cancer cells to doxorubicin, and relationships between the effect of the drug-efflux pump P-gp. Biol. Pharm. Bull. **37**(12), 1926–1935 (2014)
24. Tekchandani, P., Kurmi, B.D., Paliwal, S.R.: Nanomedicine to deal with cancer cell biology in multi-drug resistance. Mini. Rev. Med. Chem. **17**(18), 1793–1810 (2017)
25. Otto, E., McCord, S., Tlsty, T.D.: Increased incidence of CAD gene amplification in tumorigenic rat lines as an indicator of genomic instability of neoplastic cells. J. Biol. Chem. **264**(6), 3390–3396 (1989)
26. Tipping, A.J., Mahon, F.X., Lagarde, V., Goldman, J.M., Melo, J.V.: Restoration of sensitivity to STI571 in STI571-resistant chronic myeloid leukemia cells blood. J. Am. Soc. Hematol. **98**(13), 3864–3867 (2001)
27. McKinnell, R.G., Parchment, R.E., Perantoni, A.O., Pierce, G.B.: The biological basis of cancer. Cambridge University Press (1998)
28. Kuznetsov, M., Clairambault, J., Volpert, V.: Improving cancer treatments via dynamical biophysical models. Phys. Life Rev. **39**, 1–48 (2021)

# The Immune System in Health and Disease: The Need for Personalised Longitudinal Monitoring

**Hector Zenil, Abicumaran Uthamacumaran, and Kourosh Saeb-Parsy**

## 1 Introduction

There is increasing recognition that the immune system has a critical role in the pathogenesis of most, if not all, diseases. While the role of immune system in disorders such as infections and autoimmune conditions has long-been recognised, emerging evidence indicates that the immune system fundamentally underpins how human biology responds to all internal and environmental perturbations, both as a consequence of damage or disease, and as part of the normal life events such as ageing and pregnancy. A better understanding of an individual's immune system thus has the potential to transform our ability to deliver personalised healthcare.

The original version of this chapter was revised. Incorrect Abstract in this chapter has been removed. The correction to this chapter can be found at https://doi.org/10.1007/978-3-031-04379-6_16

H. Zenil (✉)
The Alan Turing Institute, London NW1 2DB, UK
e-mail: hector.zenil@cs.ox.ac.uk

H. Zenil · K. Saeb-Parsy
The University of Cambridge, Cambridge, UK

Oxford Immune Algorithmics, Reading RG1 3EU, UK

A. Uthamacumaran
Department of Physics, Concordia University, Montreal, Canada



I. Balaz and A. Adamatzky (eds.), *Cancer, Complexity, Computation*, Emergence, Complexity and Computation 46,
https://doi.org/10.1007/978-3-031-04379-6_15

## 2 The Vital Role of the Immune System in Health and Disease

Cancer is perhaps the best and most commonly studied example of a disorder in which the contribution of the immune system is increasingly recognised. The role of T lymphocytes [1], B lymphocytes [2], Natural Killer cells [3], neutrophils [4], macrophages [5] and other immune cells in the pathogenesis of a range of cancers is now firmly established. This knowledge is also being leveraged to generate new therapies that harness and manipulate the immune response to eradicate tumour cells [6]. The contribution of the immune system is now also increasingly recognised in cardiovascular diseases including myocardial infarction [7], in metabolic disorders including type 1 [8] and type 2 [9] diabetes, in neurodegenerative diseases such as Alzheimer's dementia, Parkinson's disease and Huntington's disease [10], and in mental disorders such as depression [11]. The immune consequences of dysregulation of bacteria in the gastrointestinal tract, or 'microbiome', are now understood to make important contributions to a range of diseases including autism, multiple sclerosis, asthma, liver disease, metabolic disorders, irritable bowel syndrome, diabetes, and rheumatoid arthritis [12].

The immune system is also critical for the response to injuries, from trivial falls to major accidents [13]. During injury and disease, the immune system is intimately involved in the perception of pain and its consequences for repair and regeneration [14]. Even in the absence of any recognisable injury or disease, the immune system has a key role in orchestrating 'normal' homeostasis in humans. There are also important differences in immune responses in men and women which are believed to underpin the different susceptibility of the sexes to disease [15]. The immune system changes with age and 'Inflammaging' is a recent concept in which chronic, low-grade, sterile inflammation develops and contributes to the pathogenesis of age-related diseases [16].

The immune system is important in regulating sleep dynamics and the consequences of sleep disturbances [17]. The immune system also plays a central role in the health consequences of obesity [18]. During pregnancy, maternal immune activation or dysfunction, such as precipitated by obesity, asthma, autoimmune disease, infection or psychosocial stress, is thought to lead to potential long-term consequences for the foetus, including predisposition to neurodevelopmental disorders [19].

## 3 Monitoring the Immune System: Challenges

It is clear from this evidence that a better understanding of the perturbations in an individual's immune system can potentially transform our ability to determine risks of disease more accurately, develop more effective preventative interventions, make earlier diagnoses and develop superior treatments. There is increasing recognition that the immune system is highly variable between individuals and that discrete 'immunophenotypes' exist that are likely better descriptors of how the immune system of individuals determine their health [20]. The immune system, however, is exceptionally complex and incorporates a very large number of cells, proteins, epigenetic factors, and small molecule mediators that interact in a very complicated—and only partially understood—dynamic network. This complexity may initially seem to be a compelling deterrent from attempting to decipher the 'personalised' associations between changes in the immune system and diseases for individuals. This is because—the logic would argue—the technical, logistic and economic challenges are likely to prevent 'deep immune characterisation' of individuals outside of research settings for decades. However, such a pessimistic rationale incorrectly assumes that personalised immune characterisation is entirely dependent on complex and costly diagnostic tests.

The simplest 'survey' of an individual's immune system is a routinely performed blood test, known as the Complete (or Full) Blood Count (CBC or FBC). The CBC enumerates the number and frequency of cells in the blood, including white blood cells and their main subtypes (neutrophils, lymphocytes, monocytes, eosinophils and basophils), red blood cells and platelets. The CBC is the most commonly performed medical diagnostic test performed worldwide (estimated more that 1 billion tests per year globally) and is incorporated as part of the vast majority of blood tests performed.

The utility of CBC in diagnosis and monitoring of disease progression and response to treatment is well established. However, our assessment of the results is fundamentally at odds with its use as a personalised measure of the individual's immune system. This is because the results of CBC tests are invariably categorised in a binary manner as 'normal' or 'abnormal' with reference to population norms. However, these population normal are surprisingly broad: The 'normal' total white blood cell (WBC) range is typically 4.0–11.0 million cells per microlitre of blood. The equivalent population normal reference values for neutrophils and lymphocytes are 2.0–7.5 and 1.0–4.5 million cells per microlitre of blood. This binary system of classification of CBC results implies that some immune cells in an individual's blood can increase by 300–400% in number, but such a huge change could, or should, be safely 'ignored' because they still remain within the population 'normal' reference range.

This system of analysis, taught and practised in every healthcare system in the world today, represents an astonishingly crude approach to understanding health and disease. Ironically, the medical education system ingrains in all medical professionals an acceptance of such a crude binary analysis. Yet, even school children, if left to ponder under their own deductive reasoning, are likely to assign significance to such wide variations in blood cell numbers: If, as we know to be the case, immune cells are so integral in health and disease, how can it possibly be safe to ignore a 400% rise in their number? Not attempting to capture this data, and failing to integrate the data with other known health parameters, is a colossal loss of opportunity to benefit the society.

## 4 Monitoring the Immune System: Opportunities

There is emerging evidence in support of the assertion that variations in immune cells within the 'normal' range can have important health implications. In one study, the total WBC count in 'healthy' individuals was found to be predictive of risk of short-term and long-term death, even when known risk factors (age, smoking, diabetes, high blood pressure, ethnicity and blood cholesterol levels) were controlled for [21]. In this study of >194,000 individuals, those with a 'normal' WBC of 8.65–10.05 were almost $3\times$ more likely to die within 6 months than those with a 'normal' WBC of 5.35–6.25. In another recent seminal study, the immune system of healthy individuals (as assessed by the number of a subset of their lymphocytes), was a better predictor of risk of death than age, leading to the development of the concept of Immune Age [22, 23].

These studies provide compelling motivation for a paradigm-shift in the monitoring of the immune system from a population-based approach to a personalised one. To better understand disease, and to enable and accelerate personalised diagnosis, monitoring and treatment, we must redefine norms with respect to a patient's individual physiology, and seek to detect and monitor perturbations from that individuals' personal baseline. This individualised approached, placed within the broader context of population data and reference values, has the opportunity to bridge the gap between population and personalised healthcare. To achieve this, we need to gather data about an individual's immune system, not only when disease has clinically manifested, but also during health.

## 5 Longitudinal Analysis and Causal Pattern

The understanding of complex adaptive diseases like cancers necessitate the interdisciplinary paradigm of complexity science. By merging AI and experimental cancer research, we can approach cancer cybernetics as complex dynamical systems. Tumours are complex ecosystems composed of the dynamic interactions between malignant (neoplastic) cells, healthy host cells, mesenchymal cells, endothelial cells, extracellular matrices, and infiltrated immune cells which collectively maintain inflammatory signals to nourish and regulate the tumour microenvironment (TME) [24, 25]. The signaling networks of inflammatory-immune cells regulate cancer progression and metastatic invasion by remodelling the TME, promoting immune escape (tolerance), and conferring therapy resistance [24, 25]. The signaling interactions between cancer-immune cell networks provides a vast source of liquid-biopsy biomarkers for the diagnostic screening and prognostic prediction of complex diseases like cancer in precision medicine. Longitudinal data analysis is thence required to study the complex dynamics across these evolving networks. Prior to discussing the importance of longitudinal cancer data analysis from patients-derived liquid-biopsies, the essential properties of cancer-immune network dynamics and immunotherapies must be clarified.

Cancer immunotherapy attempts to stimulate the immune system to target antigens on cancer cells to destroy them. The central advantage of immunotherapies is their personalized (patient-specific) targeted approach. The immune cells can be collected from the patient's blood, modified to activate a cytotoxic response towards a tumour-specific antigen(s), and reinjected back to the patient [26]. Despite the availability of these immunotherapies for certain types of cancers, approximately only one-third of the patients benefit from the therapies [27, 28]. Identification of potent and robust clinically targetable biomarkers for effective immunotherapies remains a challenge due to the dynamic molecular complexity and intratumoral heterogeneity in aggressive cancers [27, 28]. The molecular profiling of tumour ecosystems based on their antigenicity, tumour heterogeneity, and immunogenicity are required for effective personalized cancer immunotherapies and improve clinical outcomes [29]. Furthermore, how therapy affects immune cell counts over time is key for predicting therapy response and treatment management [27]. Liquid biopsies such as blood test-based detection of immuno-markers, immune cell count fluctuations, and blood cytogenetics provide a non-invasive approach to molecular profiling complex diseases like cancer [30]. Along with the quantification of soluble biomarkers in blood plasma/serum, the identification of systemic immune cell populations that correlate with therapeutic responses could provide promising clinical targets, which include the relative abundance of CD8+ and CD4+T-cells, the expression of immune checkpoints in peripheral blood cells, and the relative numbers of immunosuppressive cells such as regulatory T-cells and myeloid-derived suppressor cells [27]. These biomarkers exhibit fluctuations and oscillations over time. As such, longitudinal analysis of blood test-derived immune markers are critical for monitoring and regulating therapeutic efficacy and disease progression in cancer patients.

To illustrate, complete blood count parameters (CBC) were demonstrated as non-invasively accessible patient-specific biomarkers capable of predicting immunotherapy response in a study within 986 advanced-stage lung cancer patients [31]. A significantly decreasing absolute lymphocyte count (ALC), significantly increasing absolute neutrophil count (ANC) and fluctuating platelet count (PLT) levels were indicative of metastatic disease progression in the patients [31]. Further, single-cell RNA sequencing (scRNA-Seq) of patient-derived peripheral blood extractions revealed that circulating immune-cell phenotype dynamics are robust biomarkers for predicting therapy response and prognosis [26, 30]. In general, patients baring tumours with an immunosuppressive phenotype (e.g., expressing high levels of PD-L1) exhibit a lower immune cell count prior to treatment. The studies concluded that patients with a low peripheral blood mononuclear cells (PBMC) abundance would most benefit from anti–PD-1 immune therapy [30]. Further, an increased IFN signaling, and differentiation of T-cells was associated to an increased anti-PD-1 immunotherapy response in these patients [30]. Combination therapy may be more appropriate in personalized precision medicine, tailored the molecular heterogeneity and complexity of the patient's tumour.

Molecular profiling of the tumour is thence used to identify patient-specific therapeutic targets. Multi-omics approaches on the tumour biopsy when applicable in addition to the non-invasive liquid biopsies provide optimal information for quality patient-care and effective clinical decision-making. However, liquid biopsies-based measures are better suited and easily accessible for longitudinal analyses. We must also account for differences in the molecular phenotypes of cancers within pediatric patients and adult patients when assessing therapy response [32]. As children undergo immune system development in the early ages of their life, their inexperienced immune systems implies significant differences in immune responses compared to adults (i.e., reduced memory cell populations related to less antigenic exposures) [32]. Hence, pediatric cancers may require less specific immunotherapeutic agents which simultaneously create a pro-inflammatory environment and activate immune cells [32]. The infiltrating immune cells create a chronic inflammatory and immunosuppressive TME that restricts immunotherapeutic approaches [32, 33]. Pediatric solid tumours further exhibit significant myeloid and stromal cell infiltration [32]. Both pediatric and adult TME display a complex interactome of immune-cells secreted signaling pathways including chronic nuclear factor-κB (NF-κB), chemokines, cytokines, and growth factors [32, 33].

## 6 From Dynamic Cellomics to Multidimensional Omics

Cancer treatment modalities and targeted therapies take weeks if not months to observe clinically relevant effects in most cancer patients. As such, the efficacy of our current anti-cancer treatments requires longitudinal data analysis. Longitudinal studies reveal fluctuations and oscillatory patterns in blood count parameters such as immune cell counts are potent indicators of metastatic cancer progression and therapy response in cancer patients [34, 31]. Further, as stated, cancers are complex dynamical systems. That is, cancer signaling, gene expression, protein secretion, and cellular communication (e.g., immune network dynamics) are dynamic processes which may exhibit collective behaviours. As such, the concentration of key cellular biomarkers and proteins may exhibit oscillations which vary over time and require a long-term time-series analysis for effective clinical decision-making. For example, the blood sera expression of inflammatory signals such as the cytokine TNFα and C-reactive protein (CRP) exhibit ultradian oscillations which vary throughout the day and require a longitudinal (time-series) analysis to understand their complex signaling dynamics within cancer patients [35, 36].

In dynamical systems theory, such signaling oscillations are the precursors to complex dynamics and causal structures. When oscillations are perturbed, they may bifurcate towards complex, chaotic attractors. Attractors are self-organized causal patterns (set of points or states) towards which the trajectories of a complex dynamical system tend to evolve in state-space [37, 38]. Attractors govern the temporal behaviors of the complex system while the complexity of an attractor measures the control predictability and irregularity of the complex system's behaviors [37, 38]. Chaotic attractors, also known as strange attractors, are the most complex of these causal patterns making the system exhibit unstable and irregular dynamics

[38]. Studies have shown that cancer cell fates correspond to unstable, nonequilibrium attractors on the Waddington's epigenetic landscape [39, 40]. With longitudinal cancer data analysis, we can approach the causal information dynamics across cancer-immune networks as attractors and verify whether strange attractor states emerge in the biomarker/signaling oscillations.

Age-related deterioration and diminishment of immune system responses leads to co-morbidities and mortality, including complex diseases such as cancer [22, 23]. Longitudinal tracking can assess the gradual age-related changes in immune processes as dynamical systems. The merging of AI algorithms and longitudinal blood data analyses have shown capable of inferring attractor dynamics as predictors of disease emergence. To illustrate, in a longitudinal study, Alpert et al. [22, 23] used multi-omics profiling to analyze personalized and population-level immune system dynamics of 135 healthy adult patients over a 9-year period. They defined a metric, immune aging (IMM-AGE) score, by use of trajectory inference algorithms on immune cell profiles, as a causal measure of the patient's immune homeostasis and health prediction. Blood sample measurements consisted of cellular phenotyping by single-cell mass cytometry (CyTOF) to assess relative abundance of immune cell populations and their fluctuations over time, assessment of cytokine responses by cytokine-stimulation assays, and whole-blood gene expression profiling. A moderate rise in immune cells-mediated circulating inflammatory mediators were observed in patient-derived peripheral blood samples with ageing [22, 23]. This process is often referred to as inflammaging and is central to the predisposition and development of age-related chronic diseases, including cancers [22, 23]. Further, there are dynamic inter-individual variability in the degree of inflammaging due to genetics, environmental interactions, epigenetic factors (e.g., methylation), etc. to consider in personalized precision medicine.

Alpert et al. [22, 23] screened 73 immune cell subsets (phenotypes) and calculated the group level and individual level longitudinal slopes in frequency (population abundance). To study the global attractor dynamics of these immune profiles, they applied Principal Component Analysis (PCA) on the CyTOF data. The two main principal components correlated with the frequencies of many immune-senescence associated phenotypes such as such as CD28–CD8+ T-cells, naive CD4+CD8+ T-cells, and memory B cells [22, 23]. The PCA trajectory length per individual with respect to these immunosenescence-related cell-subsets exhibited a shorter trajectory in younger patients. Immune cells in older adults converged toward steady-state levels in a baseline-dependent manner to yield fixed-point attractors. Further, by applying the diffusion pseudo-time algorithm (DPT), a type of dimensionality reduction technique, the cellular frequency profiles changed along a trajectory reflecting the patient's immune age. In agreement with the PCA immune trajectories, the individual slopes of older adults along the DPT trajectory were significantly positive, as opposed to those calculated for the young adults. Key cytokine responses that significantly changed along the trajectory were also identified in the study (Alpert et al. 2018).

The longitudinal blood analysis by Alpert et al. [22, 23] demonstrated the IMM-AGE score can well-predict age-related cardiovascular disease incidence by identifying a candidate gene set whose expression was significantly correlated along the patient-derived immune profile trajectories. The cellular frequency dynamics along the trajectory showed the downregulation of genes predominantly expressed by naive CD8+T cells, naive CD4+T cells and B cells, whereas upregulated genes were predominantly expressed by cytotoxic NK cells and terminally differentiated effector memory CD8+T cells. The IMM-AGE score was shown to predict mortality in older adults by >500-fold significance than the epigenetic clock (DNA-methylation metric). Further, these findings suggested that like the epigenetic landscape proposed by Waddington to describe cell fate dynamics during differentiation processes as attractors, we can describe the cellular composition and cell fate dynamics of the immune system as traversing an immunological attractor landscape. The immune-profile variation (trajectory) observed may be a path through this attractor landscape corresponding to the processes observed in aging and aging-related diseases, with interindividual variability in the rate of progression along the trajectory. The trajectory trend toward a minimum (i.e., an attractor) in the landscape representing the patient's immune homeostasis. Populations with longevity may exhibit slower pseudo-time progression rates along the trajectory of immune cell dynamics. The presence of more complex attractors may be indicative of complex immune diseases such as cancers [40, 39]. The immunological landscape may well apply to predicting causal dynamics in response to cancer immunotherapy in patients through their blood-derived immune variation along a principal curve (PCA trajectory) or DPT curve. These attractor dynamics inferring approaches may pave the identification of clinically-relevant biomarkers and targeted therapies for immune-system related morbidities such as cancers [22, 23]. Further, they demonstrate that longitudinal, systems-level cancer data analysis may allow a robust quantitative description of disease prognosis and therapy response prediction.

Similarly, in another longitudinal systems analysis, the immune system variation within individuals were studied in n = 99 middle-aged healthy patients (aged 50–65), sampled every third month over a year's time-period to identify causal biomarkers of cardiovascular disease risk [41]. Inter-individual variation and characterization of extreme immune cell phenotypes along a principal component curve were measured by fluctuations in between various immune cell populations and a diverse panel of plasma proteins. The most variable individuals exhibited markers of poor metabolic health suggestive of a causal link between immunologic homeostasis and health progression. To perform the immune cell compositional frequency analysis, Lakshmikanth et al. [41] calculated the Aitchison's distance matrix, based on relative abundances of 53 non-overlapping immune cell populations, and visualized samples by multidimensional scaling (MDS), where each sample was positioned on its pair-wise distance to all other samples. Then, a PCA trajectory was used to analyze the variation (i.e., attractor dynamics) in immune cell composition.

In agreement with the previously discussed study by Alpert et al. [22, 23], the global immune dynamics revealed a positive correlation with CD4+ and CD8+T-cells, and a negative correlation with memory B-cell populations [41]. Using a

Bayesian variable selection method, they identified likelihood of plasma protein markers associated with the cell population fluctuations which resulted in the inference of an immune regulatory network of 226 significant associations [41]. In specific, a high immune cell variability was found to be associated with blood biomarkers of poor metabolic health. The study found a positive correlation between multiple markers of poor metabolic health, such as plasma Apo-B, Troponin T, Triglycerides (TG), and LDL cholesterol, and a negative relationship between immune variability over time (mean Aitchison's distance), as well as metabolic biomarkers such as HDL, and Apo-A1 [41]. The total white blood cell count, red blood cell count, and neutrophil count were also positively associated with high immune variability over time. Chemokines such as CCL18 expressed by macrophages were further causally linked with the predisposition to coronary artery disease in the patients.

The longitudinal study demonstrates that longitudinal analysis of patient-specific immune system profiles can be modelled as attractor dynamics underlying networks of causal relationships amidst the biomarkers (i.e., cells and proteins) [41]. However, the study was limited to the assumed notion of stable attractors points representing the immune profile trajectories/variations. The stable steady-state frequency toward which an immune cell subset converged with age were assumed to be fixed-point attractors, the simplest of attractors in dynamical systems. However, in principle, the high dimensional data points of the immunological landscape can possibly converge towards unstable, complex attractors such as strange attractors [42, 43]. Mathematical models of immune-cancer cell population growth-invasion dynamics suggest the emergence of chaotic (strange) attractors as indicators of cancer adaptiveness, resilience, and recurrence/relapse [44–47]. However, these computational predictions of lack experimental validation. As will be discussed, Algorithmic Information Dynamics (AID) provides a robust solution to overcome the challenge of causality inference in network dynamics and is emerging as a powerful detection tool for reconstructing the complex dynamics/attractors in state-space signaling networks [48].

## 7 From Statistical AI to AGI for Causation in Medicine

Machine learning (ML) algorithms are emerging as a promising statistical AI approach for clinical data analysis. For example, liquid-biopsy based cancer detection and screening via characterization of patient-derived EVs/exosomes [49, 50]. In another example, Chen et al. [16] combined a single-molecule localization and multiplex super-resolution imaging technique called DNA-PAINT to immunostain patient-derived exosomes by four specific exosomal surface biomarkers: HER2, GPC-1, EpCAM, and EGFR. Linear discriminant analysis (LDA), a type of ML classifier, enhanced by cross-validation was used to distinguish cancer patients from healthy individuals using exosomes extracted from patient-derived blood samples. The ML algorithm was shown capable of profiling cancer EVs extracted from pancreatic and breast cancer patients from those of healthy patients with a 100% accuracy [51].

More recently, Deep Learning-based spectroscopic analysis of liquid-biopsy derived exosomes has shown > 90% sensitivity and accuracy in cancer detection [52]. Exosomes are a subset of lipid-bound EVs found in the 300–100 nm size range circulating in patient blood sera. Shin et al. [52] trained a residual neural network (Resnet)-based deep learning model, with the surface-enhanced Raman spectra (SERS) of cancer exosomes and healthy plasma cell line exosomes, which the algorithm classified with a 95% accuracy. When the Deep Learning algorithm was assessed on the SERS signals of 43 patients-derived exosomes, including stage I and II lung cancer patients, the algorithm predicted lung cancer with an AUC of 0.912 for the whole cohort and stage I patients with an AUC of 0.910. These findings strongly demonstrate that the pairing of liquid-biopsy derived cancer EVs with AI may pave early-stage cancer detection with high sensitivity and specificity [52].

These arguments strongly warrant the need for pairing longitudinal cancer data analysis with AI to understand cancer-immune causal networks. Further they highlight its clinical importance in early cancer diagnosis, screening, therapeutic response assessment, and therapy management. The longitudinal studies further demonstrate that network science and attractor landscapes provide novel complex systems tools to study the evolving attractor states of immune cell subsets and identify their causal roles in cancer cell fate regulation (cybernetics). Network medicine allows the integration of multi-omics sequencing data with graph theoretic algorithms to infer signaling relationships between cells, genes, and proteins [53, 54]. For example, technologies that combine a pooled CRISPR screen with scRNA-seq, such as Perturb-seq, were shown capable of identifying the molecular transcriptional networks controlling cancer progression and steering cancer cell fate decisions [55]. ML algorithms have widely been adopted for reconstructing complex cell communication networks such as gene regulatory networks and protein–protein interaction networks [55]. However, most of our currently predominant approaches in network medicine rely on correlation metrics. Correlation does not imply causation. As such, amidst all efforts from AI to understand cancer signaling networks as dynamical systems, Algorithmic Information Dynamics (AID) has been demonstrated as a robust AGI (Artificial General Intelligence) approach to longitudinal dynamic analysis to infer causal relationships (attractor dynamics).

## 8 Algorithmic Information Dynamics

To systematically move away from 'sick' care to health care, where the purpose is the sustainability of health and extended quality of life (longevity), new methods are required to discover each person's own personalised baselines and patterns and to produce personalised medical knowledge over time based on the individual own history both when ill but also when normal. The application of Artificial Intelligence to learn from longitudinal data and help build individual health profiles represents an opportunity. Current machine learning approaches are very limited to finding simple statistical patterns and deal poorly with causal discovery often generating spurious correlations [56, 57]. For this reason, the application of responsible, intuitive, and causal machine intelligence [58] is required and has been proven to find patterns in cancer pathways [48], in genetic regulatory networks that produce proteins [48], and in deciphering the structure of the DNA [59] where the founders of Oxford Immune Algorithmics have made their greatest contributions. This new type of new AI, rooted in causal models, is traceable and more explainable compared to black-box approaches (such as deep neural networks) as more transparency is needed in medicine for a responsible application.

Algorithmic Information Dynamics (AID) is emerging as an AI approach for inferring causal information dynamics across evolving networks through perturbation analysis in software space [48]. AID provides a framework of measures from algorithmic complexity, network science, and information theory to infer the properties and causal dynamics of complex networks. According to AID, the immune system can be seen as an error-correcting code while cancer is treated as an information-theoretic problem [42, 43]. The causal relationships with interactions of the evolving networks (e.g., signaling proteins, genes, metabolic, epigenetic, etc.) can be projected as attractor state-dynamics on an epigenetic landscape (i.e., the state space projection) or an immune landscape when considering the high- dimensional phase space of cancer-immune dynamics [60, 42, 43]. Then, unhealthy states of the dynamical immune system can be defined as strange attractors on the landscape/state-space, while healthy states are considered as stable fixed-point attractors [42, 43].

AID provides a set of tools to approximate the Algorithmic (Kolmogorov) complexity of these complex graph networks and control them via perturbation analysis in software space. Perturbation analysis can be as simple as the removal of an edge or node from the complex network. For example, a gene regulatory network can be represented by a set of strings or vector of arrays (a tensor) in binary (at a multiplicity of cutoff values). The algorithmic information content of this tensor can then be described by classical measures such as Shannon entropy H(s) or Kolmogorov complexity K(s). The Kolmogorov complexity, K(s), also known as K-complexity or algorithmic complexity quantifies the shortest bits of a string or computer program required to describe a dataset (graph network) [61]. However, K(s) is a more robust tool than Shannon entropy to measure the complex dynamics of networks and distinguish them from randomness (i.e., low complexity). Although Shannon entropy can quantify the amount of information in a complex network, it fails to reconcile the causal relationships within the network. Furthermore, K(s) does not depend on a choice of probability distribution like Shannon entropy does. Therefore, it is more robust than Shannon entropy in measuring the complexity of graph networks, such as the longitudinal cancer-immune networks reconstructed from blood-analysis. The graph (Shannon) entropy is to correlation what the Kolmogorov complexity is to causation. Unlike graph entropy, graph Kolmogorov complexity is a robust measure of causal information dynamics in complex networks [48]. The algorithmic complexity of a complex network distinguishes a process as a cause or randomness [48].

Longitudinal cancer data analysis by AI and algorithms may pave robust clinical approaches for the early detection, prognostic screening, therapeutic response prediction, and therapy management of complex diseases like cancer in precision medicine. The dynamics of blood-immune cell composition and cell frequencies in response to therapies can be inferred as a metric of immunological health via longitudinal patient-blood analysis. Longitudinal variability/fluctuations in immune cell counts and plasma proteins were used to identify co-regulated immune system components within the patients. The longitudinal analysis of their gene expression patterns and biomarker fluctuations over time allow us to study the causal information dynamics (i.e., attractor inference) in evolving disease-immune networks. AID is an AI toolkit for screening complex attractor dynamics in complex processes such as cancer cell fate trajectories (on the Waddington-immunological landscape). The complexity measures provided by AID allow us to detect the complex attractor states steering cellular decision-making during disease progression and therapy response via perturbation analysis in software space. The complexity of the attractor(s) corresponds to the irregularity of the system's behaviors and dynamics. Estimating the Kolmogorov-complexity (K(G)) of the complex evolving graph networks steering cancer cell fate decision-making provides a robust screening tool to identify the critical signals (epigenetic regulations, genes, proteins, etc.) regulating the complex attractor dynamics and hence, governing the adaptive behaviors in cancer ecosystems. As such, they may pave the cell fate reprogramming of complex adaptive diseases like cancers.

## 9 Conclusion

The key for better and personalised understanding of the role of the immune system over time, that is in health and disease. The routine collection of longitudinal data that allows changes in the immune system to be tracked over time and be correlated with internal and external factors, including symptoms and sign of disease is required to move towards causal analysis and model-driven precision medicine. The delivery of this promise requires three pillars: (1) accessibility, the democratisation to access to blood testing, so that they can be performed regularly, conveniently, accurately, and affordable; (2) the collection of all relevant physiological and clinical data, such clinical symptoms, and treatments; and (3) the ability to collate, integrate and analyse these data using state-of-the-art responsible model-driven AI to generate novel insights. If performed in an ethical and sustainable manner, such an approach has the potential to propel our ability to reduce disease and suffering from incremental progress towards vast transformation.

## References

1. Philip, M., Schietinger, A.: CD8(+) T cell differentiation and dysfunction in cancer. Nat. Rev. Immunol. (2021)

2. Sharonov, G.V., et al.: B cells, plasma cells and antibody repertoires in the tumour microenvironment. Nat Rev Immunol **20**(5), 294–307 (2020)
3. Huntington, N.D., Cursons, J., Rautela, J.: The cancer-natural killer cell immunity cycle. Nat. Rev. Cancer **20**(8), 437–454 (2020)
4. Hedrick, C.C., Malanchi, I.: Neutrophils in cancer: heterogeneous and multifaceted. Nat. Rev. Immunol. (2021)
5. DeNardo, D.G., Ruffell, B.: Macrophages as regulators of tumour immunity and immunotherapy. Nat. Rev. Immunol. **19**(6), 369–382 (2019)
6. Kraehenbuehl, L., et al.: Enhancing immunotherapy in cancer by targeting emerging immunomodulatory pathways. Nat Rev Clin Oncol. (2021)
7. Epelman, S., Liu, P.P., Mann, D.L.: Role of innate and adaptive immune mechanisms in cardiac injury and repair. Nat. Rev. Immunol. **15**(2), 117–129 (2015)
8. Ilonen, J., Lempainen, J., Veijola, R.: The heterogeneous pathogenesis of type 1 diabetes mellitus. Nat. Rev. Endocrinol. **15**(11), 635–650 (2019)
9. Lackey, D.E., Olefsky, J.M.: Regulation of metabolism by the innate immune system. Nat. Rev. Endocrinol. **12**(1), 15–28 (2016)
10. Heneka, M.T., Kummer, M.P., Latz, E.: Innate immune activation in neurodegenerative disease. Nat. Rev. Immunol. **14**(7), 463–477 (2014)
11. Miller, A.H., Raison, C.L.: The role of inflammation in depression: from evolutionary imperative to modern treatment target. Nat. Rev. Immunol. **16**(1), 22–34 (2016)
12. Levy, M., et al.: Dysbiosis and the immune system. Nat. Rev. Immunol. **17**(4), 219–232 (2017)
13. Tidball, J.G.: Regulation of muscle growth and regeneration by the immune system. Nat Rev Immunol **17**(3), 165–178 (2017)
14. Baral, P., Udit, S., Chiu, I.M.: Pain and immunity: implications for host defence. Nat. Rev. Immunol **19**(7), 433–447 (2019)
15. Klein, S.L., Flanagan, K.L.: Sex differences in immune responses. Nat. Rev. Immunol. **16**(10), 626–638 (2016)
16. Franceschi, C., et al.: Inflammaging: a new immune-metabolic viewpoint for age-related diseases. Nat. Rev. Endocrinol. **14**(10), 576–590 (2018)
17. Irwin, M.R.: Sleep and inflammation: partners in sickness and in health. Nat Rev Immunol **19**(11), 702–715 (2019)
18. Larabee, C.M., Neely, O.C., Domingos, A.I.: Obesity: a neuroimmunometabolic perspective. Nat. Rev. Endocrinol. **16**(1), 30–43 (2020)
19. Han, V.X., et al.: Maternal immune activation and neuroinflammation in human neurodevelopmental disorders. Nat. Rev. Neurol. **17**(9), 564–579 (2021)
20. Brodin, P., Davis, M.M.: Human immune system variation. Nat. Rev. Immunol. **17**(1), 21–29 (2017)
21. Shah, A.D., et al.: White cell count in the normal range and short-term and long-term mortality: international comparisons of electronic health record cohorts in England and New Zealand. BMJ Open **7**(2), e013100 (2017)
22. Alpert, A., Pickman, Y., Leipold, M., Rosenberg-Hasson, Y., Ji, X., Gaujoux, R., Rabani, H., Starosvetsky, E., Kveler, K., Schaffert, S., Furman, D., Caspi, O., Rosenschein, U., Khatri, P., Dekker, C.L., Maecker, H.T., Davis, M.M., Shen-Orr, S.S.: A clinically meaningful metric of immune age derived from high-dimensional longitudinal monitoring. Nat. Med. **25**(3), 487–495 (2019). https://doi.org/10.1038/s41591-019-0381-y
23. Alpert, A., et al.: A clinically meaningful metric of immune age derived from high-dimensional longitudinal monitoring. Nat. Med. **25**(3), 487–495 (2019)
24. Gonzalez, H., Hagerling, C., Werb, Z.: Roles of the immune system in cancer: from tumor initiation to metastatic progression. Genes. Dev. **32**(19–20), 1267–1284 (2018). https://doi.org/10.1101/gad.314617.118
25. Zhao, H., Wu, L., Yan, G., et al.: Inflammation and tumor progression: signaling pathways and targeted intervention. Sig Transduct Target Ther. **6**, 263 (2021). https://doi.org/10.1038/s41392-021-00658-5

26. Waldman, A.D., Fritz, J.M., Lenardo, M.J.: A guide to cancer immunotherapy: from T cell basic science to clinical practice. Nat Rev Immunol **20**, 651–668 (2020). https://doi.org/10.1038/s41577-020-0306-5
27. Hernandez, C., Arasanz, H., Chocarro, L., Bocanegra, A., Zuazo, M., Fernandez-Hinojal, G., Blanco, E., Vera, R., Escors, D., Kochan, G.: Systemic blood immune cell populations as biomarkers for the outcome of immune checkpoint inhibitor therapies. Int. J. Mol. Sci. **21**(7), 2411 (2020). https://doi.org/10.3390/ijms21072411
28. Ventola, C.L.: Cancer immunotherapy, part 2: efficacy, safety, and other clinical considerations. P & T: Peer-Reviewed J. Formulary Manag. **42**(7), 452–463 (2017)
29. Beatty, G.L., Gladney, W.L.: Immune escape mechanisms as a guide for cancer immunotherapy. Clin. Cancer Res. Off. J. Am. Assoc. Cancer Res. **21**(4), 687–692 (2015). https://doi.org/10.1158/1078-0432.CCR-14-1860
30. Griffiths, J.I., Wallet, P., Pflieger, L.T., Stenehjem, D., Liu, X., Cosgrove, P.A., Leggett, N.A., McQuerry, J.A., Shrestha, G., Rossetti, M., Sunga, G., Moos, P.J., Adler, F.R., Chang, J.T., Sharma, S., Bild, A.H.: Circulating immune cell phenotype dynamics reflect the strength of tumor-immune cell interactions in patients during immunotherapy. PNAS **117**(27), 16072–16082 (2020). https://doi.org/10.1073/pnas.1918937117
31. Rojko, L., Megyesfalvi, Z., Czibula, E., Reiniger, L., Teglasi, V., Szegedi, Z., Szallasi, Z., Dome, B., Moldvay, J.: Longitudinal analysis of complete blood count parameters in advanced-stage lung cancer patients. Thoracic Cancer **11**(11), 3193–3204 (2020). https://doi.org/10.1111/1759-7714.13642
32. Hutzen, B., Ghonime, M., Lee, J., Mardis, E.R., Wang, R., Lee, D.A., Cairo, M.S., Roberts, R.D., Cripe, T.P., Cassady, K.A.: Immunotherapeutic Challenges for Pediatric Cancers. Molecular Therapy Oncolytics **15**, 38–48 (2019). https://doi.org/10.1016/j.omto.2019.08.005
33. Su, C., Zhang, J., Yarden, Y., et al.: The key roles of cancer stem cell-derived extracellular vesicles. Sig. Transduct. Target Ther. **6**, 109 (2021). https://doi.org/10.1038/s41392-021-00499-2
34. Madondo, M.T., Tuyaerts, S., Turnbull, B.B., et al.: Variability in CRP, regulatory T cells and effector T cells over time in gynaecological cancer patients: a study of potential oscillatory behaviour and correlations. J. Transl. Med. **12**, 179 (2014). https://doi.org/10.1186/1479-5876-12-179
35. Tian, B., Nowak, D.E., Brasier, A.R.: A TNF-induced gene expression program under oscillatory NF-κB control. BMC Genomics **6**, 137 (2005)
36. Coventry, B.J., Ashdown, M.L., Quinn, M.A., Markovic, S.N., Yatomi-Clarke, S.L., Robinson, A.P.: CRP identifies homeostatic immune oscillations in cancer patients: a potential treatment targeting tool? J. Transl. Med. **7**, 102 (2009). https://doi.org/10.1186/1479-5876-7-102
37. Janson, N.: Non-linear dynamics of Biological systems. Contemp. Phys. **53**, 137–168 (2012)
38. Strogatz, S.H.: Nonlinear Dynamics and Chaos: With Applications to Physics, Biology, Chemistry, and Engineering. Westview Press, Boulder, CO (2015)
39. Fang, X., Kruse, K., Lu, T., Wang, J.: Nonequilibrium physics in biology. Rev. Mod. Phys. 91(4) (2019)
40. Huang, S., Ernberg, I., Kauffman, S.: Cancer attractors: a systems view of tumors from a gene network dynamics and developmental perspective. Semin. Cell Dev. Biol. 20, 869–876 (2009)
41. Lakshmikanth, T., Muhammad, S.A., Olin, A., Chen, Y., Mikes, J., Fagerberg, L., Gummesson, A., Bergström, G., Uhlen, M., Brodin, P.: Human immune system variation during 1 year. Cell Rep. **32**(3), 107923 (2020). ISSN 2211-1247. https://doi.org/10.1016/j.celrep.2020.107923, https://www.sciencedirect.com/science/article/pii/S2211124720309049
42. Zenil, H. Kiani, N.A., Tegnér, J.: Low-algorithmic-complexity entropy-deceiving graphs. Phys. Rev. E **96**(1), 012308 (2017)
43. Zenil, H., Schmidt, A., Tegnér, J.: Causality, information, and biological computation: an algorithmic software approach to life, disease, and the immune system. In: Walker, S.I., Davies, P.C.W., Ellis, G.F.R. (eds.) From Matter To Life: Information To Causality, pp. 244–280. Cambridge University Press, New York (2017). https://doi.org/10.1017/9781316584200.011

44. Heltberg, M.L., Krishna, S., Jensen, M.H.: On chaotic dynamics in transcription factors and the associated effects in differential gene regulation. Nat. Commun. **10**, 71 (2019)
45. Itik, M., Banks, S.P.: Chaos in a three-dimensional cancer model. Int. J. Bifurcation Chaos **20**(1), 71–79 (2010)
46. Khajanchi, S., Perc, M., Ghosh, D.: The influence of time delay in a chaotic cancer model. Chaos **28**(10), 103101 (2018)
47. Letellier, C., et al.: What can be learned from a chaotic cancer model? J. Theor. Biol. **322**, 7–16 (2013)
48. Zenil, H., Kiani, N.A., Marabita, F, Deng, Y., Elias, S., Schmidt, A., Ball, G., Tegnér, J.: An algorithmic information calculus for causal discovery and reprogramming systems. iScience 19, 1160–1172 (2019)
49. Ko, J., Bhagwat, N., Yee, S.S., Ortiz, N., Sahmoud, A., Black, T., et al.: Combining machine learning and nanofluidic technology to diagnose pancreatic cancer using exosomes. ACS Nano **11**(11), 11182–11193 (2017)
50. Liu, L.; Chen, X.; Petinrin, O.O.; Zhang, W.; Rahaman, S.; Tang, Z.-R.; Wong, K.-C. Machine learning protocols in early cancer detection based on liquid biopsy: a survey. Life 11, 638 (2021). https://doi.org/10.3390/life11070638
51. Chen, C., Zong, S., Liu, Y., Wang, Z., Zhang, Y., Chen, B., Cui, Y.: Profiling of exosomal biomarkers for accurate cancer identification: combining DNA-paint with machine- learning-based classification. Small **15**(43), e1901014 (2019). https://doi.org/10.1002/smll.201901014
52. Shin, H., Oh, S., Hong, S., Kang, M., Kang, D., Ji, Y.G., Choi, B.H., Kang, K.W., Jeong, H., Park, Y., Hong, S., Kim, H.K., Choi, Y.: Early-stage lung cancer diagnosis by deep learning-based spectroscopic analysis of circulating exosomes. ACS Nano **14**(5), 5435–5444 (2020). https://doi.org/10.1021/acsnano.9b09119
53. Barabasi, A.-L., Oltvai, Z.N.: Network biology: understanding the cell's functional organization. Nat. Rev. Gen. **5**, 101–113 (2004)
54. Newman, M.E.J.: The structure and function of complex networks. SIAM Rev. **45**(2), 167–256 (2003)
55. Shi, H., Yan, K.K., Ding, L., Qian, C., Chi, H., Yu, J.: Network approaches for dissecting the immune system. iScience **23**(8), 101354 (2020). https://doi.org/10.1016/j.isci.2020.101354.
56. Zenil, H., Kiani, N.A., Tegnér, J.: A review of graph and network complexity from an algorithmic information perspective. Entropy **20**(8), 551 (2018a). https://doi.org/10.3390/e20080551
57. Zenil, H., et al.: A decomposition method for global evaluation of shannon entropy and local estimations of algorithmic complexity. Entropy **20**, 605 (2018b)
58. Hernández-Orozco, S., Kiani, N.A., Zenil, H.: Algorithmically probable mutations reproduce aspects of evolution, such as convergence rate, genetic memory, and modularity. Royal Soc. Open Sci. **5**, 180399 (2018)
59. Zenil, H., Minary, P.: Training-free measures based on algorithmic probability identify high nucleosome occupancy in DNA sequences. Nucleic Acids Res. gkz750 (2019)
60. Bindea, G., Mlecnik, B., Tosolini, M., Kirilovsky, A., Waldner, M., Obenauf, A.C., et al.: Spatiotemporal dynamics of intratumoral immune cells reveal the immune landscape in human cancer. Immunity **39**(4), 782–795 (2013). https://doi.org/10.1016/j.immuni.2013.10.003
61. Zenil, H., Kiani, N.A., Tegnér, J.: Methods of information theory and algorithmic complexity for network biology. Semin. Cell Dev. Biol. **51**, 32–43 (2016). https://doi.org/10.1016/j.semcdb.2016.01.011

# Correction to: The Immune System in Health and Disease: The Need for Personalised Longitudinal Monitoring

**Hector Zenil, Abicumaran Uthamacumaran, and Kourosh Saeb-Parsy**

## Correction to:
## Chapter "The Immune System in Health and Disease: The Need for Personalised Longitudinal Monitoring" in: I. Balaz and A. Adamatzky (eds.), *Cancer, Complexity, Computation*, Emergence, Complexity and Computation 46, https://doi.org/10.1007/978-3-031-04379-6_15

The original version of the book was inadvertently published with incorrect abstract and keywords for Chapter 15 in XML and book PDF. Abstract and Keywords has now been removed.

The erratum book has been updated with the change.

The updated original version of this chapter can be found at https://doi.org/10.1007/978-3-031-04379-6_15

 
I. Balaz and A. Adamatzky (eds.), *Cancer, Complexity, Computation*, Emergence, Complexity and Computation 46,
https://doi.org/10.1007/978-3-031-04379-6_16

The manufacturer's authorised representative in the EU is Springer Nature Customer Service Centre GmbH, Europaplatz 3, 69115 Heidelberg, Germany. If you have any concerns regarding our products, please contact ProductSafety@springernature.com

Printed and bound by CPI Group (UK) Ltd, Croydon, CR0 4YY
22/09/2026
02221055-0016